Impact Studies

Series Editor: Christian Koeberl

Henning Dypvik
Filippos Tsikalas
Morten Smelror
(Editors)

The Mjølnir Impact Event and its Consequences

Geology and Geophysics of a Late Jurassic/Early Cretaceous Marine Impact Event

With 166 figures, 74 in colour

ESF IMPACT

Springer

Editors
Dr. Henning Dypvik
University of Oslo
Dept. Geosciences
NO-0316 Oslo
Blindern
Norway
henning.dypvik@geo.uio.no

Dr. Filippos Tsikalas
Eni Norge AS
NO-4064 Stavanger
Norway
filippos.tsikalas@eninorge.com

Dr. Morten Smelror
Geological Survey Norway
(NGU)
NO-7491 Trondheim
Norway
morten.smelror@ngu.no

ISBN 978-3-540-88259-6 ISBN 978-3-540-88260-2 (eBook)
DOI 10.1007/978-3-540-88260-2
Springer Heidelberg Dordrecht London New York

Library of Congress Control Number: 2010931721

Cover image by Dona Jalufka

Cover design: deblik, Berlin

Printed on acid-free paper

Springer is part of Springer Science+Business Media (www.springer.com)

Preface

The study of the Mjølnir impact crater was initiated by Steinar Thor Gudlaugsson (Gudlaugsson 1993) who first had the idea that this peculiar inverted sombrero structure in the central Barents Sea had been created by an asteroid/comet impact. This "exotic" idea led to the acquisition of new geological and geophysical data at the structure and the further investigations and involvement of several scientists, as described in Sect. 1.4.

Impacts into marine environments and wet sediments have been common and important throughout the geological history of the Earth, which is covered by close to 75% of water. Only 27 of the 176 impact craters currently discovered on Earth have a marine origin, and just a couple of these have been studied in great detail. One of the important scientific features of the Mjølnir impact crater is the clear correlation between the crater and its very well preserved ejecta (Sindre Bed) found in boreholes in the Barents Sea and on land on Svalbard, and possibly in western Siberia. Furthermore, the Mjølnir impact is unique in targeting into thick successions of prolific marine source rocks for oil and gas, and thus the impact possibly resulted in an enormous post-impact fire on the paleo-Barents Sea seafloor. What are the morphological, structural and sedimentological characteristics of the Mjølnir impact crater, and what were the immediate environmental consequences of the impact event for life and later the petroleum generation? The present book outlines key features of the Mjølnir impact and sums up the results of nearly 20 years of studies of the impact crater. However, the study of the Mjølnir impact and its consequences will continue in the near future (see Sect. 1.6).

In this book the major scientific contributions of the Mjølnir impact studies are summarized, updated and presented in 10 chapters, together with a rich reference list and useful subject index. The introductory chapter gives the general setting of the different scientific involvements and sketches the hydrocarbon exploration of the region (Chap. 1). In Chap. 2, an overview of the Arctic geological setting is given, forming the foundation for the major structural, geomorphological and geophysical characteristics of the Mjølnir crater as presented in Chaps. 3 and 4. The sedimentation of the impact crater, both syn- and post-impact, is presented in Chap. 5, succeeded by the ejecta geology in Chap. 6. The mechanisms of cratering are treated in Chap. 7, but modeling and cratering mechanisms are also discussed in Chaps. 4 and 8. The generally poorly understood post-impact deformation in impact craters, is elaborated in Chap. 9, where the Mjølnir results are compared to other, well-known impact sites. The dramatic Mjølnir impact tsunami is the theme of Chap. 10, clearly accounting for the post-impact sedimentological consequences of the shallow-water target area.

In a newly submitted proposal (December 2009), we suggest to drill 6 boreholes, up to 300 m long, to further gather new, unique information on the Mjølnir impact. Financial support was requested in a combined drilling proposal submitted to the Integrated Ocean Drilling Program (IODP) and International Continental Scientific Drilling Program (ICDP), oil companies active in the Barents Sea (20 companies), and the Research Council of Norway (NRC). The drill sites have been carefully selected to cover the full variety of lithologies and stratigraphies associated with the Mjølnir impact. The project is aiming at studying the mechanisms of crater formation, ejecta generation and distribution, and the shock and seismic disturbances in the area. We are aiming at drilling the structure in 2011. The proposed drilling operations are planned to be televised, which we hope will give great public relations for the natural sciences, beneficial for both academia and industry. An international science team has been established in relation with the proposed drilling operations.

We hope and expect the current book and the planned near-future activities will take the Mjølnir research one step further and inspire to more projects within the field of marine impact cratering.

Oslo, Norway — Henning Dypvik
Stavanger, Norway — Filippos Tsikalas
Trondheim, Norway — Morten Smelror

Acknowledgements

The Mjølnir project has involved a large number of project affiliated scientists and students and other persons and organisations outside the research group. Here we will just mention a few, which have been particularly important for us to finalize this book.

We wish to thank Research Council of Norway (grant No. 154831 and Mjølnirprosjektet), SINTEF Petroleum Research (Trondheim), the Norwegian Petroleum Directorate (NPD) and the International Centre for Geohazards (ICG) for their support.

In relation with the shallow drilling of the Mjølnir structure in 1998, The Norwegian Petroleum Directorate and the crew and shipowner of Bucentauer should be thanked as should the petroleum companies Phillips Petroleum, Saga Petroleum, Norsk Hydro, and Statoil for their financial support.

The ESF Impact program should be mentioned in particular for arranging inspiring meetings and discussions keeping the impact studies and research inspiration alive in the first years.

Eni Norge and Eni E&P are acknowlegded for providing the time and resources for Filippos Tsikalas to the study.

We are especially indepted to editor, inspirator and reviewer Christian Koeberl and last, but not least reviewer Philippe Claeys for significant improvements to the manuscript. Synnøve Knivsland is thanked for doing a great job in making the first compilation of figures and text and Christian Sætre helped in the proofreading.

Acknowledgements

Contents

Contributors

Henning Dypvik Department of Geoscience, University of Oslo, NO-0316 Oslo, Norway, henning.dypvik@geo.uio.no

Olav Eldholm Department of Earth Science, University of Bergen, NO-5007 Bergen, Norway, Olav.Eldholm@geo.uib.no

Jan Inge Faleide Department of Geosciences, University of Oslo, NO-0316 Oslo, Norway, j.i.faleide@geo.uio.no

Roy H. Gabrielsen Department of Geoscience, University of Oslo, NO-0316 Oslo, Norway, roy.gabrielsen@geo.uio.no

Sylfest Glimsdal Norwegian Geotechnical Institute, NO-0806 Oslo, Norway; International Centre Geohazards (ICG) at Norwegian Geotechnical Institute, Oslo, Norway, sylfest.glimsdal@ngi.no

Steinar Thor Gudlaugsson Iceland GeoSurvey, Reykjavik, Iceland, stg@os.is

Hans P. Langtangen Department of Informatics, University of Oslo, NO-0316 Oslo, Norway; Simula Research Laboratory, NO-1325 Lysaker, Norway, hpl@simula.no

Atle Mørk SINTEF Petroleum Research, Trondheim, Norway; Department of Geology and Mineral Resources Engineering, NTNU, NO-7491 Trondheim, Norway, Atle.mork@sintef.no

Geir K. Pedersen International Centre Geohazards (ICG) at Norwegian Geotechnical Institute, NO-0806 Oslo, Norway; Mechanics Division, Department of Mathematics, University of Oslo, NO-0316 Oslo, Norway, geirkp@math.uio.no

Valery Shuvalov Institute of the Dynamics of Geospheres, Russian Academy of Sciences, Moscow, Russia, shuvalov@idg.chph.ra.ru

Morten Smelror Geological Survey of Norway, NO-7491 Trondheim, Norway, morten.smelror@ngu.no

Trond Torsvik Physics of Geological Processes, University of Oslo, NO-0316 Oslo, Norway, t.h.torsvik@fys.uio.no

Filippos Tsikalas Eni Norge AS, NO-4064 Stavanger, Norway; Department of Geosciences, University of Oslo, NO-0316 Oslo, Norway, filippos.tsikalas@eninorge.com

Stephanie C. Werner Physics of Geological Processes, University of Oslo, NO-0316 Oslo, Norway, stephanie.werner@fys.uio.no

Chapter 1
Introduction

Henning Dypvik, Morten Smelror, Atle Mørk, and Filippos Tsikalas

1.1 Background

Impact cratering is one of the fundamental processes in our planetary system and an important factor in forming the lithosphere of the Earth and the planets. The active surface processes on Earth, e.g. weathering, erosion, plate tectonics, and volcanism change the Earth's surface continuously. Therefore only a modest number of crater structures have been preserved and discovered on the surface of the Earth, compared to what can be seen on the less disturbed surfaces on the Moon and Mars (Melosh 1989; French 1998; Montanari and Koeberl 2000; Koeberl 2007) (Fig. 1.1).

So far only 176 impact structures have been recognized in the Earth Database, an apparent crater density of about 1/3,000,000 (176/509,600,000) km^2, in contrast the crater density at Moon, which is at least 3,000 times higher (Beals and Halliday 1967). If we look at the global distribution of land and sea (1:3) in combination with the 27 marine impact structures detected so far on Earth, it appears that less than 6% of the total number of possible marine impacts have been discovered (compared to the roughly more than 400 expected). This rough calculation is based on the present number of impacts on land (here called subaerial) (149, i.e. 149 *3 = 447) (Dypvik et al. 2004a). The discrepancy, with a too low number of marine impacts discovered, is mainly the result of our limited knowledge of present submarine crater locations, the ocean water breakdown of impactors, along with the burial of marine craters by post-impact sediments, plate tectonics and the young ages of the ocean floors, and the limited geophysical information from the oceans and shelf seas. Many more marine impact structures should be expected and will probably be found in the future (Dypvik and Jansa 2003; Dypvik et al. 2004a).

The Mjølnir impact crater in the Barents Sea (Fig. 1.2) was recognized in 1993 and included in the Earth Impact Database of 1996; based on the discoveries of impact-related geological and geochemical features, such as shocked quartz, Ir-enrichments, possible glass remnants, fragments of nickel-rich iron oxides, in

H. Dypvik (✉)
Department of Geosciences, University of Oslo, NO-0316 Oslo, Norway
e-mail: henning.dypvik@geo.uio.no

H. Dypvik et al. (eds.), *The Mjølnir Impact Event and its Consequences*,
Impact Studies, DOI 10.1007/978-3-540-88260-2_1, © Springer-Verlag Berlin Heidelberg 2010

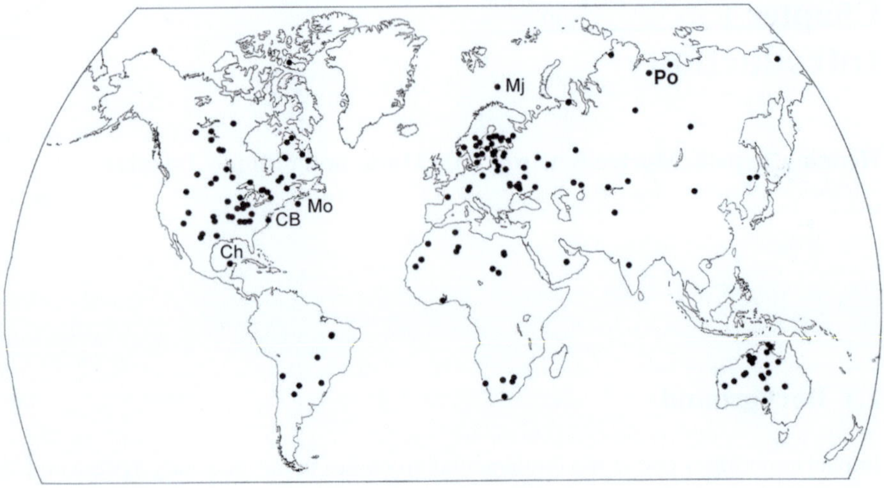

Fig. 1.1 Locations of the present known impact craters on the Earth. Ch= Chicxulub, CB = Chesapeake Bay, Mo = Montagnais, Mj = Mjølnir (Modified from French 1998)

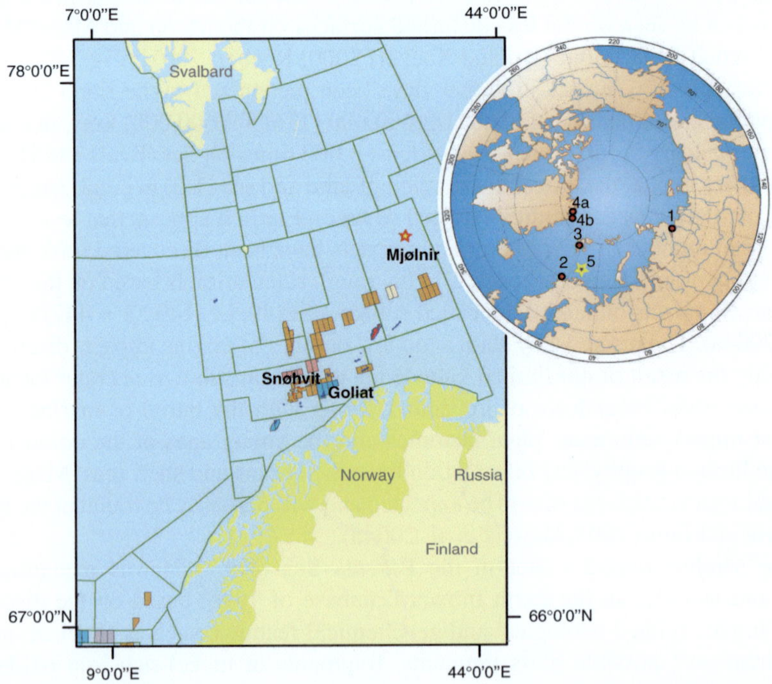

Fig. 1.2 Map with location of the Mjølnir crater, license areas in the western Barents Sea and locations of the Snøhvit and Goliat fields. *Inset* a circum-Arctic map, showing the positions of the Mjølnir crater (*5*). Nordvik in western Siberia (*1*), well 7120/12-1 off Troms (*2*), and studied onshore sections on Svalbard (*3*) and Greenland (*4*) (Modified from NPD website 2009)

addition to the convincing complex crater shape of the structure (Gudlaugsson 1993; Dypvik et al. 1996).

The marine impacts of Chixculub, Chesapeake Bay, Montagnais, and Mjølnir are found in submarine settings and consequently are difficult and expensive to study in detail. However, marine geophysical investigations along with several recent drilling campaigns have disclosed lots on information about these important events. On the marine shelves, seismic investigations can give us 3-dimensional impressions of the subsea structures, which may be a great advantage in crater research. Marine impacts, presently exposed on land in subaerial positions under shallow burial, such as, e.g., the Lockne and Kärdla craters are more easy accessible for direct study and have also contributed to our understanding of marine cratering and related processes (e.g., Jansa et al. 1989; Gudlaugsson 1993; Pilkington et al. 1995; Dypvik et al.1996; Lindström et al. 1996; Morgan and Warner 1999; Ormö and Lindström 2000; Suuroja et al. 2002; Poag et al. 2004).

The impact of an asteroid or a comet results in instantaneous generation of shock waves that penetrate the target area and attenuate into the target environment. The shock waves affect the target lithologies by vaporizing, melting, and shattering both the projectile and target rocks. The passage of the impact-induced shock wave leads to development of extremely high pressure and temperature as reflected in the characteristics of, e.g., impact lithologies, shocked minerals, and shatter cones (Melosh 1989; French 1998; Montanari and Koeberl 2000).

The processes of impact cratering can be subdivided into several phases, Kieffer and Simonds (1980) suggest 5 phases, while Melosh (1989) suggested this simpler three phase subdivision:

(a) Contact/compression
(b) Excavation
(c) Modification

The contact/compression stage starts from contact when the projectile pushes target material, compressing and accelerating it, and ends when the projectile has unloaded from high pressure (Melosh 1989). During the excavation stage the almost hemispherical shock wave propagates into the target and in combination with the following rarefaction wave moves target material. Target material is displaced within the crater and excavated as ejecta. This stage ends when the crater has been fully excavated. The modification stage includes the filling of the excavated crater (transient crater) with loose debris; large slumps etc. down the sides and the possible highs in the crater (central peak, annular ring).

Marine impacts experience different and additional effects in all of the three cratering phases compared to the impacts on land (subaerial impacts). This is due to the presence of water and water-saturated sediment-covers of varying thicknesses, in addition to submarine post-impact modification by erosion and sedimentation (Ormö and Lindström 2000; Dypvik and Jansa 2003; Dypvik et al. 2004a).

The nature of the cratering processes also depends on whether the target is crystalline or sedimentary. However, the global cratering record is biased towards crystalline, water-poor targets. Using a multi-material hydrocode, numerical simulations for Mjølnir, Shuvalov et al. (2002) have pointed out the importance of the target lithology for the cratering processes. In particular, the low-strength, water-saturated sedimentary target layer will lead to a modification and post-impact sedimentation crater stage that deviates considerably from the more typical scenarios of modification for large complex craters on land. Dypvik and Kalleson (2010) have recently recognized comparable crater filling process developments in the filling/sedimentation in marine impact craters.

The influence of water is in particular evident in the processes of vapor cloud formation, tsunami generation, and post impact sedimentation and modification. It is also reflected in the wide array of breccias and conglomerates occurring in and around the impact structures. In this compilation we will present the formation and modification of the marine Mjølnir impact structure in its Arctic geological framework, a typical marine impact crater. We will also place the Mjølnir structure in the global cratering picture, in order both to explain its formation and its regional and global significance.

1.2 Barents Sea Geology

The Svalbard and Barents Sea stratigraphy (Dallmann et al. 1999) and a review of the geological history is presented in Worsley (2008) and forms the base and main reference for the following presentation (stratigraphy in Chap. 2) (Figs. 1.3 and 1.4).

Today Precambrian rocks crop out along the western and northern part of Svalbard, in North Greenland and in Northern Norway. So far Precambrian rocks have not been reached by any of the wells drilled during the petroleum exploration of the southern and western parts of Barents Sea. The Precambrian rocks surrounding the Barents Sea region are made up of granitic and amphibolitic gneisses, which may be covering or even cross cut metasedimentary successions of sandstone, shale and conglomerate along with Vendian tillites, stromatolites, and dolomitic formations (Gee and Tebenkov 2004). Overlying the Precambrian rocks there is a succession of Cambrian to Lower Silurian marine clastics and carbonates. The Precambrian and Caledonian formations of Svalbard are referred to as the Hecla Hoek Complex (or Pre-Old Red basement). During the Caledonian orogeny the Hecla Hoek rocks were faulted, folded, thrusted, and intercepted by igneous complexes. The Hecla Hoek Complex has been estimated to be 15–20 km in thickness, spanning ages from 1,275 to 340 million years (Harland 1969, 1971; Worsley 2008; Worsley and Aga 1986; Otha 1994; Gee and Tebenkov 2004).

In the Barents Sea region the main phase of the Caledonian orogeny was followed by extensive Devonian to Permian denudation and rifting. In this period Old Red Sandstones (Devonian) along with alluvial to sabkha and desert-like deposits (Devonian to Carboniferous), followed by Permian carbonate platform deposits were deposited. The Devonian, Carboniferous, and Permian successions may locally be more than 13 km in thickness (Dallmann et al. 1999; Johnsen et al. 2001; Worsley 2008; Smelror et al. 2009).

In Late Devonian times humid conditions prevailed and a change from the underlying red Early Devonian to grey fluvial sedimentary facies can be observed

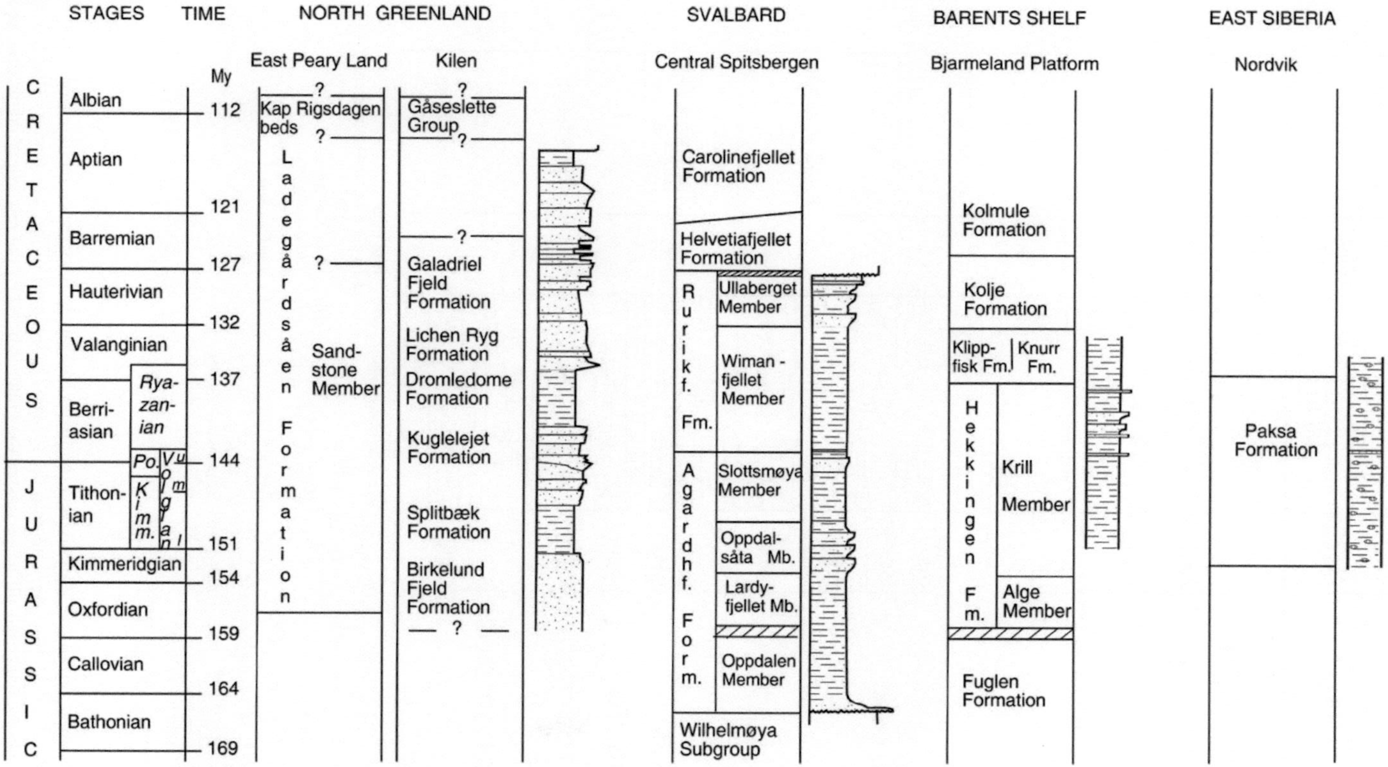

Fig. 1.3 Lithostratigraphic correlation columns for the Bathonian-Albian successions of North Greenland, Svalbard, western Barents Shelf and East Siberia (Based on Dypvik et al. 2002; Dypvik and Zakharov 2010)

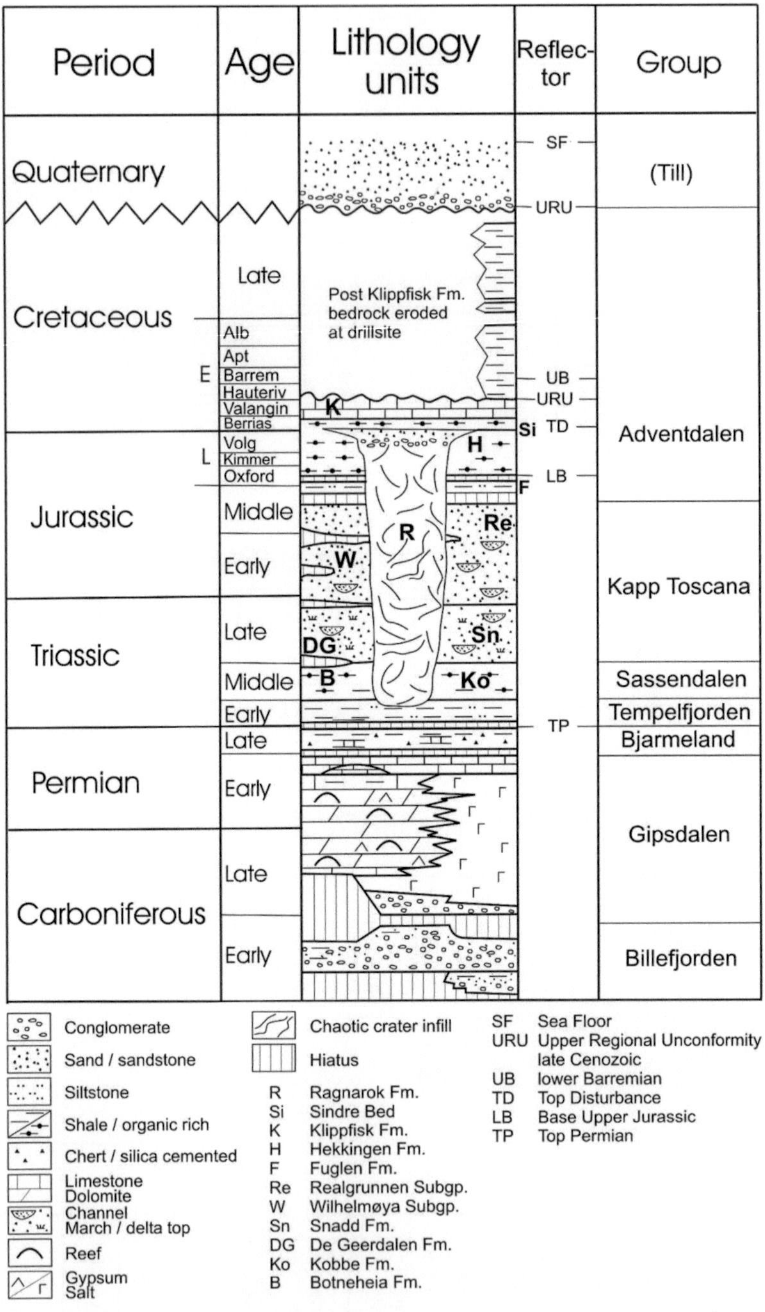

Fig. 1.4 Stratigraphy of the Barents Sea, with the Mjølnir impact crater included (R) (Dypvik et al. 2004b). The Mjølnir crater cuts into the Lower Triassic part of the sedimentary successions. During the subsequent crater filling processes the crater is filled by the reworked/redeposited beds, named the Ragnarok Formation (R)

(Fig. 1.4). Renewed extension/rifting took place in the Carboniferous, and coarse grained siliciclastics and coal beds were deposited. The Carboniferous succession is dominated by alluvial sediments succeeded by the evaporites of the Gipsdalen Group (Middle Carboniferous age). During the Late Carboniferous, Permian, and Mesozoic stable platform conditions evolved in the region, comprising Svalbard and large parts of the present Barents Shelf. In Permian time, the region was dominated by limestone, dolomite, and evaporite sedimentation grading into cherty limestones and silicified shales and siltstones. The break between the siliceous Permian beds and the much less cemented siliciclastic Triassic formations forms one of the most pronounced stratigraphical boundaries on Svalbard (Fig. 1.4).

The succeeding Lower Triassic succession consists of shales and sandstones with only moderate degree of cementation, in great contrast to the firm, siliceous Permian formations below. Consequently, the Permian/Triassic boundary is fairly well exposed/expressed on Svalbard, and can also be traced as a pronounced reflector on seismic sections in the Barents Sea.

The Mesozoic succession of the Barents Sea region (Figs. 1.3 and 1.4) represents continental to open marine environments (Worlsey 2008; Smelror et al. 2009). The succession reaches close to 3 km in thickness on Svalbard and about 6 km in thickness on the Barents Shelf. Fluvial, deltaic and coastal deposits with shifting tidal influences are found along with more open marine, shelf deposits. Sand and gravel dominate in the coastal facies, while fine grained deposition of clays and silts with varying contents of organics matter typically can be found in the mid- to outer shelf regions of the epicontinental Mesozoic sea. At times, varying dysoxic/anoxic conditions existed on the shelf, alternating with periods dominated by oceanic ventilation and storm sand deposition.

In the Late Jurassic to Early Cretaceous black and dark grey clays of the Hekkingen Formation (Oxfordian – Ryazanian) (359 m in thickness in stratotype) formed the Barents Shelf seafloor and the target area for the Mjølnir impact. The uppermost Jurassic-lowermost Cretaceous parts of these thick shale and claystone units were interrupted by the Mjølnir impact, which created the impact strata named the Ragnarok Formation (Dypvik et al. 2004b, 2006). The Sindre Bed forms an impact-derived marker horizon outside the crater rim. The impact derived units are dominated by conglomeratic and brecciated formations with a rather high content of smectite, a possible alteration product of impact glass (Dypvik and Ferrell 1998; Dypvik et al. 2004a). The material excavated consists mainly of reworked Triassic formations (Dypvik et al. 2004b).

The Hekkingen Formation is succeeded by Cretaceous limestones, marls, shales, and sandstones, of the upper Adventdalen and Nygrunnen groups (Figs. 1.3 and 1.4).

From the Late Jurassic/Early Cretaceous and until the Eocene – Oligocen transition (142–35 million years ago) sea floor spreading took place along the Nansen Ridge in the Arctic Ocean (Engen et al. 2008). The break-up of the north-east Atlantic rift system, however, started about 55 Ma (Skogseid et al. 2000) along the Mohns and Knipovich ridges, accompanied by strike-slip movements in the Fram Strait between Svalbard and North Greenland. These strike-slip movements continued to the Eocene/Oligocene transition, connecting the spreading basins of Arctic Ocean and Norwegian Greenland Sea. At this point, however, the movements of the Fram Strait shifted towards oblique extension (33.3 Ma), and a deep-water gateway opened by seafloor spreading during Oligocene-Miocene (Lawver et al. 1990; Faleide et al. 1993; Eldholm et al. 1994; Torsvik et al. 2002). Since then spreading has taken place along this major lineament, with final establishment of the present seafloor spreading regime at 9.8 Ma.

On Svalbard the Paleogene successions (up to 1,900 m in thickness), comprise continental conglomerate and sand deposits and shallow marine sandstones along with marine shales and mudstones (Fig. 1.5). Related to the opening of the Fram Strait and Norway Greenland Sea, transpressive movements took place between Svalbard and Greenland and impressive folds and fractures structures developed along the western parts of Svalbard and the Barents Shelf (Harland 1971; Nøttvedt et al. 1993; Otha 1994; Bergh and Grogan 2003; Steel et al. 1985; Bruhn and Steel 2003).

Dimakis et al. (1998) discussed the Cenozoic erosion and preglacial uplift of the Svalbard – Barents Sea region. They demonstrate a subaerial preglacial Barents Sea with uplift events and intensive erosion. The most likely mechanism for the initial uplift is thermal, possibly related to the plate tectonic opening of the Arctic.

On the land areas around the Barents Sea, e.g., North Greenland and Svalbard, outcrops of Precambrian to Paleogene rocks are found, with distributions and structural setting heavily influenced by the many stages of rifting and seafloor spreading in the Norwegian Greenland region. In North Greenland a well-developed Oxfordian to Paleogene succession rests on Upper Paleozoic and Triassic strata. The Upper Jurassic to Lower Cretaceous successions of North Greenland, as those

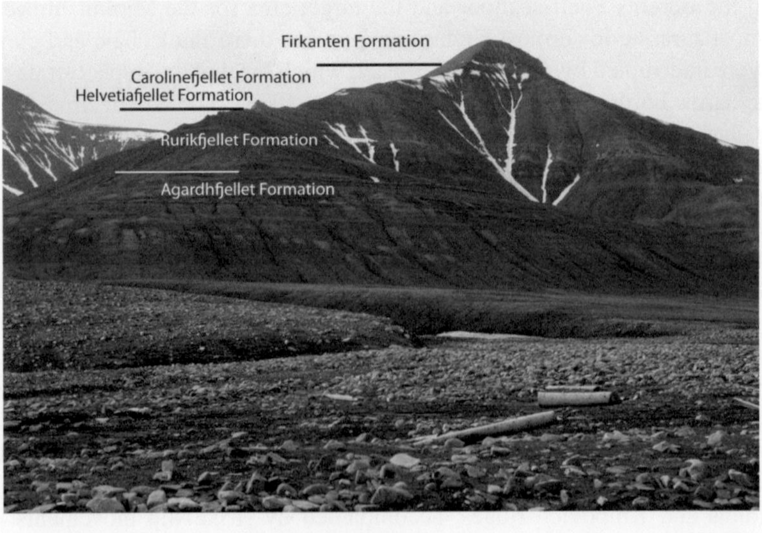

Fig. 1.5 The Mesozoic succession on the Janusfjellet; from Agardhfjellet Formation (Middle Jurassic) to the Firkanten Formation (Paleocene)

on Svalbard, are made up of black to dark grey, partly silty shales and sandstones (Dypvik et al. 1991a; Håkansson et al. 1993, 1994; Dypvik et al. 2002) time-equivalent and comparable to the Hekkingen Formation of the Barents Sea.

The general structural mode of the western Barents Sea region is characterized by northeast-southwest-trending lineaments (Fig. 1.6). A thick wedge of Upper Pliocene to Quaternary deposits, with glacial deposits and postglacial marine beds and reworked sediments, present along the western Barents Sea margin (Eidvin et al. 1993; Sættem et al. 1994; Channell et al. 1999; Vorren and Laberg 2001). The present plate tectonic setting is also reflected in the Quaternary sedimentation of the region, e.g. along the steep Barents Sea margin and into the deep basins of the Norwegian Greenland Sea, colossal slides and slumps have taken place during the Holocene (Solheim et al. 1996).

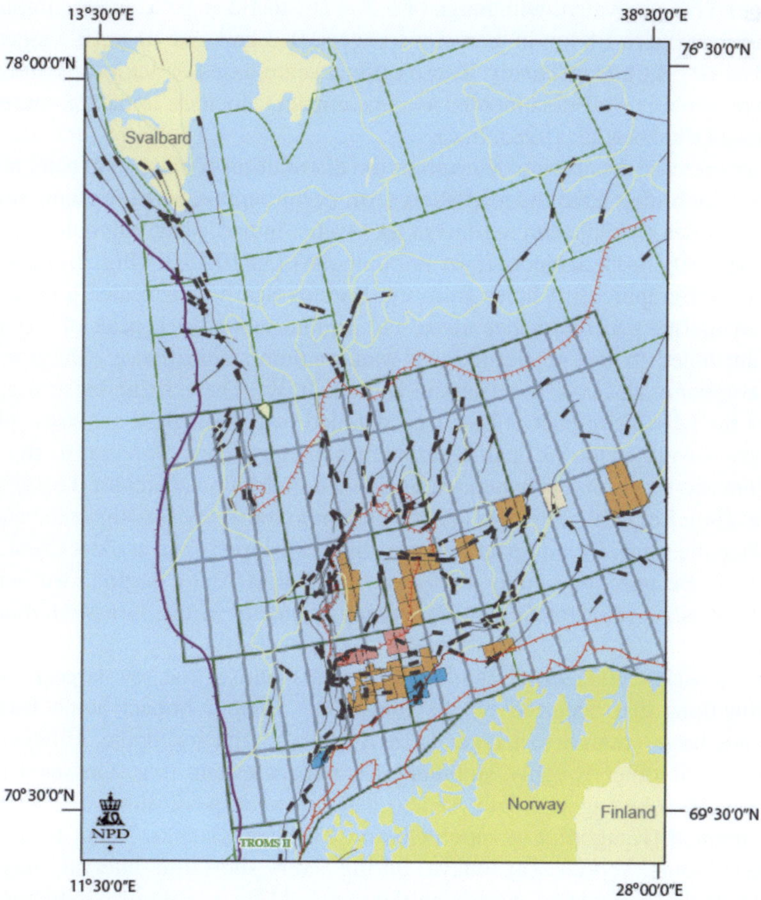

Fig. 1.6 Structural map of the Svalbard, Greenland Barents Sea (Modified from Norwegian Petroleum Directorate (NPD) website 2009)

1.3 Mjølnir Impact at Volgian/Ryazanian Boundary

Due to about 30 years of petroleum exploration in the Barents Sea, an extensive geophysical database is available from the region. At the marginal parts of the southwestern Barents Sea basins several exploration wells have been drilled. In addition, many shallow stratigraphic drillholes aiming at sub-cropping reflectors have been drilled in the more central and remote areas of the Barents Sea. Based on this information the Mjølnir structure was found and its impact origin confirmed (Gudlaugsson 1993; Dypvik et al. 1996).

The Mjølnir crater is one of the 20 largest impact structures so far discovered on Earth, ranking eight among those presently not exposed at the surface (Earth Impact Data Base 2010). In order to access the possible consequences of the impact, estimate magnitude of the impact event, i.e. the energy release, impactor sizes, and mass, have been made (Tsikalas et al. 1998a). The energy release was estimated to be in the order of 16×10^{20} J (range of $2.4–53 \times 10^{20}$ J; translating into 3.8×10^5 megatons TNT equivalent with range of 5.7×10^4 to 1.2×10^6), and the impactor's size and mass were 1.8 km in diameter (range, 0.9–3 km) and 10×10^{12} kg (range, $1.5–33 \times 10^{12}$ kg), respectively. These estimates are based on various scaling laws and on reasonably well-documented average impact velocities, impactor angles, and densities (Tsikalas et al. 1998a, b, c).

Energy release dissipation determines the distribution of ejecta and tsunami generation, which may have induced short-term perturbations/environmental stress in the Barents Sea and adjacent regions of the Arctic. In particular, palynological studies of the ~80-cm-thick ejecta-layer of borehole 7430/10-U-01 (Fig. 1.7) and other shallow stratigraphic drill holes from the Barents Sea have revealed a high abundance of marine *prasinophycae* algae and a minor abundance peak of freshwater algae attributed to the impact-induced water-column disturbance (Smelror et al. 2002; Bremer et al. 2004; Smelror and Dypvik 2006). The existing biostratigraphic age for the Mjølnir impact is based on detailed paleontological analyses, placing the impact event at the Volgian-Ryazanian boundary, corresponding to the informal "Jurassic-Cretaceous boundary", as defined in the Boreal Realm (i.e., at 142 ± 2.6 Ma) (Smelror et al. 2001a, b). A correlation to the Tethys stratigraphic subdivision place the time of the impact in the earliest Berriasian (i.e., earliest Cretaceous) (Fig. 1.4). Because our study concerns the Boreal/Arctic region, we will for practical reasons refer to Volgian-Ryazanian boundary as the Jurassic-Cretaceous boundary.

The Jurassic-Cretaceous boundary represents one of the ten largest biological extinctions that occurred on the Earth. The Mjølnir impact alone was definitely not large enough to trigger such a global extinction spike. However, the occurrence of other roughly simultaneous impact events (e.g. Gosses Bluff in Australia, Morokweng in South Africa) and the weak possibilities for additional coeval impacts (Kgagodi, Liverpool, Obolon, Puchezh-Katunki, Tabun-Khara-Obo, Upheaval Dome, Vepriai, Zapadnaya) during a very short time-interval, may be of importance and capable of surpassing the threshold for a biological extinction at a global scale (data from Earth Impact Data Base 2009).

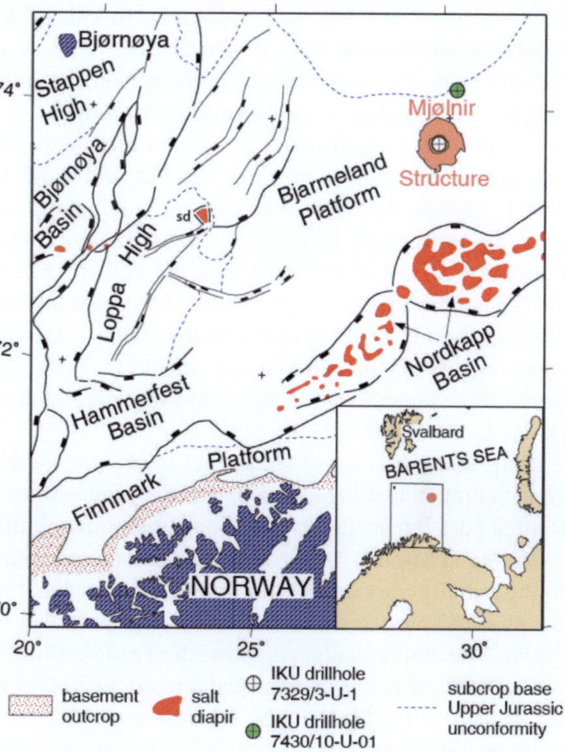

Fig. 1.7 Structural lineaments of the Barents Sea and the location of the Mjølnir crater and studied boreholes in the Barents Sea. sd= salt diapir (Based on Tsikalas et al. 1998b)

The ejected impact material of marine impacts is normally widely distributed. This will in particular be the case for ejecta from a marine impact due to the evaporation/cloud effect of the water which make it more violent compared to a similar size subaerial impact (Melosh 1989; Shuvalov and Dypvik 2004; Poag et al. 2004). In such cases both the sea-water and the water from the water-saturated sediments will take part in the formation of the vapor cloud. The water is in addition active in the later marine transportation of the ejecta.

The ejecta consist mainly of rock fragments from the target area and bolide, meteorite materials and spherules, shocked minerals, Ni-rich spinels and soot. It will have varying composition and distribution, controlled by several factors such as bolide and target area composition, sedimentary environment, mode of emplacement, timing and Earth rotation. The amount of ejecta will generally decrease away from the crater, and with increased crater size the ejecta may be dispersed over larger distances at a regional and even global scale (Shuvalov and Dypvik 2004). Consequently the ejecta forms a unique stratigraphic marker layer.

Ejecta deposits from the Mjølnir impact event (i.e., the Sindre Bed) have been recognized in the Barents Sea, on Svalbard, and possibly in Siberia (Nordvik Peninsula) (Zakharov et al. 1993; Smelror et al. 2001a; Dypvik and Zacharov 2010) at distances just 30 km from the crater (Barents Sea), via 600 km (Svalbard) and to more than 2,300 km away (Nordvik, Siberia) (Figs. 1.2 and 1.7). Furthermore, numerical simulations have shown that the presence of water at the Mjølnir impact did not have any major direct influence on the initial, first phase crater-forming process but became crucial during the subsequent crater infilling, by back-rushing water-resurge, as well as for the ejecta distribution and tsunami development and propagation (Shuvalov et al. 2002; Shuvalov and Dypvik 2004; Glimsdal et al. 2007). During the high pressure and high temperature conditions at the target area, the uppermost 100–200 m of the organic-rich claystones of the Hekkingen Formation were hit and may have caught fire. On average, the claystones of the Hekkingen Formation contain more than 8% total organic carbon (TOC), and serve as an important source rock for hydrocarbon in the Barents Sea. The effects of pressure, heat, tsunami, currents, and ejecta on such a target may consequently have provided a valuable marker horizon for the regional correlation of the Arctic region (Dypvik et al. 2008a).

Syn- and post-impact sediments reach considerable thicknesses in the Mjølnir area. Sediment loading above the primary impact structure may have resulted in substantial post-impact deformation and structural modification. Integrated geophysical modeling at Mjølnir demonstrates a close correspondence of geophysical anomalies to the radially-varying distribution of structural and morphological units, as well as the physical-property distributions (Tsikalas et al. 1998a, b, c; Tsikalas and Faleide 2004). The impact-induced substratum suffered differential compaction, triggered by a considerable overburden that altered the impact crater morphology and geometry. Indeed, at the Mjølnir site the deposits of brecciated periphery were compacted more than the denser central crater formations, resulting in a central high that maybe stood taller that the surrounding platform (Fig. 1.8). Details in the structural evolution and sedimentation around the central peak are discussed in Chap. 8. Post-impact modifications may have obscured or blurred many marine impact craters, which has caused their complex identification. The quantification of post-impact effects may be more difficult in the subaerial impact record compared to the submarine situation.

1.4 The Investigation History of Mjølnir

For more than 100 years, onshore geological exploration has been taking place in the onshore areas bordering the Barents Sea (i.e., Western Siberia, Novaya Zemlya, Franz Josef Land, Kola Peninsula, Bjørnøya, Hopen, Svalbard, and Greenland) (Fig. 1.2). During the last 30 years, extensive geophysical and geological investigations have been carried out in the Barents Sea (Fig. 1.7). This activity, including seismic analyses, gravimetric measurements, magnetic surveys, and drilling of stratigraphic and exploration wells, clearly picked up in late seventies when the petroleum industry threw their eyes on the region. Until today (2010) more than 70 exploration wells have been drilled in the Norwegian part of the Barents Sea, the first well being spudded on June 1. 1980. The exploration activity has shifted

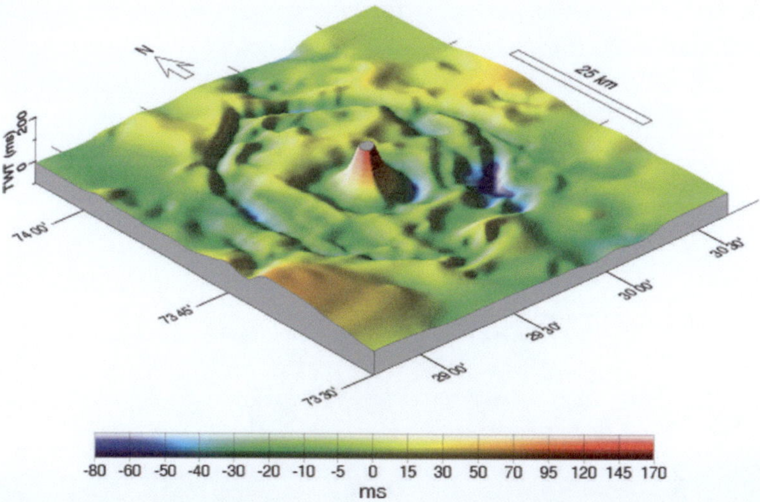

Fig. 1.8 The classical, base Barremain reflector, seismic model of the Mjølnir structure (Dypvik et al. 1996)

through the years, but in the latest years the activity has increased, in particular triggered by high oil and gas prices, the Statoil-headed exploitation of the Snøhvit Field, and the recent oil discovery of the Goliat Field. In addition to the exploration wells, several shallow stratigraphic coreholes have been drilled with the aim to penetrate sub-cropping seismic reflectors and sample high-quality cores from the different Paleozoic to Paleogene formations (Figs. 1.7 and 1.9). These programs have been run by IKU (Continental Shelf Institute Norway)/SINTEF Petroleum Research in close cooperation with the Norwegian Petroleum Directorate. Shallow and deep well information is available from the Norwegian part of the Barents Sea. Along with the opened geophysical information, this formed the foundation for our first Mjølnir studies.

Based on available seismic lines along with gravimetric and magnetic information, Steinar Gudlaugsson in 1993 launched the innovative idea that the domal structure on the Bjarmeland Platform was an impact crater (Gudlaugsson 1993). Prior to that, both salt- or mud-dome and volcanic explanations had been presented for this shallow structure that is buried beneath 50–800 m of younger sediments and about 350 m of water (Figs. 1.7 and 1.10). Gudlaugsson claimed the structure to be about 40 km in diameter and with an appearance resembling a typical complex crater as defined by Melosh (1989). Gudlaugsson's idea soon got great support and triggered new and more detailed geological analyses on available geophysical data and corematerial.

In 1993 Steinar Gudlaugsson contacted Henning Dypvik, who had 20 years of research experience from the Jurassic-Cretaceous strata on Svalbard and Barents Sea. A multidisciplinary group was put together and the detailed search for impact

Fig. 1.9 The drillship Bucentaur RS was used to drill boreholes 7430/10-U-01 adjacent to the Mjølnir crater and 7329/03-U-01 within the crater

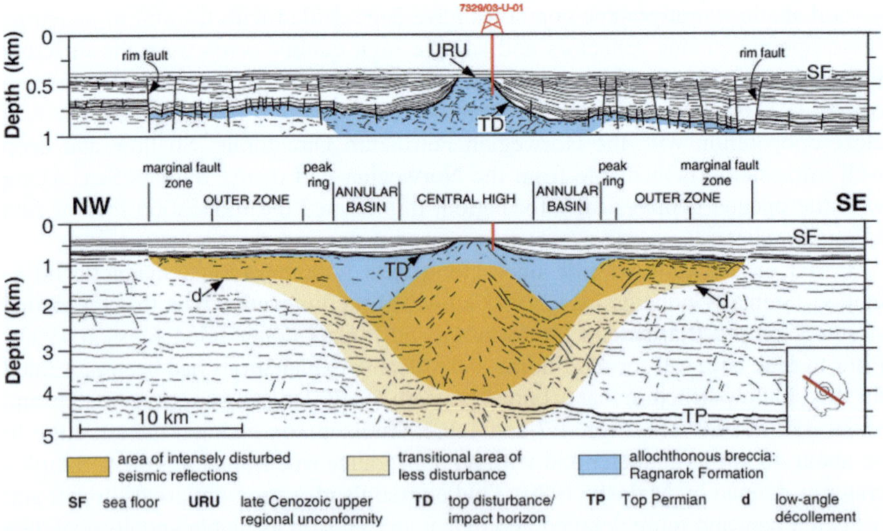

Fig. 1.10 A generalized seismic line across the Mjølnir crater (Modified from Tsikalas 1998a, b)

evidences started on available core and cuttings material from the region. Of particular interest were the well-dated shallow, stratigraphic drillcores of IKU/SINTEF and in particular from borehole 7430/10-U-01, which included a well preserved core across the Jurassic/Cretaceous boundary (location map in Fig. 1.7). After 3 years of tedious search and thousands of analyses, the first grains of shocked minerals and samples with enrichments of Ir were discovered in the 7430-core (Dypvik et al. 1996; Langenhorst and Dypvik 1996; Dypvik and Attrep 1999). This formed the geological evidence and confirmation of Gudlaugsson's suggestion. During this time Filippos Tsikalas had started his PhD studies and detailed analyses of all available geophysical information from the area.This work was performed in close cooperation with the supervisors Jan Inge Faleide and Steinar T. Gudlaugsson. Tsikalas finalized his PhD in 1997, and has since continued the geophysical analyses of the structure. His and Jan Inge Faleide's interest have in particular focused on the subsidence and uplift history of the structure along with its asymmetric structural developments, underlining the oblique impact configuration. Based on these analyses, detailed geometrical information of the Mjølnir structure has been achieved.

The geochemical and mineralogical studies, in combination with analysis of the sedimentological ejecta developments and crater filling sediments have continued since 1993, coordinated by Henning Dypvik. Detailed sedimentological and petrological studies showed the shocked quartz and Ir carrying beds of well 7430 to reflect deposition from suspension currents, most likely derived from the crater rim (Dypvik et al. 1996). The sedimentological, mineralogical, and geochemical studies have recognized possible altered impact glass and Mjølnir ejecta in Svalbard and several Barents Sea cores (Dypvik et al. 2004c; Dypvik and Ferrell 1998; Dypvik et al. 2003). Palaeontological studies have mainly been performed by Morten Smelror, Jenø Nagy, Jorunn Os Vigran, Merethe A. Bremer and Simon R. A. Kelly (Smelror et al. 2001a, 2002; Bremer et al. 2004; Smelror and Dypvik 2006). The first dating of the impact event, and the correlation of seismic lines and the impact structures formed natural key geological and geophysical information in understanding the impact evolution (Figs. 1.10 and 1.11). Macro- and micropaleontological along with palynological analyses confirm an age at the Volgian-Ryazanian boundary, i.e., 142 ± 2.6 Ma, of the impact event (Smelror et al. 2001a). In addition the palynological discoveries of algal blooms of disaster species just after impact, in concert with geochemical enrichments made it possible to trace the effects of impact in even wider areas around the Mjølnir structure (Smelror and Dypvik 2005, 2006; Dypvik et al. 2006). During some short spring weeks in 1998 we were able to collect sufficient industrial support and sponsorship to hire the drillship Bucentaur for a week operation, allowing us to drill the so-called Mjølnir crater drillhole (7329/03-U-01) (Figs 1.7, 1.9 and 1.10). The 7329/03-U-01 borehole is located on the central peak, below 350 m of water. The drill-site was selected for maximum stratigraphical depth of penetration at a place with only 54 m of overburden. The operation was generously supported by the Norwegian Petroleum Directorate, Statoil ASA, Norsk Hydro ASA, Saga Petroleum ASA and Phillips Petroleum, the ship-owner Seateam and the operator IKU/SINTEF Petroleum Research (Fig. 1.9). The 121 m long core

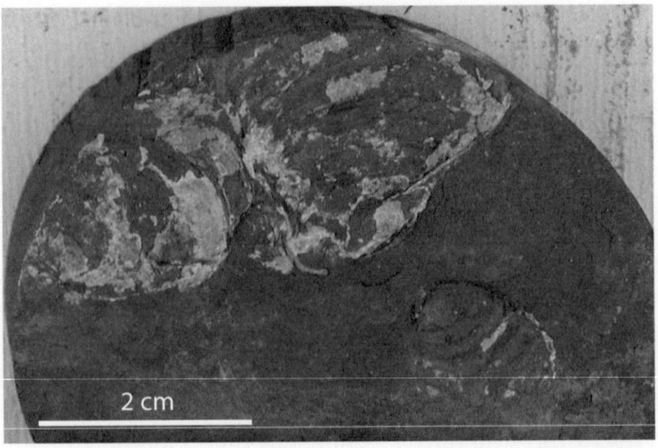

Fig. 1.11 Macrofossil (*Buchia* sp.) from core the Mjølnir Crater 7329/03-U-01, core-level 73.71 m

has since been studied in great detail and much new information has been added to our knowledge about the structure. It is the only core from inside the crater and consequently a firm evidence for the impact origin.

Rumor has it that still in the summer of 1998, while drilling the Mjølnir core on the central peak, betting was going on among doubtful geologists not believing the impact hypothesis. Some hard-core geologists and geophysicists, still at that time, preferred other explanations (e.g., volcanism, salt diapirism, liquefaction). The first papers from these core studies were presented in the ESF (European Science Foundation) supported workshop of the successful IMPACT program, which was arranged in Longyearbyen the fall of 2001 (Dypvik et al. 2004a; Smelror et al. 2001c).

Based on the paleontological analyses of Mjølnir crater core samples the dating of the impact was further confined (Smelror et al. 2001a). Through the master thesis of Pål Sandbakken (Sandbakken 2002) on core material from borehole 7329/03-U-01, additional pressure relations of the impact were disclosed (Sandbakken et al. 2005). Detailed discussions of the Mjølnir core also resulted in sedimentological descriptions of the formations and mechanical interpretations of the processes active along the central peak. In addition stratigraphical descriptions and formal definitions of the syn-impact and post-impact formations were done (Dypvik et al. 2004c).

In the Mjølnir crater core, in surface samples from Svalbard and in the 7430-core soot particles have been found, expressing a story of intense, impact-induced sea-floor fires. Potential petroleum source rocks within the crater, an about 5–10 km^2 area of the Barents Sea, were locally put on fire by the Mjølnir bolide (Wolbach et al. 2001; Dypvik et al. 2008b).

A major advance in our understanding of the Mjølnir Crater was achieved when Valery Shuvalov joined the research group with advanced numerical modeling of both impact mechanics and ejecta distribution. In this way, timing of various impact phases along with the first insight of impressive tsunami generation events was evident (Shuvalov et al. 2002; Shuvalov and Dypvik 2004). The simulations also displayed asymmetrical ejecta distribution, along with an interesting downrange

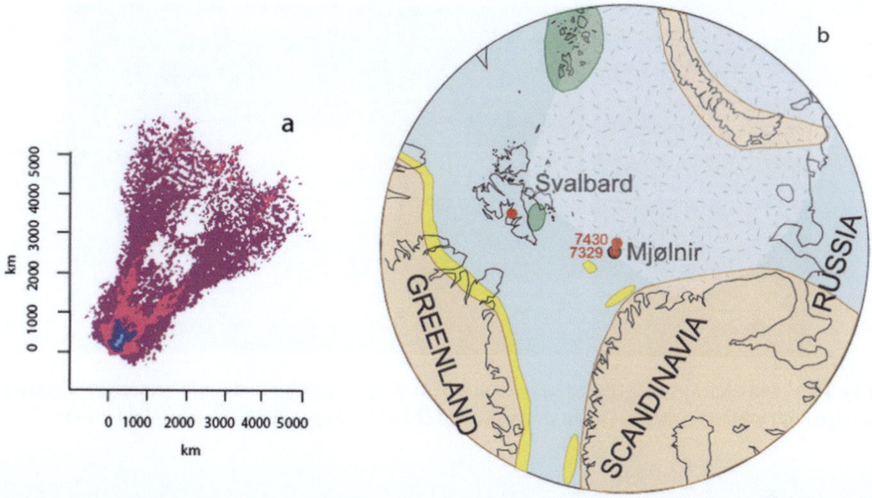

Fig. 1.12 Map showing the NE-trending ejecta distribution of the Mjølnir impact; **(a)** display the areal distribution of ejecta from the impactor and in **(b)** the compareable distribution (hatched) in the paleogeographic reconstruction is shown (Modified from Shuvalov and Dypvik 2004)

movement of the central peak during crater evolution (Fig. 1.12). The recent direct involvement of structural geologist Roy H. Gabrielsen has demonstrated how the different mechanical phases in the crater development in combination with such numerical information can explain, in this case, some of the mechanisms in the crater-fill processes.

During and immediate after marine impacts, tsunamis will be generated. Tsunami-generated deposits (tsunamites) are well known from the Chicxulub impact in the Mexican Gulf area (Smit 1999; Claeys et al. 2002). The Eltanin impact event tsunami in the Pacific has been modeled by Asphaug and Ward (2002) and Korycansky et al. (2003). The palaeogeographical reconstructions of the Barents Sea area shows that an extensive epicontinental sea covered the area at the time of impact and tsunamis must have been formed. The initial tsunami was also evident in the first simulations of Shuvalov et al. (2002). These simulations formed the starting point for PhD-student Sylfest Glimsdal and his supervisors Geir Pedersen, Hans Petter Langtangen, Shuvalov, and Dypvik for modeling the Mjølnir tsunami (Glimsdal et al. 2005, 2007). Simulation of tsunami generation in relation to marine impacts is a new and important topic, which has not been carried out in great detail before. Large amounts of new, basic knowledge were gained in the study of Glimsdal et al. (2007) and formed the basis of Glimsdal's PhD thesis defended in June 2007. This work was continued in the PhD studies of Rolv Bredesen, focusing on the tsunami runup problematics.

The colossal starting heights of the first tsunami (more than 200 m), along with its fast advancement across the paleo-Barents Sea are evident. Series of tsunamis were formed and had obvious influence on the Barents Sea sedimentation, and

Fig. 1.13 Possible ejecta bed may be present in phosphates near the Jurassic- Cretaceous boundary in Nordvik, western Siberia; **(a)** an overview and **(b)** a close up. (Photos: Victor Zakharov)

the marine life of paleo-Barents Sea region. The Mjølnir ejecta distribution is asymmetrical around the structure, as evident from both the sedimentological investigations (Dypvik et al. 2006) and in the numerical modeling (Shuvalov and Dypvik 2004) (Fig. 1.12). This asymmetrical ejecta distribution and environmental disturbances are in great contrast to the modeled, symmetrical tsunami wave propagation (Glimsdal et al. 2007). Ejecta material traced as Ir-enrichments, along with some Ni-rich iron-oxides have been found in the Barents Sea, on Svalbard and along the bolide direction of movement, towards the North East (Dypvik et al. 2006; Robin et al. 2001).

Zakharov et al. (1993) found extremely high Ir-enrichments in time-equivalent sedimentary beds from north-central Siberia (Nordvik) (Figs. 1.2, 1.3 and 1.13). Recent Siberian investigations of new, comparable samples from Nordvik have been carried out by Zacharov, Kyte and Dypvik. Correlatable stratigraphical sections from Siberia have been analyzed mineralogically and geochemically and compared with the Barents Sea and Svalbard sections. The stratigraphical developments are comparable, but the Ir-anomaly of Zacharov et al. (1993) has not been confirmed (Dypvik and Zacharov 2010; Koeberl, personal communication).

1.5 The Search for Oil and Gas in the Barents Sea

According to the Norwegian Petroleum Directorate website (2006) totally about 0.2 billion Sm^3 o.e. (Standard cubic meters oil equivalents) of extractable oil and gas (mainly gas) has been identified on the Norwegian side in the Barents Sea. Another estimated 1 billion Sm^3 o.e. unidentified oil and gas are probably present. On the Russian side of the border the numbers are of another dimension, 2 billion Sm^3 o.e. are discovered and up to 15 billion Sm^3 o.e. may be undiscovered (USGS 2000).

Petroleum exploration started in the Norwegian part of the Barents Sea in the seventies, and the first well, 7120/12-1 in the Troms I area, was spudded 1. June 1980 (Fig. 1.2). The semi-submersible installation Treasure Seeker drilled to TD at 3,573 m (1/6/80 to 12/10/80). Based on seismic interpretations and regional geological data, the well location was selected to test possible sandstone reservoirs of Middle Jurassic, Early Jurassic, and Late Triassic ages. Post-Jurassic sediments were not considered prospective, due to lack of closure and/or reservoir rocks. The results showed traces of hydrocarbons in thin sandstone reservoirs of both Early Cretaceous and Late Triassic formations.

Since then, more than 70 wells have been drilled in 39 production licenses, all in the southern part of the Norwegian section in the Barents Sea. A number of these wells yielded minor and medium sized gas discoveries. The Snøhvit Field is currently in production (July 2009), while the Goliat Field is under evaluation (Fig. 1.2). The Upper Jurassic-lowermost Cretaceous black to dark grey, organic rich shales of the Hekkingen Formation and possible the Middle Triassic Botneheia and Steinkobbe formations are the main source rocks in this part of the Barents Sea. The Stø and Tubåen formations of the Kapp Toscana Group form possible reservoirs (Figs. 1.3 and 1.4). In the Barents Sea, Paleozoic carbonates are also possible reservoir targets. The main reservoir proven so far are, however, Middle and Upper Jurassic sandstones found in the giant Stokhman Field.

The Goliat discovery (48 km from Norwegian shores) was made in 2000, an ENI discovery of 100 million bbl (barrels) oil (Fig. 1.2). It is a rather small field compared to others on the Norwegian shelf, but recent information has doubled the size of the discovery. The Goliat structure is a faulted structural closure in the crest part of a major northeast-southwest trending rollover anticline situated in the southeastern part of the Hammerfest Basin. The Kapp Toscana Group (Tubåen Formation) of Early Jurassic to Late Triassic age is the main reservoir. In addition to oil, the Goliat discovery also includes oil-bearing gas.

When the proto-Atlantic Ocean rifted open in Paleogene times, areas along the rift were uplifted and eroded with decreasing erosional effects, eastwards, away from the rift zone, along the western side of the Barents Shelf. The erosion resulted in released burial pressure, cracking of rocks and extensive leakage of oil and gas out of the originally deeply buried traps close to the rift. Gas expansion after pressure release forced the oil out of the traps. This explanation has been given as the major reason for the presences of almost only gas in the Snøhvit field and other areas in the Hammerfest Basin (Stewart et al. 1995; Nyland et al. 1992).

This uplift and pressure release effect was less effective in the eastern regions towards Russian territories, possibly explaining the much larger amount of hydrocarbons discovered in the Russian sector. The giant Russian Stokhman Field is an excellent example, 3.2 million Sm^3 o.e. in one field, one of the largest gas fields in the world.

1.6 Future Mjølnir Studies

During 2002 and 2003, ICDP (International Continental Scientific Drilling Program) and IODP (Integrated Ocean Drilling Program) were approached with applications for scientific drilling of the Mjølnir Crater. The ICDP application received a positive recommendation, while the IODP application was not recommended full proposal. Based on this mixed response, and founded on discussions and recommendations from several scientists, we submitted workshop proposals (ICDP, IODP, ESF (European Science Foundation Magellan Workshop), Statoil ASA, Hydro ASA) in spring 2006. The aim was to obtain financial support for a meeting to discuss marine cratering and its consequences, aiming at a future scientific drilling of the Mjølnir impact structure (Mjølnir Scientific Drilling Project). The workshop was approved and financially supported by ICDP, ESF (Magellan Workshop), Statoil ASA, and Norsk Hydro ASA, and arranged in Longyearbyen, Svalbard, during September 2007. The main objective of the workshop was to shed new light on one of the basic geological processes on the Earth; the mechanism and consequences of marine impact cratering. The following scientific sub-goals are of particular importance:

1. The Mjølnir crater is one of the very few cases where a source-crater and ejecta-layer (shock effects, geochemistry, paleontology) correlation has been established and correlated directly. A drilling project will deepen our understanding of this relation and make it applicable to other locations where the relations are not that obvious.
2. Through the proposed Mjølnir drilling a better stratigraphic control of the target and impact-induced lithologies will be reached. Combined with the entire spectrum of seismic reflection profiles better constraints on the amount of excavated (allochthonous) breccia volume, structural uplift, gravitational collapse, and infilling will be reached.
3. The passage of the impact-induced shock wave leads to development of extreme high pressure and temperature. The proposed drilling aims to resolve the pressure and temperature distribution occurring during the Mjølnir impact. This work should, initially, include detailed geochemical (inorganic/organic) and mineralogical analyses (optical, electron-microscopical) of shocked mineral grains along with authigenic formations. When comparing pressure/temperature distribution to theoretical models, a testable framework for increased understanding of impact cratering physics can be achieved.
4. The nature of the cratering processes depends on whether the target is crystalline or sedimentary. However, the existing global cratering record used to represent typical impact craters is biased toward crystalline, water-poor targets. Using a multi-material hydrocode, numerical simulations for Mjølnir (Shuvalov et al. 2002) have documented the importance of the target lithology for the cratering processes.
5. Magnetic modeling shows that dispersed melts are most probably located at the periphery of the structure (Tsikalas et al. 1998b). Therefore, the proposed Mjølnir drilling will contribute to the understanding of the actual impact processes by providing, for the first time in impact crater research, a direct calibration to both empirical relationships and numerical simulations.
6. Impact energy released dissipation determines the distribution of ejecta and tsunami, that may have induced short-term perturbations/environmental stress

in the Barents Sea and adjacent regions in the Arctic. These environmental effects can be studied in greater details.

7. The analysis will also involve high-precision dating of the impact event. This will provide the correlation of the Mjølnir crater to other time-equivalent (Jurassic-Cretaceous boundary) impact events, such as the 22-km-diameter Gosses Bluff crater in Australia (Milton et al. 1996) and the 70-km-diameter Morokweng crater in South Africa (Koeberl et al. 1997a; Henkel et al. 2002).
8. At the Jurassic-Cretaceous boundary, one of the five largest biological extinctions during Earth history occurred. The Mjølnir impact alone was definitely not large enough to trigger such a global extinction spike. The presently known three impact events from a short time-interval near the Jurassic-Cretaceous boundary (according to the impact data base it may be up to eleven), may be of great importance and capable to provide the threshold for biological extinctions at a global scale.
9. Ejecta material derived from an impact event may be found close to the crater. With increased crater size the ejecta are dispersed over increasingly larger distances at a regional, and even global, scale. The ejecta forms an unique stratigraphic marker and could be a valuable aid for Arctic correlation across one of the most heavily discussed Stage boundaries in the Northern Hemisphere.
10. The new project can supply information for quantification of post-impact modifications due to sediment loading. Such information is totally absent from the terrestrial impact record. Post-impact sediments can reach considerable thicknesses, and sediment loading above the primary impact relief may result in substantial post-impact deformation and structural modification. The proposed drilling, core analyses, and well-log measurements will contribute to the assessment of petrophysical properties within the impact structure. It will add to an understanding of post-impact operating compaction-processes and evolution, and to the identification and recognition of marine impact craters on sedimentary targets.

These ten technical and scientific targets and suggestions can be reached by drilling five to six, up 300 m deep core holes around the Mjølnir crater. During the workshop in Longyearbyen the following main scientific goal for such a drilling project was established; to map and understand ejecta formation and distribution, coupled with in situ disturbance of sediments due to seismic, shock waves, or erosion by displaced water near the crater. The cores will also be accompanied by sophisticated simulation models of ejecta formation and distribution in a marine environment (Dypvik et al. 2008a).

We will highlight the fact that this project aims at an integrated co-operation between the ICDP/IODP/ECORD/ESF organizations. This cooperation is important and needed due to the location of the Mjølnir crater in the central Barents Sea, and also because a program of obtaining high-quality cores from the crater and the surrounding ejecta will result in solving key scientific questions addressed by these programs and organizations.

1.7 Etymology

The Mjølnir impact structure has been named after the hammer of Thor, the god of thunder, in Nordic mythology. Thor was responsible for making thunder and lightning and he could even throw the hammer as a projectile at items he would destroy. The name Mjølnir was consequently suggested and applied as a name of this spectacular depression structure (Gudlaugsson 1993).

The Ragnarok Formation was chosen as a name of the main impactite and syn/post impact sedimentary formation in order to continue the mythology style of nomenclature (Dypvik et al. 2004c). Ragnarok means Armageddon in Norse or a catastrophe of devastating consequences, the complete collapse of the Earth. Although the Mjølnir impact was not of such a magnitude, the name seems well suited for the chaotic sediments as it characterizes the violent processes that lead to their deposition. The name Ragnarok Formation was used informally for a couple of years and formally acknowledged as the name of the crater infill material by the Norwegian Committee on Stratigraphy in 2004. Both the Ragnarok Formation and its lateral extension of ejecta, named the Sindre Bed (see Chap. 7), are included within the Adventdalen Group of Mørk et al. (1999) (see Chap. 2).

Sindre was selected as the name of the Mjølnir ejecta layer. Sindre is an old Nordic male's name, sounding like and forming the possible etymological root of the English word cinder. Sindre was the dwarf and blacksmith who made Thor's hammer; Mjølnir. Forging the hammer involved lots of sparks, fire, and glowing melts, which is exactly the meaning of the name Sindre, a proper name for air-borne ejecta from the Mjølnir crater.

Chapter 2
Geological Framework

Henning Dypvik, Morten Smelror, Atle Mørk, Filippos Tsikalas, Jan Inge Faleide, Stephanie C. Werner, and Trond H. Torsvik

2.1 Plate Tectonic Evolution of the Arctic

The Arctic Ocean comprises two main deep subocean basins, the Amerasia and Eurasia basins, separated by the elongate Lomonosov Ridge (Fig. 2.1). In a plate tectonic framework the Eurasia Basin is linked to the Norwegian-Greenland Sea and the Atlantic through the northernmost part of the Eurasia-North America plate boundary (Fig. 2.2). The plate boundary comprises two mid-ocean ridges, the Knipovich Ridge in the Norwegian-Greenland Sea and the Gakkel Ridge in the Eurasia Basin, linked by transform faults and oblique spreading segments in the Fram Strait (Engen et al. 2008). The passive margins flanking the plate boundary feature two marginal plateaus, the Morris Jesup Rise and the Yermak Plateau.

The Norwegian-Greenland Sea and the Eurasia Basin developed in a three-plate setting as Cretaceous-Paleocene rifts propagated northwards on either side of Greenland (e.g., Talwani and Eldholm 1972). By earliest Eocene time, seafloor spreading was established in both basins (Talwani and Eldholm 1972; Vogt et al. 1979), while relative motions between the Eurasia and Greenland plates were accommodated by a megashear region, the De Geer Zone, in the emergent ocean basin between Svalbard and Greenland (Harland 1969; Faleide et al. 1993, 2008). At the Eocene-Oligocene transition, Greenland became part of the North American plate and transform faulting in the Greenland Sea was replaced by oblique rifting and seafloor speading (Talwani and Eldholm 1972).

The opening of the Eurasia Basin split off a segment of the Eurasian continental margin that subsequently became the Lomonosov Ridge continental sliver (e.g., Grantz et al. 2001). The inner part of the Yermak Plateau is continental while its outer part and the Morris Jesup Rise have been attributed to voluminous volcanism at the Eurasia-Greenland-North America extinct triple junction (e.g., Jackson et al. 1984). Continental remnants in the Norwegian-Greenland Sea include the

---

H. Dypvik (✉)
Department of Geosciences, University of Oslo, NO-0316 Oslo, Norway
e-mail: henning.dypvik@geo.uio.no

H. Dypvik et al. (eds.), *The Mjølnir Impact Event and its Consequences*,
Impact Studies, DOI 10.1007/978-3-540-88260-2_2,

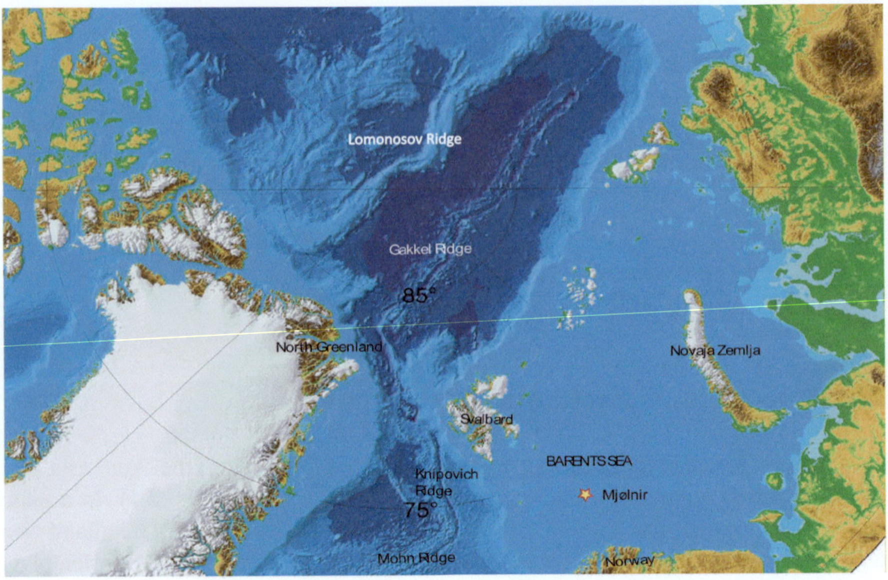

Fig. 2.1 Map of the Arctic, showing the location of the Mjølnir impact (*star*) between Svalbard and the mainland Norway. Modified from Jakobsson et al. (2003)

Hovgård Ridge microcontinent (Myhre et al. 1982) and, possibly, the Greenland Ridge (Tsikalas et al. 2002b; Døssing et al. 2008).

2.2 Mesozoic Stratigraphy and Depositional Environments of the Arctic

The Barents Shelf consists of several kilometers thick Paleozoic to Cenozoic sedimentary successions. From seismic lines crossing the Mjølnir crater it is evident that the impact disturbed strata down to, but not including the Upper Paleozoic strata (Figs. 1.4 and 1.10).

The Upper Palaeozoic succession (Larssen et al. 2005; Worsley 2008) comprises basal coarse clastics grading into Carboniferous carbonates and evaporites. Carbonate deposition with thick reef complexes continued into the Permian. Towards the Late Permian the northward drift of the land areas and the fusing of Laurasia and Siberia by closing the Uralian Sea, sealed the contact with the southern latitude seas. On this northern margin of the thereby formed Pangea supercontinent the northern ocean embayment changed from shallow to deep shelf setting within a cool climate. The abundant silica sponges resulted in cherty sediments forming the uppermost Permian succession. The top of this unit is visible as a pronounced seismic reflector throughout the Barents Shelf, and is seen as a marked ledge in the landscape on Svalbard. The reflector does not appear to be disturbed by the Mjølnir impact in the target area (see Chaps. 3 and 4).

Fig. 2.2 Tectonic map of the North Atlantic and the Barents Sea. The Mjølnir impact crater is located on the Bjarmeland Platform in the Barents Sea. The map outlines the Jan Mayen Fracture Zone in *dark red* (EJMFZ, WJMFZ) and the North Atlantic continental margin (HFZ, BFZ, MMH). Modified from Faleide et al. (2008)

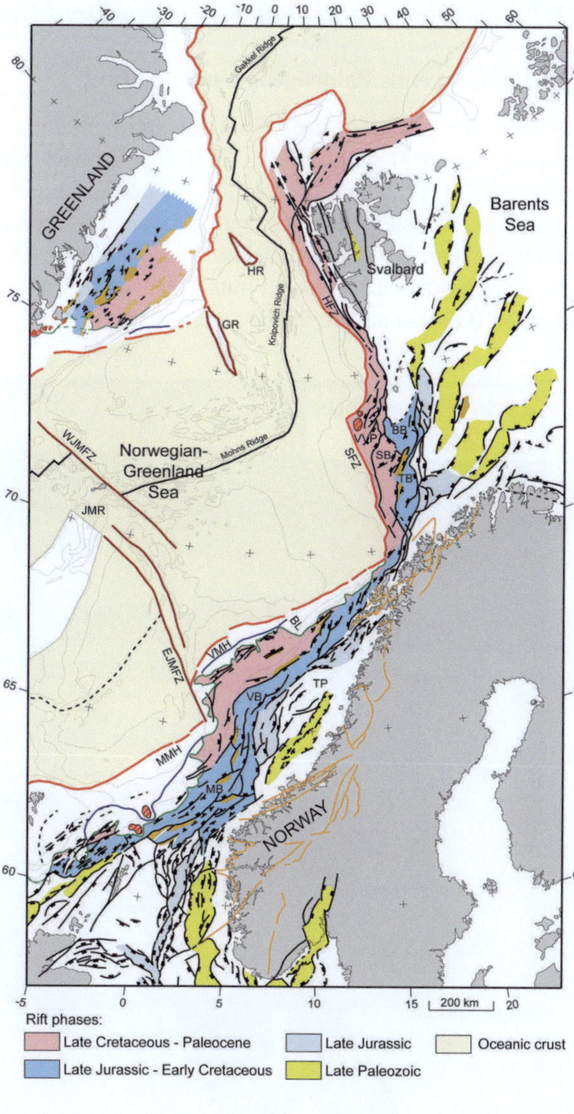

This Palaeozoic-Mesozoic boundary reflects the marked change from the silica cemented cherts to overlying clastic Triassic shales. The Lower and Middle Triassic succession, the Sassendalen Group is shale dominated, but also contains several coarsening upward sequences ending with sandstones landwards and towards the upper regressive part of each sequence (Mørk et al. 1999; Worsley 2008). Palynomorphs and re-deposited sediments from this group are found in the Mjølnir crater infill.

The Upper Triassic and Lower Jurassic succession (Kapp Toscana Group) is sandstone dominated. It represents shallow shelf to deltaic depositional environments (Worsley 2008; Riis et al. 2008). Re-deposited sandstone fragments from this group are abundant in the Mjølnir crater core.

2.2.1 Geological and Palaeogeographical Setting

2.2.1.1 Cretaceous Palaeogeographic Setting

A 142 million year plate reconstruction shows that the Barents Sea region was centered on the 55°N parallel (Fig. 2.3). The impact palaeolatitude is calculated to 56.4°N, at a time when Greenland also bordered and defined the western margin of the Barents Sea. Thus, and with relevance to tsunami modelling of the impact (compare Chap. 10), the distance to Greenland (ca. 300 km) was approximately the same as to Northern Norway (Finnmark), where waves as high as 100 m have been estimated (Glimsdal et al. 2007).

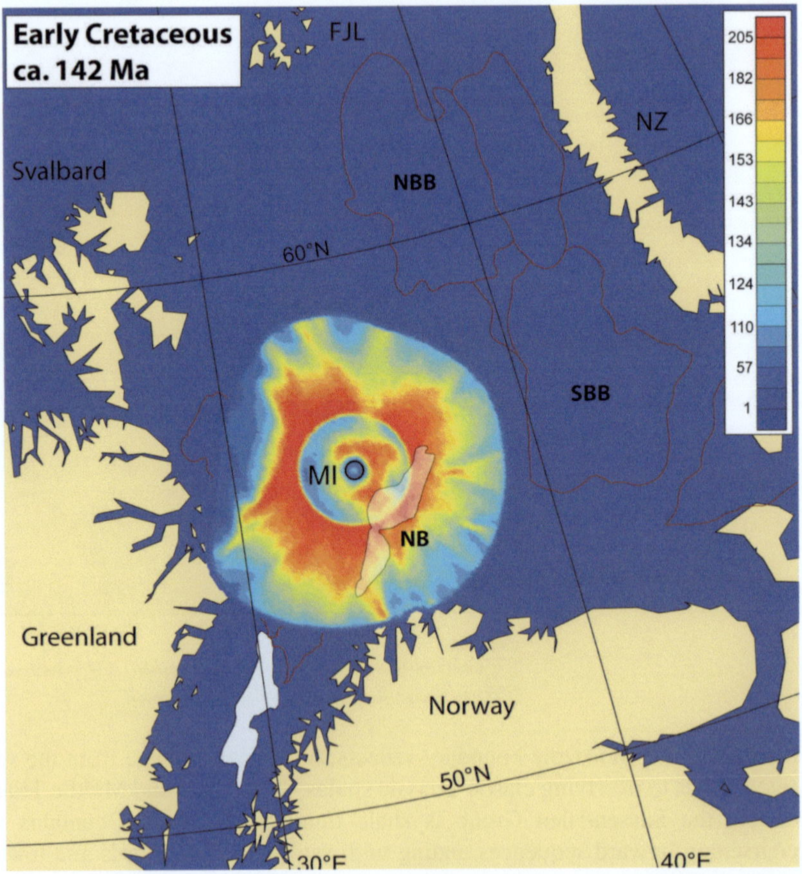

Fig. 2.3 The palaeogeographic setting of the Mjølnir impact site reconstructed to the Lower Cretaceous, at ca.142 Ma. The main differences with respect to the current setting are the palaeolatitude of 56.4°N and the then young and narrow Atlantic Ocean. The estimated tsunami wave height (after Glimsdal et al. 2007) and sedimentary basin outlines are draped on the reconstruction. FJL = Franz Josef Land, NZ = Novaya Zemlya, NBB = North Barents Basin, SBB = South Barents Basin, NB = Nordkapp basin, MI = Mjølnir impact crater

2.2.1.2 The Barents Sea in Time and Space

The Barents Sea realm has drifted northward over the past 300 million years, crossing tropic and sub-tropic latitudes as confirmed in the sedimentary record (Fig. 2.4). Fauna and sedimentary facies are sensitive to climate changes and to latitude. Distribution of evaporites, coal and certain carbonates are strongly latitudinally dependent. Evaporites are mostly deposited in arid sub-tropical regions, whilst coal is formed in wet equatorial regions or the northerly and southerly wet belts. This is well exemplified for the Barents Sea realm: 300 million years ago the Barents Sea

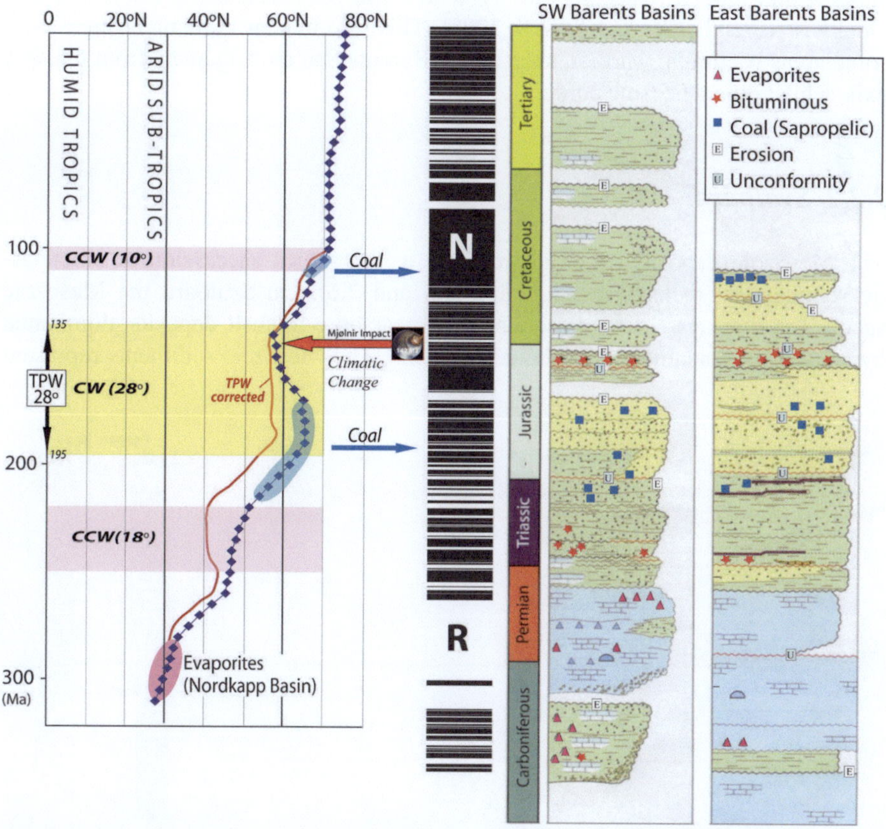

Fig. 2.4 The Barents Sea in space and time. Along with the stratigraphic column for the southwestern Barents Sea and the eastern Barents Sea basins, the magnetic polarity, the latitudinal drift of the Mjølnir impact site is shown. The drift curve is derived from a global apparent polar wander path with (*red curve*) and without (*blue curve*) corrections for true polar wander (TPW; Torsvik et al. 2008, Steinberger and Torsvik 2008). The TPW corrected curve shows a general northward movement of the Barents Sea realm (with respect to the mantle) but the flat Jurassic-Early Cretaceous section of the curve show that the apparent Mid-Late Jurassic southward movement (with respect to the spin-axis) is an artefact of TPW. The episodes of true polar wander are marked in *yellow* for clockwise (CW) and *reddish* for counterclockwise (CCW) motion

was at sub-tropical latitudes, a palaeolatitude that fits well with the occurrences of Late Carboniferous-Early Permian evaporites in the Nordkapp Basin. Subsequently the Barents Sea drifted northward, and by late Triassic to Jurassic time, we find coal witness that the region had entered the northern wet belt. We also show that the Barents Sea region drifted somewhat southward during Mid-Late Jurassic before continued northward drift (incidentally coinciding with the Mjølnir impact event) with coal once again appearing in the East Barents deposits (Fig. 2.4). These back-and-forth N–S movements (ca. five degrees in total) led to a climatic change, not because of "continental drift" but due to true polar wander (TPW). TPW is the rotation of the entire solid Earth's outer shell with respect to the spin-axis, and during Jurassic and Early Cretaceous time (195–135 Ma) the entire Earth rotated ca. 28° clockwise (Steinberger and Torsvik 2008). This lead to slow climatic changes with some areas becoming warmer, such as the Barents Sea moving away from the spin axis, whilst others became colder.

2.2.2 Svalbard

The Mesozoic deposits on Svalbard form a 3-km-thick succession of siliciclastic sedimentary rocks (Figs. 1.3, 1.4, 2.5 and 2.6). On Svalbard the Mesozoic succession displays repeated sequences of nearshore to shelf deposits, dominated by coastal progradation and deltaic sediments. The Mesozoic sediments represent

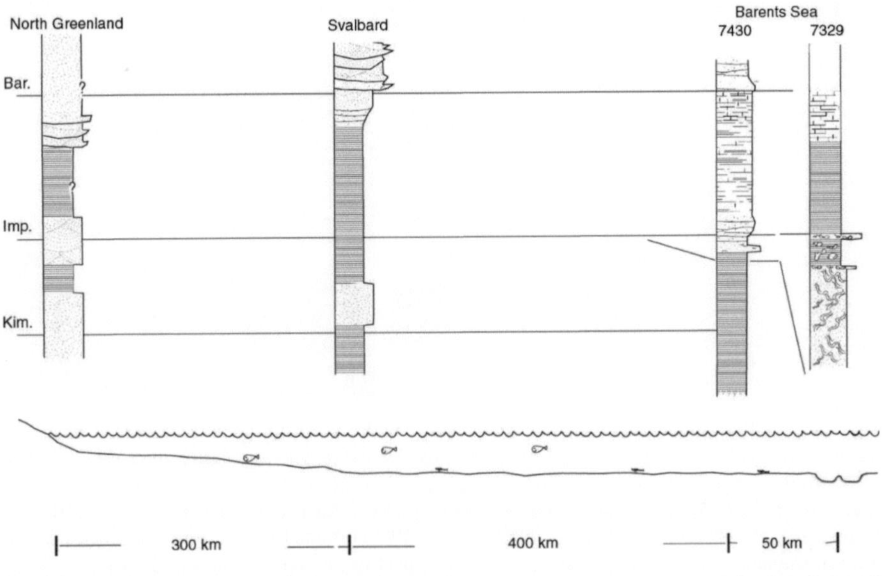

Fig. 2.5 Generalized stratigraphic correlations between Barents Sea, Svalbard and North Greenland close to the time of the Mjølnir impact. Bar. = Barremian, Imp. = Time of impact, Kim. = Kimmeridgian

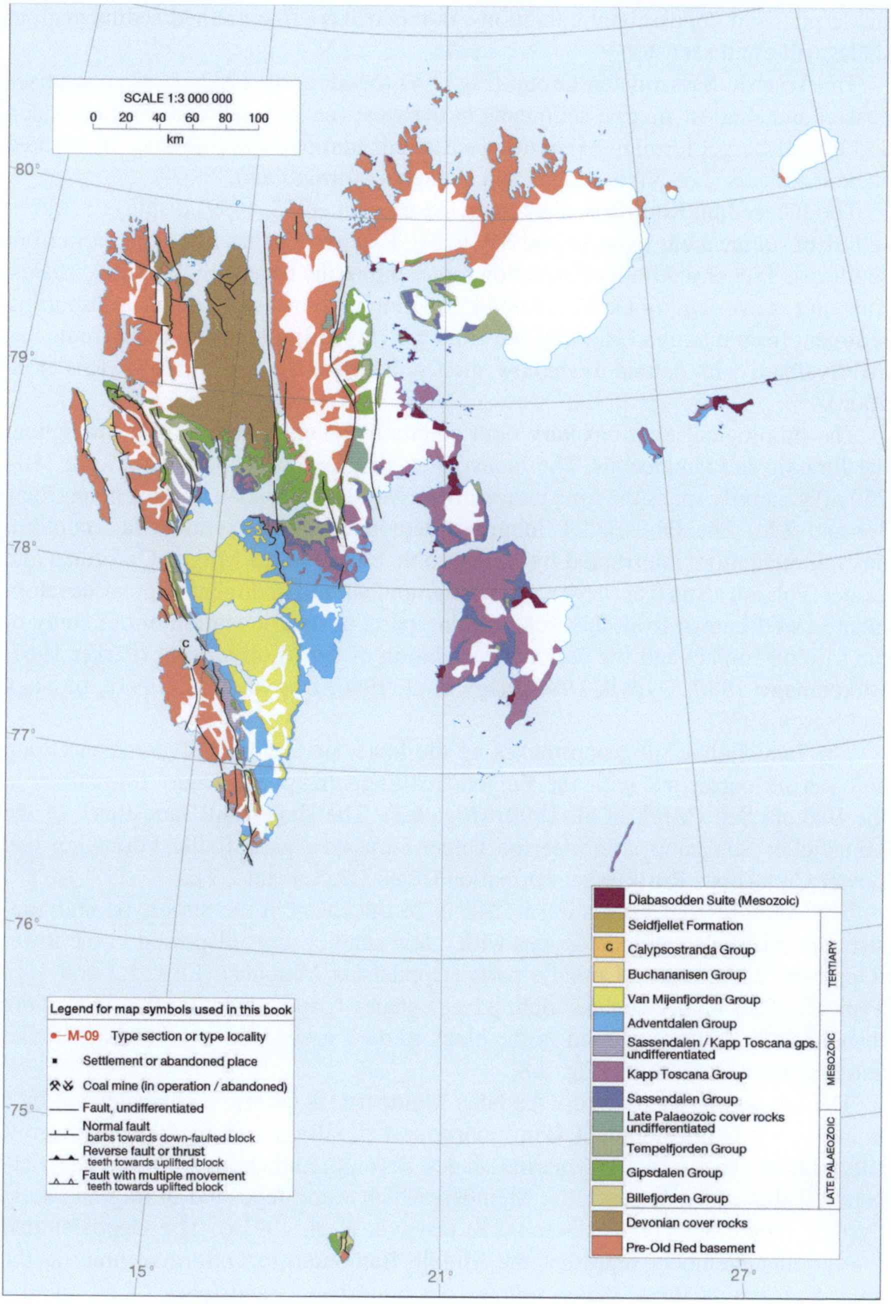

Fig. 2.6 Geological map of Svalbard (Dallmann et al. 1999)

stable platform depositional conditions, dominated by fine-grained sedimentation, shales, silt-, and sandstones.

The Triassic Sassendalen Group (Fig. 1.4) (Mørk et al. 1982, 1999) comprises coastal and shallow marine sediments in the west (i.e., Vardebukta, Tvillingodden and Bravaisberget formations) grading eastwards into open marine shale dominated shelf sediments (i.e., Vikinghøgda and Botneheia formations).

The succeeding Kapp Toscana Group (Mørk et al. 1982, 1999) displays a shift in sediment source areas from W and SW to SE. Extensive deltas filled the basin from southeast, Tschermakfjellet Formation representing the base with prodeltaic shales. This unit is overlain by De Geerdalen Formation with extensive deltas transporting sediment from mainland Norway and from the Urals filling most of the Barents Sea and Svalbard with deltaic to shallow shelf sediments (Mørk et al. 1982; Riis et al. 2008).

The lithological relations vary both qualitatively and quantitatively throughout the Jurassic and Cretaceous. The Janusfjellet Subgroup (thickness variations; 280–750 m) generally spans the time interval from Middle Bathonian to Barremian (Figs. 1.4 and 2.5). The Janusfjellet Subgroup demonstrates an overall, wide, continuous sedimentation, interrupted by stratigraphic breaks in the Middle Oxfordian and Upper Volgian (Smelror 1994). Its stratigraphical and sedimentological development is well known from the area, forming parts of the foundation of our study of the Mjølnir impact and the regional correlation of the Mjølnir event (Parker 1967; Birkenmajer 1980; Dypvik 1980; Nagy et al. 1990; Dypvik et al. 1991a, b; Nagy and Basov 1998)

The Janusfjellet Subgroup makes up the lower part of the Adventdalen Group and can be correlated with the Fuglen/Hekkingen/Klippfisk/Knurr formations in the Barents Sea (Mørk et al. 1999) (Fig. 1.3). The shales and sandstones of the Janusfjellet Subgroup comprise the Upper Jurassic Agardhfjellet Formation and Lower Cretaceous Rurikfjellet Formation (Figs. 1.3, 1.4 and 2.7).

The Agardhfjellet Formation is 242 m in thickness in the stratotype and consists of dark shales and claystones with a few sandier units dispersed in the lower (Oppdalen Member) and middle parts (Oppdalsåta Member) (Figs. 1.3 and 1.4). Typically also highly organic rich, papery shales (paper shales), often with more than 10 wt% TOC, are found in the black to dark grey shale units of Lardyfjellet and Slottsmøya members (Fig. 2.8).

The lowermost sandy unit, Oppdalen Member (10–60 m), is characterized by a fining upwards development from poorly sorted, silty sandstones into dark grey, silty shales. These shallow marine shales develop further into the organic rich, paper shales of the Lardyfjellet Member, which were deposited during anoxic to dysoxic conditions (Figs. 1.3 and 2.8) (Dypvik et al. 1991a). The Oppdalen and Lardyfjellet members represent the Middle Bathonian to Oxfordian units in the area. Succeeding these shales follows the bioturbated sandstones of the overlying Oppdalsåta Member. These sandstones and sandy shales are, on the average 28 m in thickness, mainly forming coarsening upwards successions of Oxfordian to Kimmeridgian age. The Oppdalsåta Member has been interpreted by Dypvik et al. (1991b) to represent storm deposits, formed in the wide, epicontinental

Fig. 2.7 The Janusfjellet Subgroup at Wimanfjellet, Svalbard. The Janusfjellet Subgroup, Agardhfjellet Formation, Rurikfjellet Formation, and Helvetiafjellet Formation are marked (Photo: Jenö Nagy)

Fig. 2.8 Close up (photo about 30 cm in height) photos of paper shales from Agardhfjellet Formation, Wimanfjellet

paleo-Barents Sea. The vivid sea floor life returned to the region immediately after the stormy period of the Oppdalsåta Member and disturbing these sandy formations completely. Consequently, the coarsening upwards Oppdalsåta sandstones are characterized by homogeneous appearance, due to severe post-storm bioturbation. The Oppdalsåta Member is today rich in parallel and current oriented belemnites, but poor in sedimentary structures other than the signs of thorough bioturbation.

The uppermost part of the Jurassic succession is made up of the dark grey to black, organic rich shales of the Slottsmøya Member. It often reaches more than

100 m in thickness and consists dominantly of paper shales, with varying degrees of fissile ability.

In central Spitsbergen the Agardhfjellet Formation is succeeded by the Myklegardfjellet Bed (mainly Ryazanian in age), which is 0–11 m in thickness, and forms the base of the succeeding Rurikfjellet Formation. The Myklegardfjellet Beds consists of soft, often yellowish to greenish plastic clays commonly rich in dolomite, pyrite and sporadic altered glauconites present in separate layers. The Myklegardfjellet Bed deposits may represent the transgressive start of a flooding event, marking the transition from shallow shelf to relatively deep mid-shelf conditions in the succeeding Wimanfjellet Member. The depositional development continues into the regressive prodeltaic deposits of the uppermost part of the Rurikfjellet Formation (Figs. 1.3 and 1.4) (Dypvik et al. 1991a).

The Rurikfjellet Formation (Wimanfjellet and Ullaberget members) often reaches more than 200 m in thickness and is composed of dark grey, commonly silty shales. Increasing amounts of siltstone and sandstone occur in its upper part, commonly in upwards coarsening units. The lowermost part of the Rurikfjellet Formation, the Myklegardfjellet Bed (Fig. 2.9), is succeeded by the very fine claystones of the Wimanfjellet Member. They form the finest grained unit of the

Fig. 2.9 The Myklegardfjellet Bed is located in between the Slottsmøya Member (*below*) and the Wimanfjellet Member above at Glitrefjellet (Reindalen, Svalbard). The possible Mjølnir impact level is located at person's hand

Fig. 2.10 The Ullaberget Member of the Rurikfjellet Formation at Aasgaardfjellet, with a few sandstone beds is seen in the upper photo (persons for scale). The lower photo is a close up of a hummocky cross stratification unit (HCS) found in these beds

Janusfjellet Subgroup, representing a relative deep shelf setting at maximum flooding stage. The smectitic claystones of the Wimanfjellet Member are succeeded by the general coarsening upwards successions of the Ullaberget Member, 155 m in thickness in the stratotype (Fig. 2.10). In contrast to the mid-shelf facies of the Wimanfjellet Member the Ullaberget Member represents prodeltaic depositional environments, sedimentologically related to the succeeding deltaic deposits of the Helvetiafjellet Formation.

The Barremian Helvetiafjellet Formation varies from 40 to 155 m in thickness (Figs. 1.3 and 2.11). It consists in the lower part of the coarse sandstones of the pronounced Festningen Member (up to 16 m in thickness), which is covered by coal-bearing successions. The Helvetiafjellet Formation represents complex transgressive interacting fluvial, delta plain, mouth bar, barrier bar, tidal estuary and transgressive sheet sandstone facies succeeding a period of relative sea level fall (Gjelberg and Steel 1995; Mørk et al. 1999; Midtkandal et al. 2007). The Festningen Member is a fluvial dominated delta complex prograding from the northwest. It correlates with the lower part of the Kolmule Formation of the Barents Sea. The Helvetiafjellet Formation displays an overall transgressive trend (Gjelberg and Steel 1995), but contains numerous punctuated regressive pulses within its general transgressive development.

The Aptian–Albian Carolinefjellet Formation has large thickness variations, from 190 m in the northern areas to more than 1,200 m in the southeastern part of Svalbard. The Carolinefjellet Formation consists of alternating shallow marine

Fig. 2.11 The Helvetiafjellet Formation, the Festningen Sandstone Member as it appears at Lardyfjellet in the overview photo to the *left* (sandstone unit about 6 m high). Well-developed cross-bedding in the sandstones of the same locality is shown in the *right* photo

shales and sandstones, reflecting prodeltaic to distal marine depositional conditions. It correlates with the upper part of the Kolmule Formation in the Barents Sea (Fig. 1.3).

The Cenozoic beds of Svalbard span the time interval from Late Paleocene to Oligocene, and represent cyclic deposition in a foreland depression (Dallmann et al. 1999). The succession consists of intermixed continental and marine clastics. A thick pile of more than 2,500 m of Cenozoic sediments covered large parts of the central and southern Svalbard. Maturation studies of organic matter indicate that an even 1,000 m thicker original Tertiary succession has been present in Svalbard (Manum and Throndsen 1978).

2.2.3 Barents Sea

In the Barents Sea the Lower Triassic sediments show moderate deep-shelf facies consisting of mixed shales and sandstones assigned to the Havert Formation (Worsley et al. 1988) (Fig. 1.4). It is correlated with the Vardebukta Formation and lower part of the Vikinghøgda formations on Svalbard (Mørk et al. 1999). This unit is succeeded by the sandstone, siltstone and shales of the Klappmyss Formation (Worsley et al. 1988) correlated to the Tvillingodden Formation and upper part of Vikinghøgda Formation on Svalbard (Mørk et al. 1999).

The Middle Triassic succession comprises claystones and sandstones of the Kobbe Formation (Worsley et al. 1988), however in central part of the Barents Shelf, it is replaced with the organic-rich Steinkobbe Formation (Mørk and Elvebakk 1999). The Kobbe Formation dominates on the eastern part of the basin and is an equivalent of the Bravaisberget Formation on western Svalbard. The time equivalent Steinkobbe Formation is an equivalent of the Botneheia Formation of central and eastern part of Svalbard (Mørk et al. 1999).

The Late Triassic uplift of the area north of Svalbard and the Norwegian–Kola land areas resulted in dominantly sand deposition in the paleo-Barents Sea. The

Snadd Formation mainly got its sediment supply from the south and southeast and form extensive delta front to delta top deposits, equivalent to the De Geerdalen Formation of Svalbard (Worsley 2008; Riis et al. 2008). Both on Svalbard and in the Barents Sea the uppermost Triassic and Lower/Middle Jurassic succession consists of extensively reworked sandstones; i.e. the Wilhelmøya (on Svalbard) and Realgrunnen (in the Barents Sea) subgroups, respectively. In the Barents Sea, it is formed by the deltaic to shallow marine Fruholmen Formation and the overlying shallow marine to coastal Stø Formation (Worsley et al. 1988; Worsley 2008).

The Mesozoic sediments of the Barents Sea, the Adventdalen Group in particular, can be recognized and match with similar deposits in Svalbard (Figs. 1.3, 2.5 and 2.6). These correlations are as follow; the Fuglen Formation correspond to the Oppdalen and Lardyfjellet members of the Agardhfjellet Formation, the Hekkingen Formation can be correlated to the Oppdalsåta and Slottsmøya members, while Klippfisk and Knurr formations are equivalent to the lowermost part of the Rurikfjellet formation (Wimanfjellet Member). The Kolje Formation can be correlated to the Ullaberget Member of Svalbard. The Kolmule Formation of the Barents Sea can be match the Helvetiafjellet and Carolinefjellet formations of Svalbard (Figs. 1.3, 1.4 and 2.5).

In the southeastern Barents Sea the Fuglen Formation varies between 20 and 50 m in thickness, but it reaches more than 200 m in the Troms III area (Worsley et al. 1988). The formation is of Late Bathonian to Middle Oxfordian age. It consists of mudstones and minor limestones and represents marine shelf deposits.

In the Barents Sea the Hekkingen Formation, which is of Late Oxfordian to Ryazanian age, has a thickness varying between 110 and 400 m. It consists of mostly dark grey to black shales, with some silt and sandstone beds dispersed. The Alge and Krill members form the lower and upper parts of the Hekkingen Formation, respectively. Both these units are dominated by dark grey, to black organic rich shales, which is reflected in the names of the two members (Figs. 1.3 and 2.12). In the Barents Sea exploration wells the Fuglen Formation is recognized by its characteristic log response, in particular high gamma activities (Dypvik et al. 2004). The highest gamma response is, however, found in the Alge Member of the Hekkingen Formation and correlate with increased uranium concentrations in this organic rich member with TOC values up to 12%.

The Lower Cretaceous Hekkingen Formation is succeeded by the Lower Cretaceous Knurr and Klippfisk formations. The Knurr Formation is a Volgian to Barremian unit, consisting of mudstones, sandstones and limestones. The formation represents distal marine shelf depositional conditions, with only locally restricted bottom ventilation. The Knurr Formation varies from 56 to 285 m in thickness. The time-equivalent Klippfisk Formation comprises condensed carbonate deposits preserved on structural highs and platforms (Fig. 2.13). These carbonate platform deposits have been found as far east as in the Olga Basin and it outcrops on Kong Karls Land on Svalbard (Smelror et al. 1998). The Klippfisk Formation is thin, generally only from 4 to 15 m in thickness and consists of marls and limestones of Berriasian to Hauterivian age.

Fig. 2.12 The Alge member of the Hekkingen Formation from the Barents Sea. IKU (Continental Shelf Institute Norway) core 7430/10-U-01. The finely laminated dark grey shales are seen disrupted by very thin silt laminae and a large pyrite concretion in the upper part of the core

In the Barents Sea the succeeding Kolje Formation, which is between 15 and 403 m in thickness, consists mainly of shales and mudstones representing distal, open marine depositional conditions with moderate water circulation. It forms a transgressive phase on top of the Klippfisk/Knurr formations couplet.

2.2.4 Greenland

In North Greenland, the Mesozoic Wandel Sea Basin comprises a succession with Lower Carboniferous to Paleogene sediments. This succession was deposited on the margin of the stable Greenland craton where the Caledonian and Ellesmerian orogenesis intersected. On top of these metamorphic Precambrian to Silurian formations Upper Paleozoic to Lower Mesozoic sedimentary sequences are found in an extensional setting, with four main tectonic events recognized: one in mid-Jurassic, one in mid-Cretaceous, one at the end of Cretaceous and one post-Paleocene extensional event. This last event took place after the tectonic shift of the plate boundary towards its present position between Greenland and Svalbard (Håkansson et al. 1993). Accumulation of Carboniferous and younger sediments postdate the Caledonian and Ellesmerian Orogeny. In Kilen a more than 3 km-thick succession

Fig. 2.13 The core piece shows bioturbated carbonates from the Klippfisk Formation. It was cut from IKU (Continental Shelf Institute Norway) core 7430/10-U-01

of such sediments are found, including about 1,400 m of Upper Triassic, Jurassic, and Cretaceous deposits (Fig. 1.3).

Jurassic formations are widely distributed in East Greenland, but are only found in two major sites in North Greenland: Kilen and East Peary Land. These North Greenland locations are recognized by coastal and shallow marine clays and well sorted cross-bedded sands (Håkansson et al. 1994; Heinberg and Håkanson 1994; Dypvik et al. 2002). In particular, the Lower Cretaceous sand units are well developed; as seen in the white, well sorted, sandstone formations of the Ladegårdsåen (East Peary Land) and the Lichen Ryg (Kilen) formations (Fig. 1.3). In both areas, the uppermost part of the Jurassic succession consists of dark grey, organic rich shales, commonly containing plant fragments (Fig. 2.14). The successions comprise upward coarsening, 5–30 m thick sequences, covering the Oxfordian to Valanginian time interval.

In the Late Jurassic depositional models of Dypvik et al. (2002), the North Greenland area forms the southwestern coastline of the palaeo-Barents Sea, with a possible opening towards the Canadian Sverdrup Basin (Figs. 2.15 and

Fig. 2.14 Upper Jurassic dark grey shales of the Dromledome Formation in Kilen, North Greenland. Thin siderite beds are found throughout the formation, see, e.g., one such bed just *below* the person

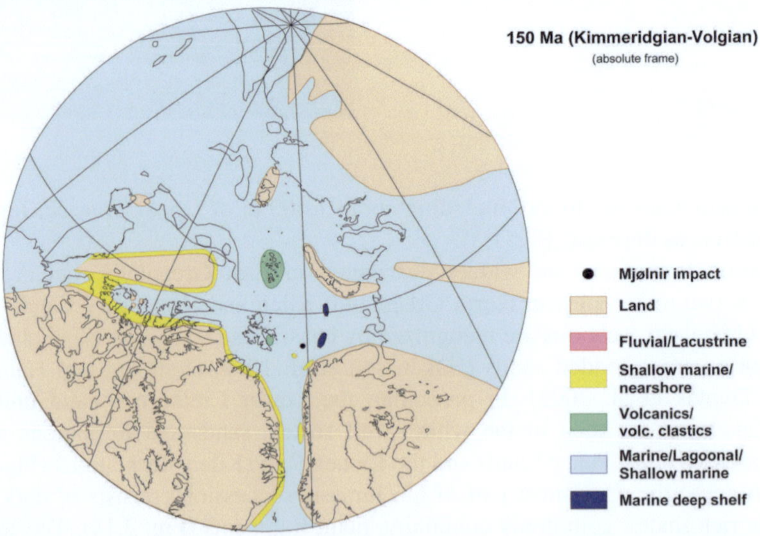

Fig. 2.15 The paleogeographic setting during the Kimmeridigan-Volgian (~150 Ma) when the Mjølnir bolide hit the paleo-Barents Sea is shown. Based on the plate reconstruction of Lawver et al. (1999)

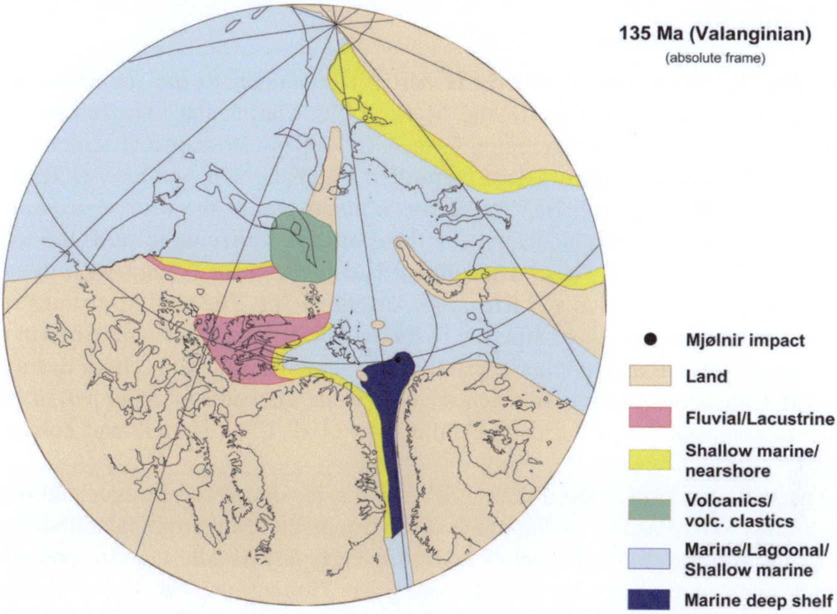

Fig. 2.16 The paleogeographic setting during the Valanginian (~135 Ma) is shown. The possible opening/deep water connection through the pre-North Atlantic are discussed. The reconstruction is based on Lawver et al. (1999)

2.16). This opening was closed in Early Cretaceous time in connection with the northerly uplift/doming of the region, most likely related to breakup in the Amerasia Basin and associated regional magmatism within an Arctic large igneous province (LIP) and formation of the Alpha Ridge (Grogan et al. 1998; Maher 2001). The doming was initiated in Valanginian time and continued through the Early Cretaceous. Valanginian to Hauterivian/Barremian regressive sequences are developed in North Greenland, but comparable units are also evident in the Sverdrup Basin (Canada) (Embry 1991) and Svalbard (Helvetiafjellet Formation).

From the early Eocene (53.3 Ma), sea floor spreading took place along the Gakkel Ridge in the Arctic Ocean (Eurasia Basin) and the Mohns Ridge in the Norwegian–Greenland Sea, accompanied by strike-slip movements between Svalbard and North Greenland (Skogseid et al. 2000; Eldholm et al. 2002; Engen et al. 2008) (Figs. 2.1 and 2.2). The strike-slip movements (and associated opening of the southern Greenland Sea) continued to the Eocene/Oligocene transition, connecting the spreading basins of Arctic Ocean and Norwegian Greenland Sea. At this point in time (33.3 Ma) the movements between Svalbard and NE Greenland shifted towards oblique extension, and the Fram Strait deep-water gateway opened by sea floor spreading in the Miocene (Lawver et al. 1990; Faleide et al. 1993; Eldholm et al. 1994; Torsvik et al. 2001; Engen et al. 2008). Since then, spreading has taken place along this major lineament with final establishment of the present seafloor spreading regime at about 10 Ma.

2.2.5 Siberia

The Triassic succession of Siberia is partly comparable to the time-equivalent Barents Sea succession (Egorov and Mørk 2000). During the Late Jurassic and Early Cretaceous the Siberian region formed a stable biogeographical ecotone with faunistic changes to a large extent controlled by changes in sea-level (Zakharov and Rogov 2003). The Siberian Arctic was a major part of the wide, epicontinental paleo-Arctic Sea, which consisted of three branches opening towards the south (Zakharov et al. 2002). These depositional basins are named, from west to east; the Pechora, West Siberian and Khatanga/Anabar basins. The Ural Mountains separated the Pechora and West Siberian basin, which in turn was separated from the Khatanga/Anabar Basin by the Central Siberian landmass. In the Pechora Basin the Jurassic-Cretaceous succession reaches a composite thickness of about 370 m (120 m Cretaceous and 250 m Jurassic) (Malinovsky et al. 1999; Dypvik and Zakharov 2010).

The sedimentation in the different Siberian basins during the Jurassic and Early Cretaceous took place in a wide range of environments from proximal alluvial, via lacustrine swamps, through shallow marine in Early and Middle Jurassic, and dominating open marine in Late Jurassic and Early Cretaceous. The Upper Jurassic, mainly marine clays, were deposited in the central regions of the basin, and are presently found as black and grey organic rich shale formations (Figs. 2.17 and 2.18). Anoxic bottom water conditions dominated in wide central regions, while coarser grained, sand sedimentation took place along the more ventilated margins of the basins. Finely laminated organic-rich, black shales, without any visible bioturbation, dominated in the basin central areas of the Siberian basins (Zakharov and Rogov 2003). Marginally, carbonate cemented sandstones are commonly found. These are often enriched in phosphates and glauconite. Generally, the amount of marine indicators increases from south to north and towards the opening to the paleo-Arctic seas (Malinovsky et al. 1999).

The prolific hydrocarbon source rock of the Bazhenov Formation in the West Siberian Basin (Volgian to Berriasian in age) is made up of black to brown, organic rich shales (Gavshin and Zakharov 1996; Zakharov et al. 1998). It was deposited during a period of 5–6 million years and it is on the average about 25–30 m in thickness. The formation represents a basinal depositional facies, covers more than 1 million km^2, and is normally buried beneath 2,000–3,000 m of younger sediments. It contains on the average weight 8% TOC and can be correlated to the Hekkingen Formation of the Barents Sea (Fig. 2.12). The Bazhenov shales contain less than 5 weight % of sand and silt, but the formation displays high concentrations of biogenic silica (Zakharov et al. 1998). In the northern, more ventilated environments of the Siberian basins towards the wide Arctic basin, several fossil species similar to the ones discovered in Greenland and Svalbard have been recognized (Zakharov 2004, personal communication).

Marine shelf sedimentation continued into the Early Cretaceous, but at that time the subsidence was reduced and consequently the basins were gradually filled up. This regressive trend dominated in the Early Cretaceous, but was interrupted by

Stage	Substage	Barentz-Kara Plate	Svalbard			Nordvik peninsula				
		Beds with foraminifera	Beds with foraminifera	Foraminiferal zones		Beds with foraminifera		Ammonite zones and subzones		Buchia zones
Berriasian	Upper	?	Assemblages with Recurvoides obskiensis, Gaudryina gerkei, Lenticulina sossipatrovae	Gaudryina aff.milleri	?	Recurvoides romanovae	Recurvoides obskiensis, Valanginella tatarica	Bojarkia mesezhnikowi		Buchia inflata
					Gaudryina rostellata					B. tolmatschewi
							Gaudryina gerkei, Ammobaculites gerkei	Surites analogus	Surites analogus	
									Surites subquadratus	Buchia jasikovi
	Middle	Recurvoides obskiensis Orientalia(?) baccula						Hectoroceras kochi	Surites praeanalogus	Buchia okensis
									Borealites constans	
									Hectoroceras kochi	Buchia unshensis
	Lower			Recurvoides obskiensis	Recurvoides obskiensis		Ammodiscus veteranus Evolutinella emeljanzevi / Evolutinella emeljanzevi	Chetaites sibiricus	Chetaites sibiricus	
									Praetollia maynci	
Volgian	Upper	Ammodiscus veteranus Evolutinella emeljanzevi	Ammodiscus veteranus Evolutinella emeljanzevi	Trochammina aff.abrupta	Calyptammina praegyroidiniformis	Trochamminoides emeljanzevi - Recurvoides praeobskiensis		Chetaites chetae		
								Craspedites taimyrensis		
								Craspedites okensis	Craspedites originalis	Buchia obliqua
							Ammodiscus veteranus		Craspedites okensis	
									Praechetaites exoticus	

Fig. 2.17 The figure shows a comparison of Siberian and Barents Sea Volgian and Berriasian detailed biostratigraphy. It is based on Dypvik and Zakharov (2010)

Fig. 2.18 Black shales exposed on the Nordvik Peninsula, West Siberia. The *white flags* showing a phosphate concretionary layer near the Jurassic/Cretaceous boundary. (Photo Victor Zakharov)

several minor transgressive episodes. In the south, deposition of anoxic, organic rich clays still took place in Early Cretaceous time, but increased ventilation in the water masses terminated the organic rich clay deposition.

2.2.6 Late Jurassic and Early Cretaceous Depositional Configuration

The Late Jurassic (Kimmeridgian – Volgian) and Early Cretaceous (Valanginian) paleogeographic reconstructions (Figs. 2.15 and 2.16) of Barents Sea, Svalbard, North Greenland and Siberia are based on our own field studies and detailed literature surveys. The plate tectonic reconstructions are founded on Lawver et al. (1999).

During the Mesozoic era, the Barents Sea region was part of an extensive epicontinental sea on the northern margin of the Pangea supercontinent, with wide and shallow branches stretching into the present Siberia, the Pechora, West Siberian and Khatanga/Anabar basins (Figs. 2.15, 2.16 and 2.17).

In this Late Jurassic epicontinental sea the average water depth was from 300 to 500 m and mainly low lying land areas surrounded the basins (Fig. 2.15). Within the distant west and northwest, the Arctic basin widened and opened towards the Panthalassa Ocean (paleo-Pacific). In the Arctic basin shallow marine, clastic sedimentation dominated during the Jurassic, as elaborated in the stratigraphical presentation above. The low relief sea floor was disrupted by a limited number of topographic highs, where coarser grained sedimentation took place. Coarse-grained sedimentation also happened along the coastlines and in the near-shore environments of this extensive basin. Eustatic sea level changes controlled the overall major depositional conditions, while heavy winds and storms triggered periods with sand deposition in this normally rather calm, epicontinental sea. Consequently, the general sedimentation in the basin was dominated by deposition of fine grained clays, with sparse amounts of sand and silt. The low relief surrounding land areas suffered rather heavy chemical weathering and dominating well-weathered dissolved and fine-grained material was transported into the basin. The Barents Sea region was at that time located in a position of about 50° N-latitude and clearly experience more chemical weathering and both warmer and more humid conditions than of today. Both along the coastal margins of the mainland of Norway and the Greenland region of that time, shallow marine and highly reworked, well weathered quartzitic sands were deposited. In periods and regions with reduced clastic sedimentation, carbonates and organic matter dominated.

Due to the wide and shallow bathymetry of this extensive Late Jurassic-earliest Cretaceous sea, the total organic production must have been substantial. In the water-masses large amounts of planktonic algae thrived and provide good living conditions for a variety of animals higher up in the food-chain. This is partly reflected in periods with high accumulation of fossil skeletal fragments from ammonites, belemnites, mollusks and marine reptiles such as ichtyosaurus and plesiosaurs, which are commonly found preserved along shallow banks and highs in this sea. The *Dorsoplanites* Beds of Svalbard is a good example of such skeletal accumulations (Fig. 2.19) (Dypvik et al. 1991a).

The sedimentation and preservation of organic matter was prolific, as shown on Svalbard by the black shales of the Janusfjellet Subgroup, in the western Barents Shelf by the Hekkingen Formation and in the Timan-Pechora region by the organic rich Bazhenov Formation. In the epicontinental sea, high algal production took place, while rich vegetation must have been present in the surrounding land areas. Large parts of this organic matter were eventually deposited on the sea floor. Due to lack of global glaciations in the Late Jurassic, oceanic circulation was reduced and only sporadic storms existed. The limited ocean-circulation resulted in rather sluggish to partly stagnant bottom water conditions on the Barents Shelf. Consequently, deposited organic matter was rarely altered or oxidized, and came to make up a large part of the bottom sediments. Some of these dark, organic rich clays (e.g., parts of the Hekkingen Formation) have been found to contain more than 20 wt% total organic carbon (Dypvik et al. 1991a; Leith et al. 1992).

The wide paleo-Arctic seaways in the Jurassic provided a good basis for establishing inter-basinal stratigraphical correlations, since comparable beds were deposited over wide areas and the deposition in the Circum-Arctic basins was mainly controlled by sea level changes (Figs. 1.3, 2.5 and 2.16) (Mørk and Smelror

Fig. 2.19 Ammonites in the Dorsoplanites Bed of the Agardhfjellet Formation at Knerten, near Wimanfjellet, on the south shore of Isfjorden, Svalbard. The ammonite is *Dorsoplanites maximus* of Middle Volgian (Photo: Hans Arne Nakrem)

2001). Consequently the Circum-Arctic geological successions, we study today contain several stratigraphical markers that can be traced across wide areas. In the Mjølnir case, this makes the search, discovery and dating of ejecta layers possible.

Figure 2.15 shows the Kimmeridgian and Volgian palaeogeographic architecture of the Arctic. The Oxfordian setting was overall transgressive. This transgressive development and widening of the epicontinental sea is evident in the Canadian Arctic (Balkwill 1978; Embry 1991) and in the Siberian basins. The evolution may be related to the Late Jurassic opening of the North Atlantic and the initial sea floor spreading. The Valanginian reconstruction (Fig. 2.16) displays regressive developments into the Early Cretaceous of the Canadian and central Arctic basin, while the deepening of the North Atlantic takes place in combination with a tectonic uplift of the northern paleo-Barents Sea.

The Late Jurassic to Early Cretaceous setting with widespread fine-grained sedimentation lasted for 15–20 million years. The succeeding Middle and Upper Cretaceous formations generally is characterized with several additional stratigraphic breaks and non-continous sequences, and more coarse grained sedimentation filling up the subsiding basins.

Early Cretaceous (Barremian-Aptian) volcanic activity was widespread on Kong Karls Land and Franz Josef Land and the offshore surroundings areas. The volcanics are represented by subalkaline tholeiites (Amundsen et al. 1998). In the Svalbard and the northern Barents Sea dolerites of Late Jurassic to Early Cretaceous ages are present (Halvorsen 1989; Halvorsen et al. 1996).

The paleogeographic setting during the Volgian-Ryazanian boundary times when the Mjølnir bolide hit the paleo-Barents Sea is shown in Fig. 2.15. The bolide approached the target site at an angle of 45° from the south/southeast (Tsikalas 2005). No doubt that the 1.5–2 km in diameter bolide created a major disturbance in this Jurassic/Cretaceous world. The effects of the impact, its related mass flows and great tsunamis are described in detail in the following chapters of this book.

During the 20–25 million years time-span from the beginning of Oxfordian to end of the Valanginian several impactors should have hit (based on periodicity and statistical estimates) the extensive paleo-Greenland –Barents Sea – Eastern Siberian epicontinental seas. Numerous tsunamis should have formed, and their traces preserved in the sedimentary record, e.g., as coarse grained and exotic deposits with sharp erosional features along the palaeo-coasts lines. In the deeper, more central parts of the basin, it may be difficult to see tsunami effects, in contrast to the shallow banks or platforms of the Arctic Sea. These deeper regions could consequently be well suited for crater search, both due to the low and rather continuous accumulation rates of fine-grained clastics and the well developed fossil floras and faunas making detailed stratigraphical correlation possible.

Chapter 3
Impact Structure and Morphology

Filippos Tsikalas, Jan Inge Faleide, Steinar Thor Gudlaugsson, and Olav Eldholm

The Mjølnir crater (Fig. 3.1) was first interpreted as an impact structure by Gudlaugsson (1993) based on its geophysical signature and overall geological setting. This inference was derived from a limited amount of multichannel seismic profiles, and regional gravity and magnetic profiles. The impact hypothesis prompted the acquisition of high-resolution seismic, gravity and magnetic profiles by the Norwegian Defence Research Establishment in 1992 and 1993. Together with the previously acquired shallow and conventional multichannel seismic profiles, these data comprise an extensive and unique geophysical database (Fig. 3.2 and Table 3.1). In addition, stratigraphic, and sedimentological information exists from two shallow boreholes, one near the center (7329/03-U-01) and another ~30 km north-northeast from the crater periphery (7430/10-U-01) (Fig. 3.2). The detailed seismic reflection correlation to the two boreholes, together with the existing regional grid of seismic profiles on the Bjarmeland Platform, have made it possible to correlate the main seismic sequence boundaries at Mjølnir to the established stratigraphic framework of the Barents Sea (Worsley et al. 1988; Gabrielsen et al. 1990; Richardsen et al. 1993; Smelror et al. 2002).

3.1 Seismic Reflection Database

The seismic reflection method is the most powerful geophysical method in sedimentary targets, where the pre-impact stratification provides a series of reference horizons for the large-scale impact induced structures to be identified and mapped, e.g., at Montagnais (Jansa et al. 1989; Pilkington et al. 1995), Manson (Keiswetter et al. 1996; Hartung and Anderson 1996), Chicxulub (Morgan and Warner 1999; Morgan et al. 2002b), and Chesapeake Bay (Poag 1996; Poag et al. 2004) craters.

F. Tsikalas (✉)
Eni Norge AS, NO-4064 Stavanger, Norway; Department of Geosciences, University of Oslo, NO-0316 Oslo, Norway
e-mail: filippos.tsikalas@eninorge.com

H. Dypvik et al. (eds.), *The Mjølnir Impact Event and its Consequences*,
Impact Studies, DOI 10.1007/978-3-540-88260-2_3, © Springer-Verlag Berlin Heidelberg 2010

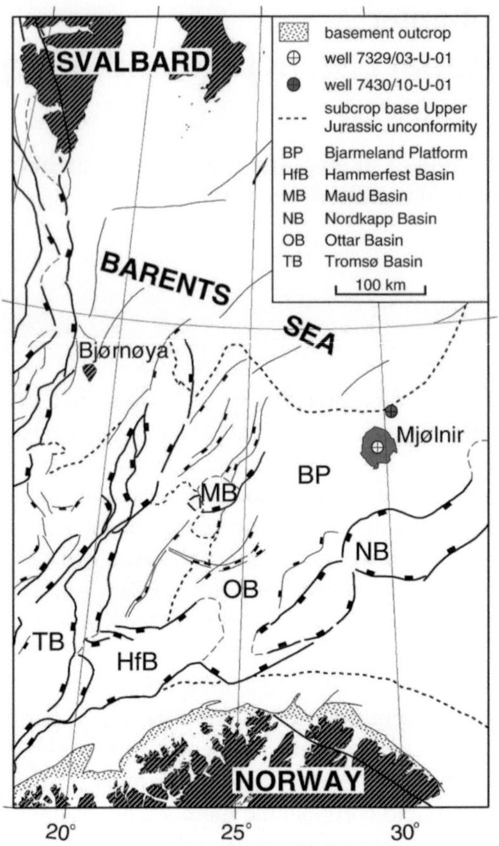

Fig. 3.1 The Mjølnir impact crater with respect to the principal Late Jurassic-Early Cretaceous structural framework in the southwestern Barents Sea, based on Faleide et al. (1993, 2008) and Gudlaugsson et al. (1998)

Most craters presently underwater have not necessarily been subaerially exposed and may be well-preserved.

The extensive Mjølnir seismic database consists of three types of reflection profiles: high-resolution single-channel, shallow multichannel, and conventional multichannel profiles (Fig. 3.2). The 872-km-long single-channel survey was carried out with a fan-like geometry in 1992 and 1993 by the Norwegian Defence Research Establishment after the impact interpretation was proposed. In 1992 a sparker source and analogue recording were used, whereas in 1993 a high-frequency sleeve-gun-array and digital recording were employed (Table 3.2). The data have a record length of 1.0 s two-way traveltime, a sampling rate of 1 ms, and retain frequencies from 70 to 500 Hz. The vertical resolution is estimated at 4–8 m, although deconvolution processing has not been performed (Figs. 3.3a, 3.4a and 3.5).

The 174 km of shallow multichannel profiles were acquired by IKU Petroleum Research using a 1.5 km long streamer with 12.5 m group-interval, and an airgun source (Table 3.2). They comprise 60-fold common-depth-point, digital seismic data recorded with 2 ms sampling rate and a frequency range of 8–180 Hz. The seismic profiles were processed by IKU Petroleum Research and image the structure to 1.0–1.5 s depth with a vertical resolution of 5–10 m (Fig. 3.5).

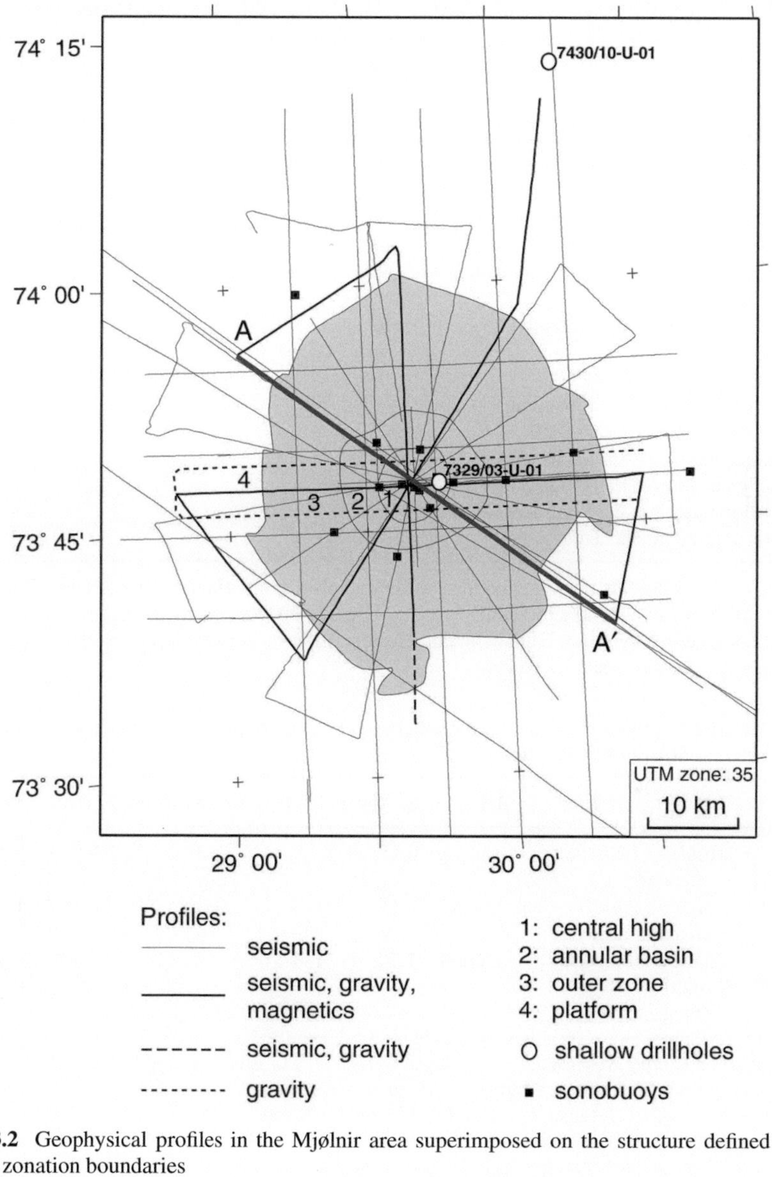

Fig. 3.2 Geophysical profiles in the Mjølnir area superimposed on the structure defined by its radial zonation boundaries

The 1,081 km of conventional multichannel profiles are part of the Norwegian Petroleum Directorate′s regional seismic grid, acquired with a variety of recording parameters and processing sequences (Figs. 3.3b and 3.4b). These profiles have also been used to establish a velocity function for the Mjølnir area, based on stacking velocity analyses (Tsikalas et al. 1998b). The velocity function is used for converting traveltimes to depths and layer thicknesses. In addition, velocity information was acquired from the interpretation of shallow refracted arrivals

Table 3.1 Type of geophysical data used in the study of Mjølnir crater

Data type[a]	Number of profiles	Profile length (km)	Number of velocity analyses
Conventional multichannel seismic reflection profiles (NPD, IKU, BGR)	20	1,081	361
Shallow multichannel reflection profiles (IKU)	4	174	128
Shallow high-resolution single-channel reflection profiles (NDRE)	23	872	
Shallow refraction profiles (sonobuoys) (NDRE)	16		
Side-scan sonar profiles (IKU)	4	174	
Marine gravity measurements along single-channel profiles (NDRE)	9	397	
Marine magnetic measurements along single-channel profiles (NDRE)	7	292	
Base station magnetic measurements at Hopen, Bjørnøya, and Tromsø (UT)			

[a]BGR = Bundesanstalt für Geowissenschaften und Rohstoffe, Hannover; GSN = Geological Survey of Norway, Trondheim; IKU = IKU Petroleum Research, Trondheim; NDRE = Norwegian Defense Research Establishment, Horten; NPD = Norwegian Petroleum Directorate, Stavanger; UT = University of Tromsø.

Table 3.2 Mjølnir crater shallow seismic reflection database depicting details on the acquisition parameters and the resulting resolution

AG	TY	SO	RF	SL (km)	RL (twt, s)	SR (ms)	FI (Hz)	VR (m)
NDRE[a] (1992)	Single-channel	Sparker	Analog	0.15	1	1	70–500	4–8
NDRE[a] (1993)	Single-channel	Sleeve-gun array	Digital	0.15	1	1	70–500	4–8
IKU[b] (1988)	Shallow multi-channel	Airgun	Digital	1.5	1–1.5	2	8–180	5–10

AG, agency; *TY*, type; *SO*, source; *RF*, recording format; *SL*, streamer length; *RL*, record length; *SR*, sampling rate; *FI*, frequency interval; *VR*, vertical resolution.
[a]NDRE = Norwegian Defense Research Establishment, Horten.
[b]IKU = SINTEF Petroleum Research, Trondheim.

recorded by sonobuoys (Fig. 3.2; Table 3.1), and from shallow multichannel profile stacking velocities (Fig. 3.5). The seismic investigation of the Mjølnir crater includes a deep and a shallow part, indirectly reflecting the differentiation into primary impact-induced, and secondary crater-influenced, deformation. This distinction also reflects the resolution of the three seismic datasets comprising single-channel, shallow multichannel, and deep multichannel profiles (Table 3.1).

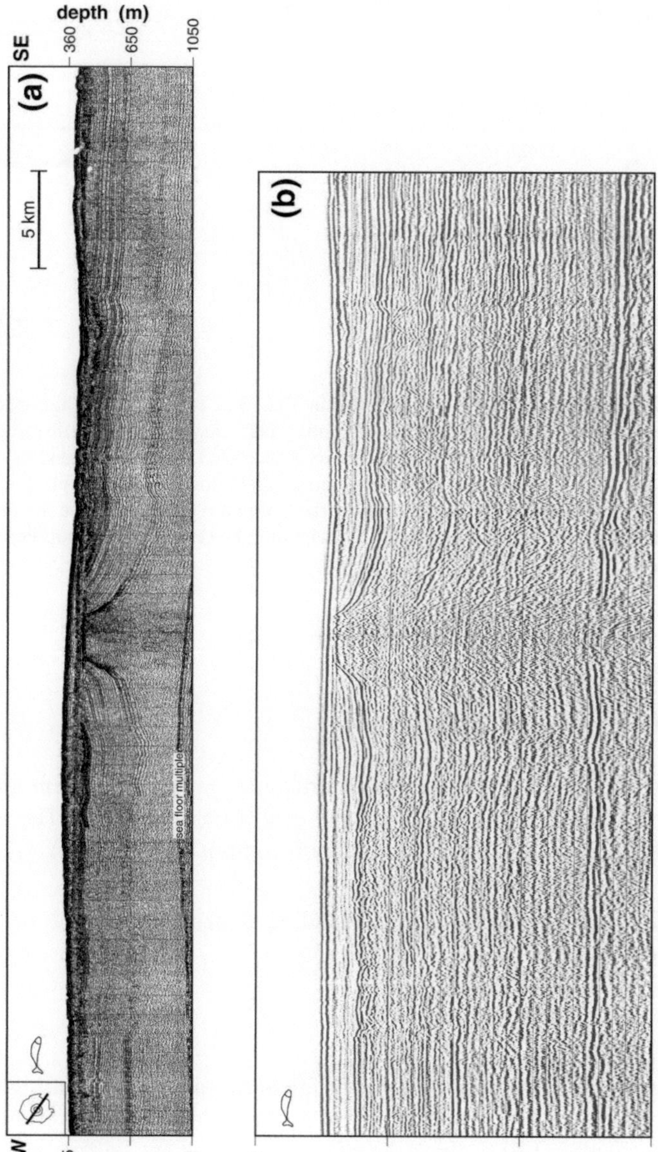

Fig. 3.3 Seismic profiles (along the same location) crossing the entire structure through the centre; **a** high-resolution single-channel profile, and **b** multi-channel profile. Confer Fig. 3.4 for detailed line-drawing interpretation

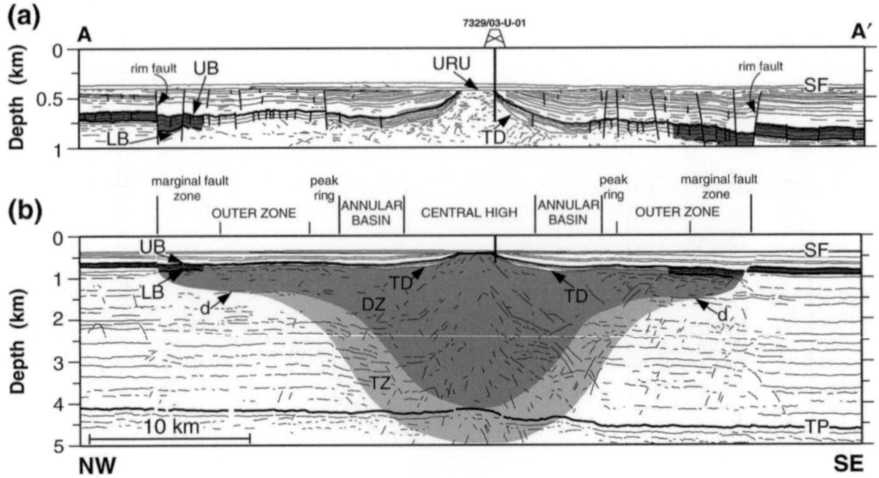

Fig. 3.4 Seismic type sections along profile AA′ in Fig. 3.2. **a** Interpreted high-resolution single-channel profile. **b** Interpreted multichannel profile. DZ, area of intensely disturbed seismic reflections; TZ, transitional area of less disturbance. Reflectors UB and LB bound the time of impact. SF, sea floor; URU; Late Cenozoic upper regional unconformity; UB (upper boundary), lower Barremian; TD (top seismic disturbance), the first continuous reflector above the disturbed seismic reflections; LB (lower boundary), upper Callovian/middle Oxfordian; TP, top Permian; d, low-angle décollement

3.2 Shallow Structure

3.2.1 Main Features

The seismic data clearly show that the major structural features of Mjølnir are typical of large complex impact structures (Tsikalas et al. 1998a, b, 2002a). The seismic profiles image the distinct radial zonation pattern, comprising (Fig. 3.4):

1) a 12-km-wide complex outer zone, including a marginal fault zone and a modestly elevated ring,
2) a 4-km-wide annular depression, and
3) an uplifted 8-km in diameter, central high

In addition, we observe distinct boundary faults forming a ~150 m high near-circular rim wall separating highly deformed strata within the crater from intact platform strata; and a 45–180-m-thick disturbed and incoherent seismic reflectivity unit caused by the impact and confined by prominent fault-blocks and post-impact strata (Figs. 3.4 and 3.6).

At the time of impact, the Bjarmeland Platform was covered by Upper Paleozoic strata, mainly carbonates and evaporites, overlain by 4–5-km-thick Mesozoic siliciclastic sediments (Figs. 3.4b and 3.7) (Gudlaugsson et al. 1998; Faleide et al.

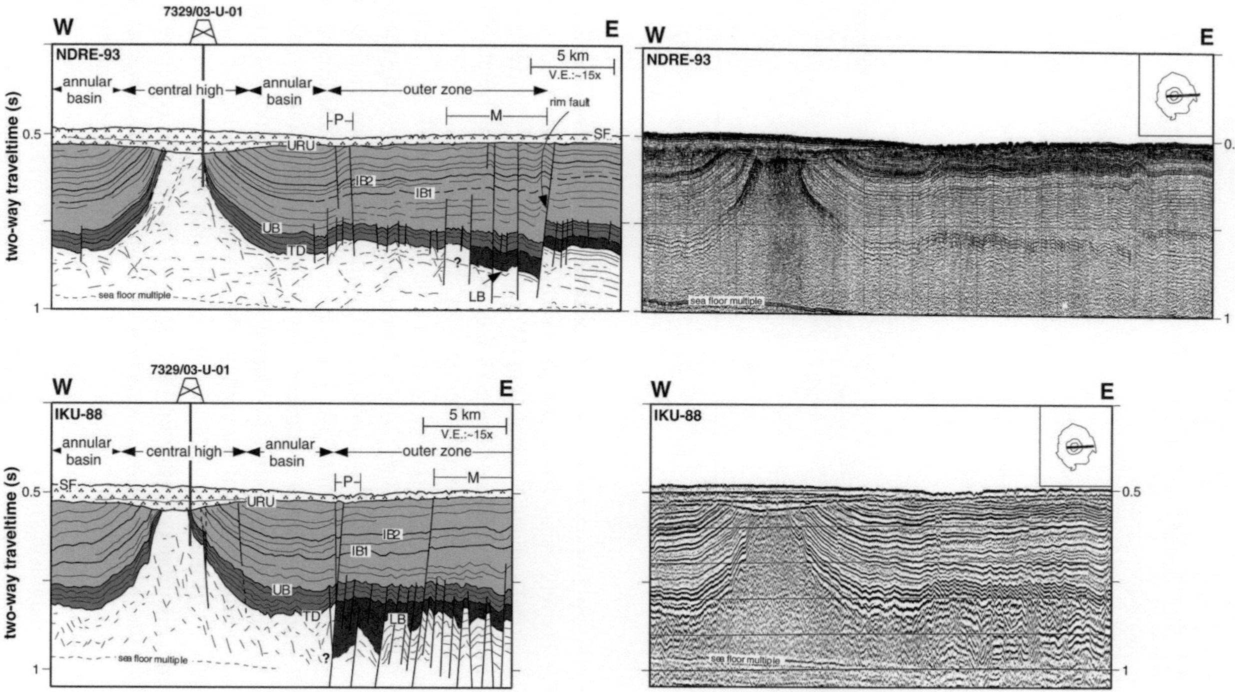

Fig. 3.5 Interpreted (*left*) and uninterpreted (*right*) examples of a single-channel (NDRE-93) and a shallow multichannel (IKU-88) profile crossing the center of the structure and borehole 7329/03-U-01. *P*, peak ring; *M*, marginal fault zone; IB1/IB2, intra-Barremian reflectors. Other annotations as in Fig. 3.4

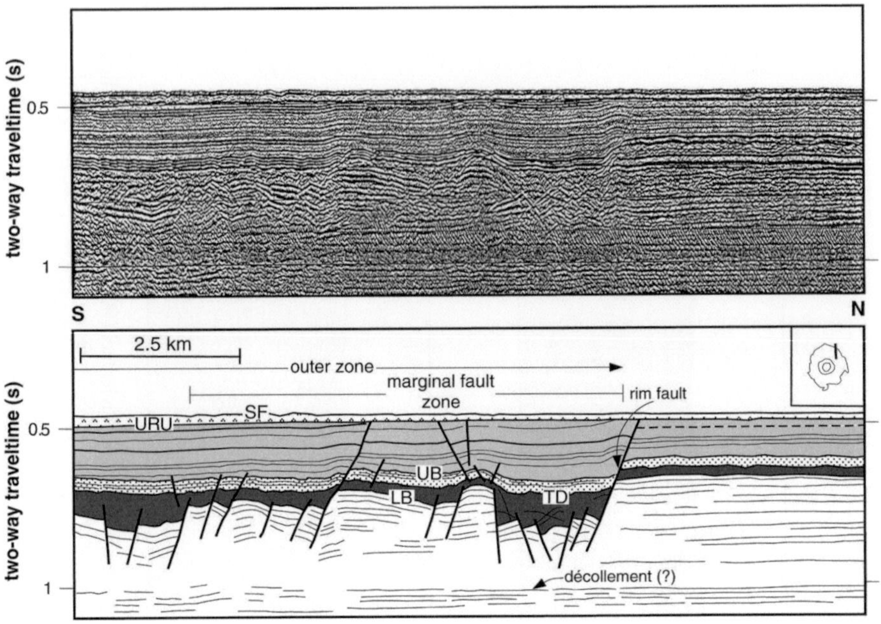

Fig. 3.6 Shallow multichannel seismic profile across the crater rim. Abbreviations in Fig. 3.4

2008). The stratigraphic impact level is well constrained by the two shallow drill-holes (Fig. 3.2) (Dypvik et al. 1996, 2004b; Tsikalas et al. 1998a; Smelror et al. 2001a) and by regional sequence correlation (Worsley et al. 1988; Gabrielsen et al. 1990; Tsikalas et al. 2002b). More aspects of the detailed seismic correlation to the two nearby shallow boreholes are provided in Sect. 3.2.2.

We interpret the impact to be constrained by a Late Ryazanian to Early Barremian limestone unit (UB, upper boundary; lower Barremian reflector defining the top of limestone unit) and a upper Callovian/middle Oxfordian unconformity (LB, lower boundary) (Figs. 3.3, 3.4, 3.5 and 3.6). The top of the seismic disturbance (TD, top disturbance; impact horizon) is located only 35–70 ms (50–100 m) below the regionally prominent lower Barremian reflector (UB), which, in contrast to reflector TD, can be confidently traced throughout the dataset (Figs. 3.3, 3.5 and 3.6). Therefore, reflector UB has been used as a marker horizon that excellently images the top of seismic disturbance, reflector TD (Fig. 3.8). Errors due to differences in resolution among the three seismic datasets were minimized by cross-over analysis, resulting in a root-mean-square deviation of ± 10 ms (± 13 m). Subsequently, the gently south-tilting regional trend was removed by fitting a second-order polynomial surface to the data using the Generic Mapping Tools (GMT) software package (Wessel and Smith 1998). Finally, reflector UB was contoured by fitting a spline surface to the residual depths (Figs. 3.9 and 3.10).

The maps of the Lower Barremian surface (Figs. 3.9 and 3.10) clearly exhibit a structure separated from the surrounding Bjarmeland Platform strata by a

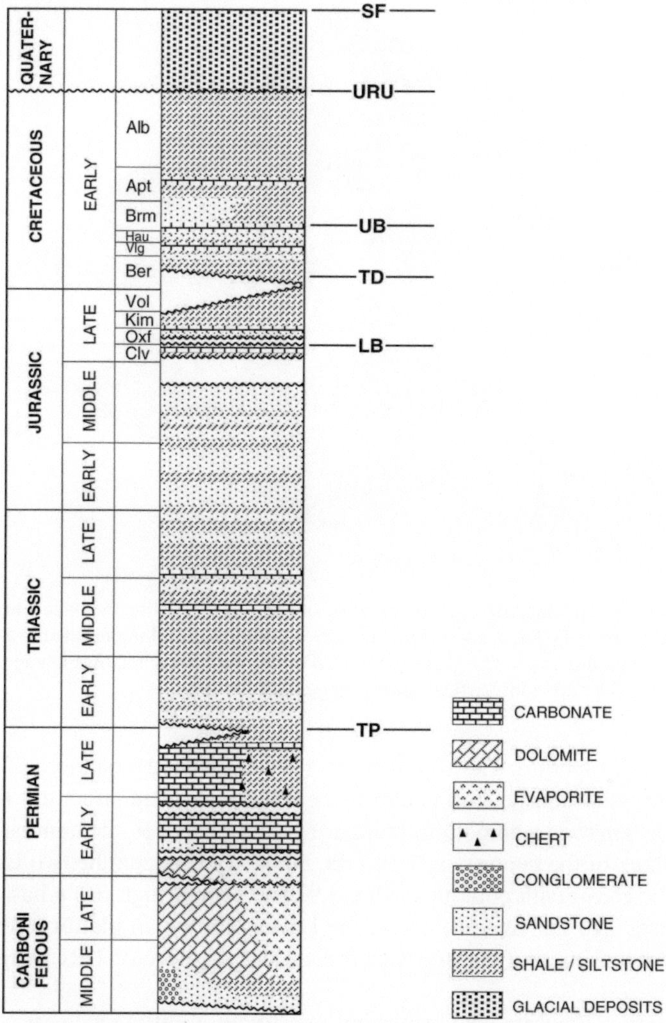

Fig. 3.7 Simplified lithostratigraphic scheme for the Bjarmeland Platform based on integration of the southwestern Barents Sea stratigraphy (Worsley et al. 1988; Gudlaugsson et al. 1998; Smelror et al. 2002), the Bjarmeland Platform sequence stratigraphy, and drillholes 7430/10-U-01 and 7329/03-U-01 (Dypvik et al. 1996; Tsikalas et al. 1998b, 2002a; Smelror et al. 2001a, b; Dypvik et al. 2004b, c). Annotations in Fig. 3.4

2.5–7 km wide, marginal fault zone. The marginal fault zone is part of a 12 km wide complex outer zone that also includes a 3–6.5 km wide intermediate area and an inner area of slightly raised relief at a radius of 8–12 km (seen most clearly in Fig. 3.10). Although irregular in shape, varying in width from 1 to 3 km, the raised relief gives the impression of a subdued ring structure, similar to those typically found in peak ring craters. For want of a better expression, we will use the term

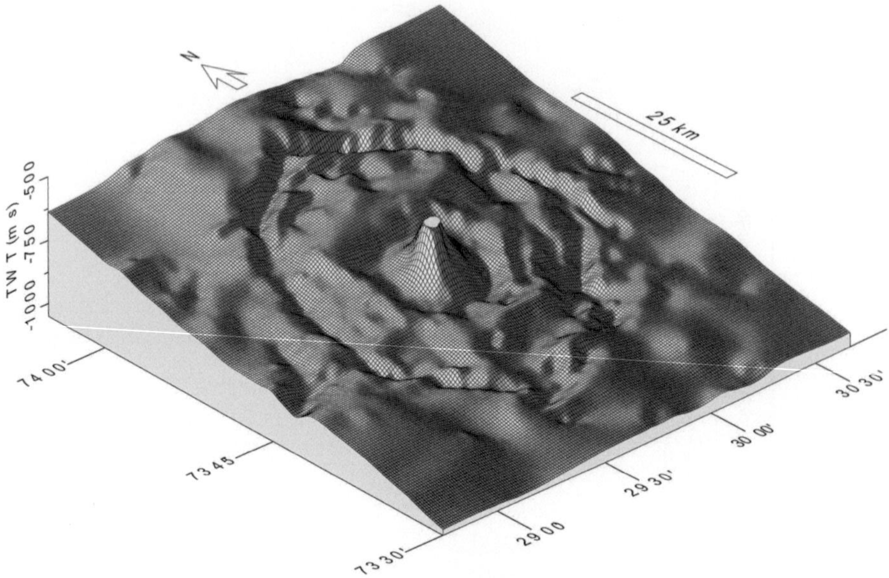

Fig. 3.8 Illuminated perspective diagram of reflector UB, approximating the impact horizon (without regional tilt removal). The view is from 30° above the horizon looking northeast (azimuth 40°); light source at azimuth 290°. The *grey* area on *top* of the central high shows where reflector UB has been truncated by erosion. Vertical scale exaggeration is ~20×

peak ring for this feature. The complex outer zone surrounds a smoother, 4 km wide annular depression that in turn encloses the most prominent structural feature, the central high. There is a smooth unfaulted transition between the annular basin and the rising flank of the central high, and the boundary between the two is arbitrarily placed at the zero-depth contour in Fig. 3.9. The central high has a basal diameter of 8 km, rises 170 ms (250 m) above the Lower Barremian platform level, and is truncated by erosion at a circular subcrop surface at reflector URU (Figs. 3.9 and 3.10).

The pattern of faulting also brings out the structural difference between Mjølnir and the surrounding platform (Figs. 3.9 and 3.10). The radial fault distribution at the level of the impact horizon (reflector TD) exhibits two maxima (Fig. 3.11); a major peak over the marginal fault zone and another over the peak ring. Furthermore, we note only minor faulting in the adjacent platform area and the unfaulted nature of the central high (Figs. 3.3, 3.5, 3.9 and 3.11).

3.2.2 Detailed Seismic Correlation to Nearby Shallow Boreholes

3.2.2.1 Borehole 7430/10-U-01

The shallow borehole 7430/10-U-01 was drilled in 1988, prior to the interpretation of Mjølnir as an impact crater (Gudlaugsson 1993), and before acquisition of the

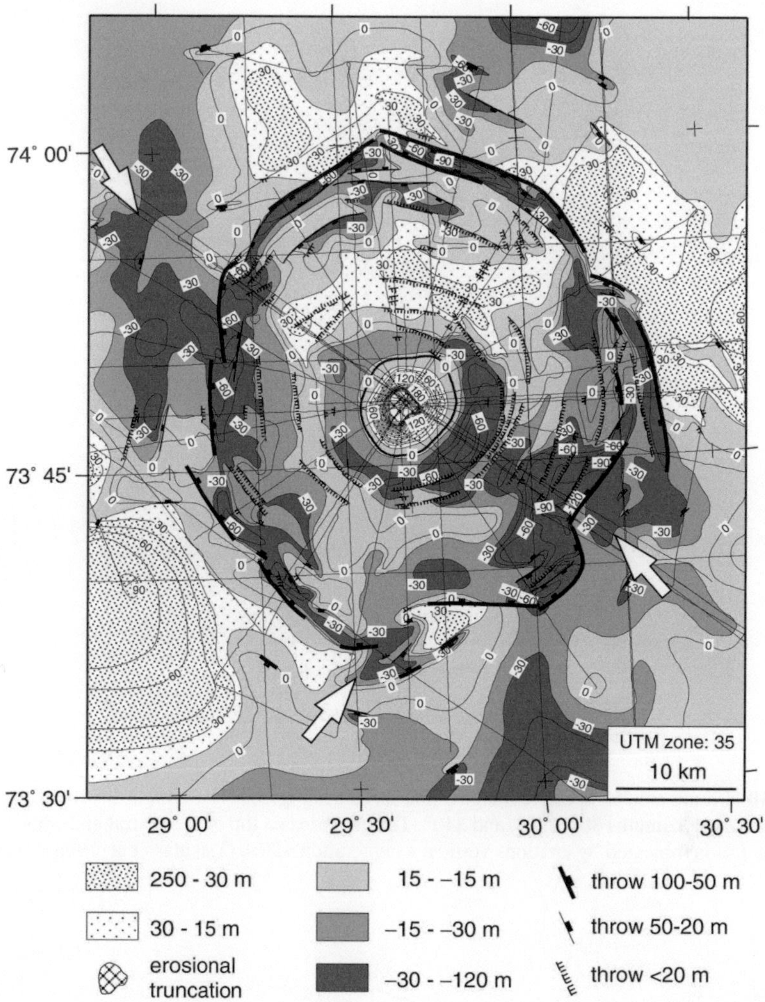

Fig. 3.9 Morphology and structure at the level of reflector UB, expressed as depth residuals with reference to a regional reflector surface dipping to the south. The regionally prominent reflector UB has been used as a marker horizon in the entire seismic reflection dataset because it is located just above and excellently images the top of seismic disturbance (TD, impact horizon), and in contrast to reflector TD can be confidently traced throughout the dataset. Therefore, the depth residuals image the impact-generated relief at the sea bottom after the end of the impact-related processes. *Arrows* point at the resurge gullies. Map is based on the entire seismic reflection database available (*thin solid lines*). Velocity used for depth conversion at the level of reflector UB is 3.0 km/s (range 2.7–3.4 km/s, Tsikalas et al. (1998b)). Contour interval 15 m

high-resolution seismic, gravity, and magnetic profiles (Tsikalas et al. 1998b, c). The borehole is located ~30 km northeast of the crater periphery (Fig. 3.2) and was drilled to a total depth of 67.6 m. A 57.1-m-long continuous core, ranging in age from Late Kimmeridgian (Late Jurassic) to Early Barremian (Early Cretaceous),

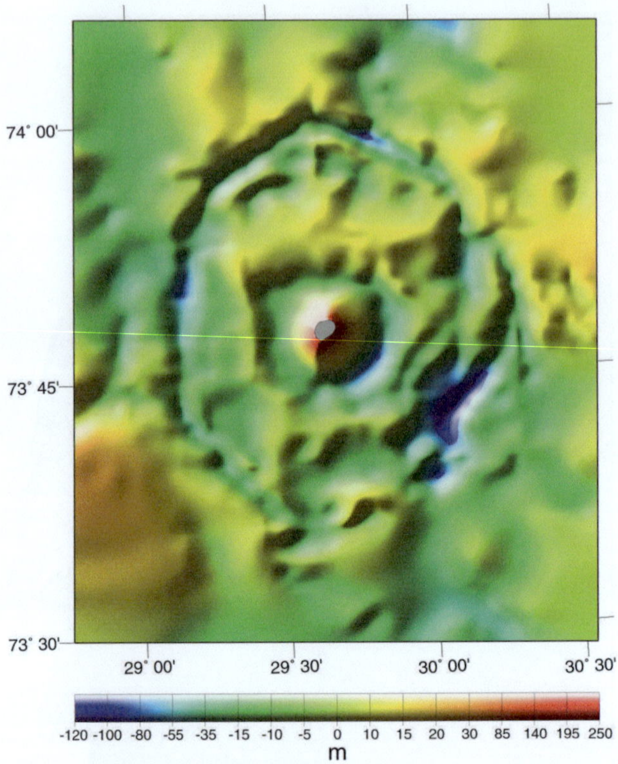

Fig. 3.10 Illuminated perspective image of the surface in Fig. 3.9. The view is directly from above; light sources at azimuths 30°, 290°, and 340°. The grey area on top of the central high shows where reflector UB is truncated by erosion. Vertical exaggeration ~20× (Tsikalas et al. 1998c)

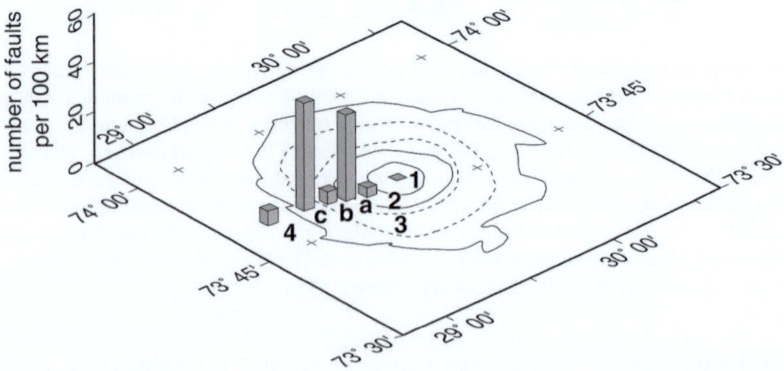

Fig. 3.11 Distribution of the number of faults per 100 km for the various zonation boundaries at Mjølnir. *1* – central high; *2* – annular basin; *3* – outer zone (*a*, peak ring; *b*, intermediate area; *c*, marginal fault zone); *4* – platform

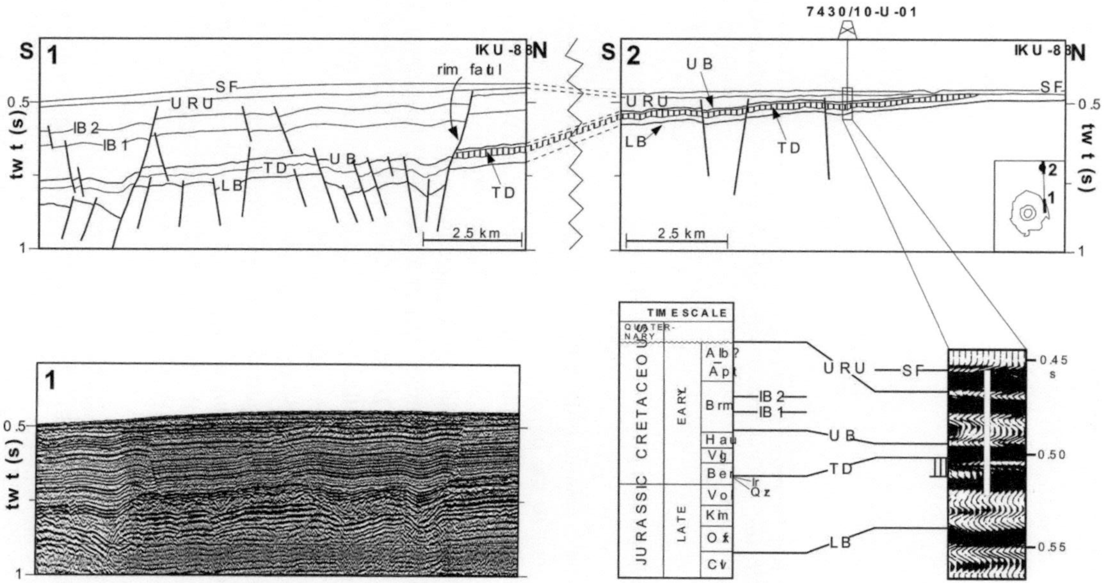

Fig. 3.12 Stratigraphic relationships between the main seismic units at Mjølnir and borehole 7430/10-U-01. *Vertical-bars* raster in profiles 1 and 2 denotes the uncertainty in the seismic tie of the impact horizon (reflector TD). Core recovery is indicated by the *white bar*. Qrz/Ir, position of shocked quartz grains and iridium peak, respectively (Dypvik et al. 1996). Other annotations as in Figs. 3.4 and 3.5

Table 3.3 Correlation of seismic reflectors with lithological units in core 7430/10-U-01. Note that interval velocities for the seismic units are based on average stacking velocities around the borehole. TD_{min}/TD_{max} (TD, top seismic disturbance) define the seismic tie bracket of impact horizon. Annotations as in Fig. 3.4

SR	DE (twt, s)	IV ($km\ s^{-1}$)	DB (m)	CU
SF	0.456		0	
		1.80		Glacial deposits
URU	0.469		12	
		2.10		Post-impact succession
UB	0.491		34	
		2.45		Post-impact succession
TD_{min}	0.501		46	
		2.30		Proximal ejecta/platform sediments
TD_{max}	0.511		57	
		2.70		Platform sediments
LB	0.540		98	

SR, seismic reflector; *DE*, two-way traveltime (twt, s) depth from sea-level; *IV*, interval velocity; *DB*, depth (m) below sea floor; *CU*, core unit.

was recovered from beneath 10.5 m of Quaternary sediments (Dypvik et al. 1996, 2004b, c).

We used continuous seismic reflection profiles to correlate the impact deformation event at Mjølnir to the borehole (Fig. 3.12; Table 3.3). Information from the borehole, combined with regional sequence correlation (Worsley et al. 1988; Gabrielsen et al. 1990; Smelror et al. 2001b, 2002) show that the impact event occurred between the Late Ryazanian- Early Barremian limestone unit (UB, upper boundary, ~130 Ma; lower Barremian reflector defining the top of limestone unit) and upper Callovian/middle Oxfordian unconformity (LB, lower boundary, ~159 Ma). As already indicated, reflector UB can be confidently traced throughout the dataset and has been used as a marker horizon to produce distinct seismic images of the crater relief (Figs. 3.9 and 3.10). Reflector LB marks the top of a thick unit of parallel reflectors, which are broken into distinct, tilted fault blocks along the crater periphery. LB is not recognized towards the central parts of the crater, where the strata reflections of the older platform change into a chaotic and diffraction-dominated seismic zone (Figs. 3.4 and 3.5).

The fault blocks are overlain by a variably thick and incoherent seismic unit LB-TD (Figs. 3.4, 3.5, 3.6 and 3.12) ascribed to a polymictic allochthonous breccia, which formed during impact and was covered by post-impact strata (Tsikalas et al. 1998a; Tsikalas and Faleide 2007). Reflector TD (top disturbance) defines the impact horizon and intersects the borehole at a level where a 0.8-m-thick core section contained impact ejecta. Within the ejecta, quartz grains showed planar fractures and recrystallized planar deformational features. An iridium peak of about 1,000 ppt was recorded approximately at the same level (Fig. 3.12) (Dypvik et al. 1996; Dypvik and Ferrell 1998; Dypvik and Attrep 1999).

The configuration of rim faults at the crater periphery and the regional northward thinning of the Upper Jurassic-Lower Cretaceous strata (LB-UB unit) are the main sources of uncertainty in the seismic tie (Fig. 3.12). However, the prominent character of reflectors LB and UB together with the smooth and continuous character of both the LB-TD and TD-UB units provide a robust correlation (Fig. 3.12; Table 3.3).

3.2.2.2 Borehole 7329/03-U-01

The study of the Mjølnir crater led to a new shallow borehole, drilled in 1998, specifically to recover both impact-related and post-impact strata. The 171-m-deep borehole was drilled at the edge of the up-dipping flank of the crater central high (Dypvik et al. 2004b, c). A 121-m-long core was recovered below 50 m of glacial cover (Fig. 3.13).

A detailed seismic stratigraphy has been established based on correlation with the high-resolution shallow multichannel profile crossing the crater center (Fig. 3.13; Table 3.4). The interval velocities for the seismic units in Table 3.4 are based on one stacking velocity analysis at the borehole location (Tsikalas et al. 2002a). Reflector TD represents the top of a unit of chaotic deposits, reflecting impact-induced deformation (Smelror et al. 2001a). Away from the crater center, the reflector defines the top of the uppermost impact-related breccia unit, the allochthonous breccia (Tsikalas et al. 1998a). This unit exhibits minor reflectivity in places but is mostly characterized by reflection-free zones. The allochthonous breccia is expected to consist of crushed fallback, collapsed crater-wall, and washback material (e.g., Melosh 1989; Turtle et al. 2005). A velocity inversion below reflector TD (Table 3.4) is possibly associated with the brecciated character of the impact strata. The allochthonous breccia is underlain by the *in situ* fractured autochthonous breccia. The autochthonous breccia comprises the tilted and internally stratified fault blocks at the periphery, but changes into disrupted and chaotic reflections, diffractions, and reflection-free zones toward the crater center. The top of the allochthonous breccia corresponds to the apparent crater floor. Theoretically the base of the allochthonous breccia defines the true crater floor (Grieve and Pesonen 1992; Turtle et al. 2005) but this level is difficult to identify on the shallow seismic reflection profiles. However, MCS profiles provided estimates of 1.5–2 km of structural uplift at the center of the structure (Tsikalas et al. 1998b). Furthermore, gravity and seismic traveltime modelling, together with the MCS profiles demonstrate that the allochthonous breccia unit exhibits a lenticular cross-sectional shape beneath the annular basin pinching out towards the central high and towards the periphery (Tsikalas et al. 1998a–c; Tsikalas et al. 1999). At the borehole location, it is, therefore, not obvious whether reflector TD defines the top of the allochthonous breccia wedge or marks a mixture of allochthonous and autochthonous breccia.

Reflector TD is overlain by a unit, TD-R3 (Fig. 3.13), which contains sediments related to the final current activity imposed by the Mjølnir impact (Smelror et al. 2001a; Tsikalas et al. 2002a). In a broad sense, the unit can be considered the topmost part of the allochthonous breccia because it is related to processes triggered by the impact event. It contains current-reworked and fine-grained debris that can be considered either as fallout or suspended sequences. Between reflector R3 and the glacial cover the post-impact succession comprises 16 m of Berriasian shales and 8 m of Upper Ryazanian-Lower Barremian carbonates (Fig. 3.13). The

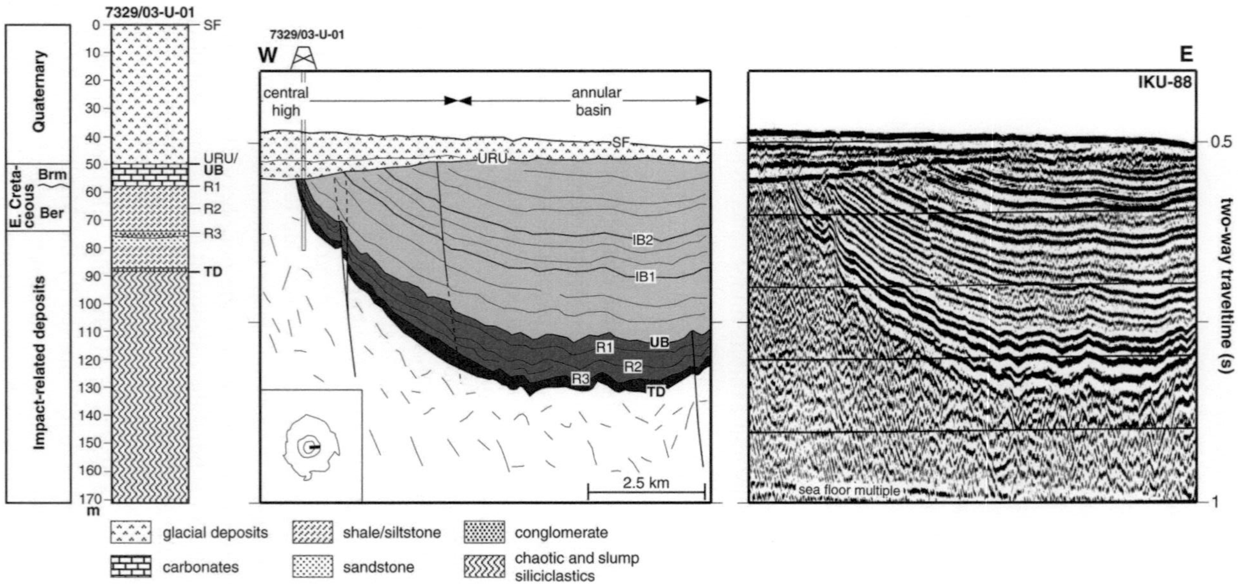

Fig. 3.13 Generalized lithostratigraphic log of borehole 7329/03-U-01 and interpreted shallow multi-channel seismic profile across the borehole. R1/R2/R3, continuous reflectors within the TD-UB unit at the crater center. Annotations as in Figs. 3.4 and 3.5

Table 3.4 Correlation of seismic reflectors with lithological units in core 7329/03-U-01. Annotations as in Fig. 3.4

SR	DE (twt, s)	IV (km s^{-1})	DB (m)	CU
SF	0.488		0	
		1.77		Glacial deposits
URU/UB	0.545		50	
		2.51		Post-impact succession
R1	0.556		57	
		2.51		Post-impact succession
R2	0.562		65	
		2.51		Post-impact succession
R3	0.570		75	
		2.59		Current reworked material
TD	0.581		89	
		2.38		Impact breccia
(Borehole bottom)	0.650		171	

SR, seismic reflector; *DE*, two-way traveltime (twt, s) depth from sea-level; *IV*, interval velocity; *DB*, depth (m) below sea floor; *CU*, core unit

regionally prominent reflector UB correlates with the top of the carbonate bed, which is truncated at the borehole location (as are all the cited reflectors) by the glacial URU unconformity (Fig. 3.13). Reflector R1 defines the base of the carbonate bed, whereas reflector R2 is an internal Berriasian reflector (Fig. 3.13; Table 3.4).

3.2.2.3 Impact Timing as Revealed from Seismic Correlation

Stratigraphic control in borehole 7430/10-U-01 indicated that the Mjølnir impact took place in the latest Jurassic to earliest Cretaceous, 141–149 Ma (Dypvik et al. 1996). New biostratigraphic evidence from the same borehole and, in particular, from borehole 7329/03-U-01 (Smelror et al. 2001b) further constrains the age of impact to the Volgian-Ryazanian boundary, i.e. 142±2.6 Ma according to the geological time scale of Gradstein et al. (1994, 1999). This age is based primarily on macrofauna recovered just at the base of the post-impact succession in the central crater core (R3-R2 unit, Fig. 3.13) represents a return to normal sedimentary conditions after the impact.

We believe that the 142±2.6 Ma age should be considered as a minimum age estimate for the impact; it may have been slightly older, because the shallow reflection profiles provide evidence for crater-influenced sedimentation (Figs. 3.5 and 3.13). In particular, the depression around the central high exhibits infilling by progressive accumulation of sediments in to the original annular basin. The depression is completely smoothed above reflector IB1 as indicated by the uniform thickness of the IB1-IB2 (intra-Barremian) unit when the crater-relief control on sedimentation was minor (Fig. 3.13) (Tsikalas et al. 1998a; Tsikalas and Faleide 2007). More importantly, we observe onlap against the high for most of the early, pre-IB1, post-impact deposits, and especially for those immediately after-impact, between reflectors TD and UB, which were penetrated by the borehole (Fig. 3.13). The increase in reflection amplitude between the strata onlapping the central high also suggests lateral changes in depositional facies (Fig. 3.5) (Tsikalas et al. 1998a). Therefore, the lateral thickness variations of the immediately post-impact sequences, especially the R3-R2 unit (Fig. 3.13), suggest that the impact took place somewhat earlier than the borehole biostratigraphic age, but most likely still within the Upper Volgian stage (144.2±2.6 to 142±2.6 Ma). Note that with the recent geological time scale of Gradstein et al. (2004) the absolute age boundaries of the Volgian/Tithonian stage are shifted to 145.5–150.8 Ma.

3.2.3 Impact-Induced Deformation

The marginal fault zone of the Mjølnir crater is characterized by rim faults separating the crater from the platform and by a series of smaller normal faults (Figs. 3.6 and 3.14). The faults bound tilted sedimentary blocks which are among the most conspicuous impact-generated structures imaged in the seismic data. Within the adjacent platform, reflector LB forms the upper boundary of an older thick unit of parallel, subcontinuous, and subhorizontal reflections that can be correlated a short distance within the Mjølnir crater. Inside the structure, reflector LB is the uppermost reflector that can be identified within the fault-blocks of the marginal fault zone (Figs. 3.6 and 3.14). It thus predates the faulting. In several places, the rim faults, and associated antithetic faults, form marginal grabens that closely follow the concentric crater rim (Figs. 3.9 and 3.10).

Most of the faults in the marginal zone were initiated sometime during the deposition of the LB-TD unit and have later been reactivated. At present, some of them cut the entire sedimentary sequence up to URU (Figs. 3.6, 3.12 and 3.14). At the level of reflector LB, the cumulative throw of the rim faults ranges from 60 ms (90 m) to 150 ms (225 m) with an average of about 100 ms (150 m). Furthermore, the LB-TD unit shows lateral thickness variations changing from a uniform and thin, 25–40 ms (35–60 m) layer outside the structure to a 30–120 ms (45–180 m) thick sedimentary unit overlying the fault-displaced, pre-impact strata in the outer zone (Figs. 3.6 and 3.14).

An impact event within the LB-TD unit is compatible with the seismic observations, in particular (Figs. 3.4, 3.6 and 3.14):

1. the thickening and the dispersed character of the seismic reflectivity of the unit inside the crater rim,
2. its confinement by horizontally layered sequences above and a faulted base below, and
3. the timing of faulting sometime during the deposition of the unit.

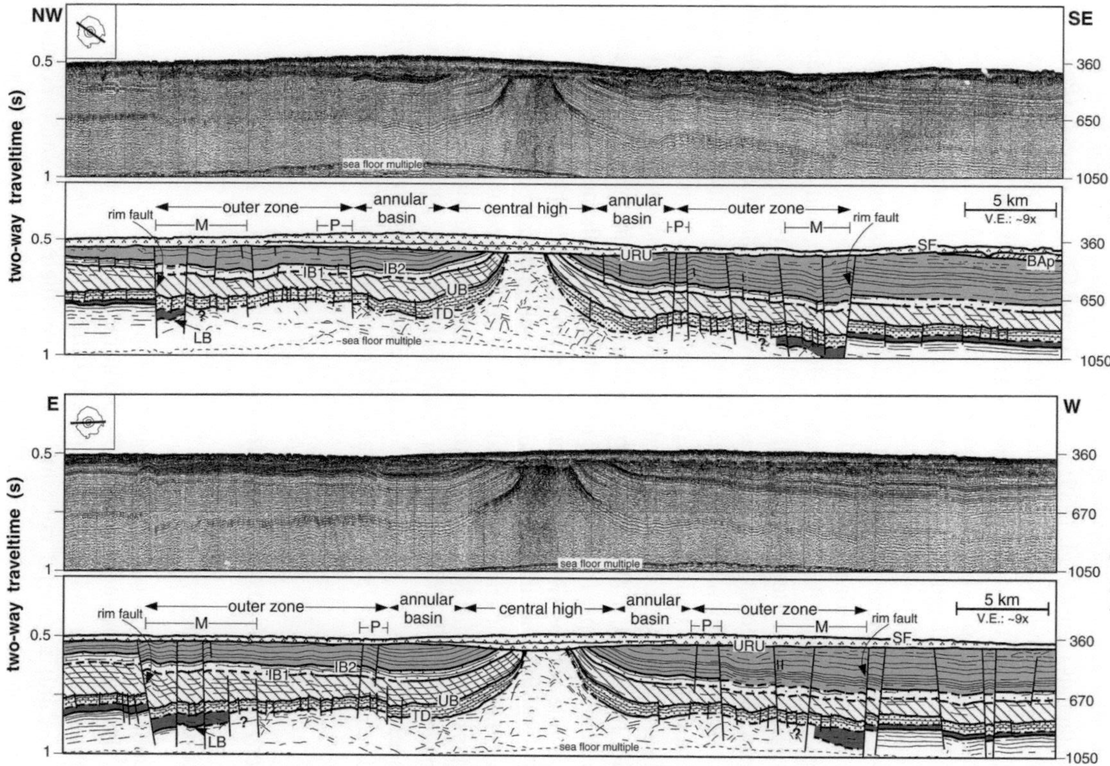

Fig. 3.14 High-resolution single-channel seismic profiles, and interpretations, crossing the entire structure through the centre. Annotations as in Figs. 3.4 and 3.5

The impact-origin of Mjølnir was confirmed by the discovery of impact ejecta near the top of this interval in drillhole 7430/10-U-01 (Dypvik et al. 1996), and slumped and brecciated sediments including shocked quartz grains and abundant planar fractures/deformation features in drillhole 7329/03-U-01 (Fig. 3.2) (Dypvik et al. 2004b, c; Sandbakken et al. 2005). We, therefore, interpret the LB-TD unit as a polymictic allochthonous breccia compareable to those described from the submarine Montagnais (Jansa et al. 1989; Jansa 1993) and Chesapeake Bay (Poag et al. 1994, 2004) impact craters. Such breccia is formed during the passage of the shock wave from the impact and the subsequent excavation and partial infilling of the crater. It is composed of fall-back and back-wash material (Melosh 1989), and may contain highly shocked and dispersed melt clasts and glass similar to suevite breccia (Stöffler et al. 1977).

The marginal fault zone is an unusual feature of the Mjølnir crater. Typical terrestrial craters are characterized by step-like terraces at the periphery. Such marginal terraces are absent from the Mjølnir crater (Figs. 3.4, 3.6, and 3.14). Instead, the prominent continuous reflectors beneath the fault-blocks (Fig. 3.6) suggest inward displacement of the blocks during collapse above an apparent low-angle décollement, without the development of terraces. The faults appear to fade out within a low coherence zone that is underlain by the prominent continuous reflectors (Fig. 3.6). Similar observations have been reported from the Chesapeake Bay impact crater (Poag 1996; Poag et al. 2004). The deep multichannel seismic profiles indeed support gravitational collapse through listric normal faulting at the periphery (Figs. 3.3b and 3.4b), while collapse is characterized by density flows farther inward (Tsikalas et al. 1998b; Shuvalov et al. 2002). A likely explanation of these unusual features is the fact that the Mjølnir crater has experienced a large degree of collapse. The estimated collapse factor for Mjølnir, i.e., the ratio of final crater diameter to transient crater diameter, is 2.5–3.0 (Gudlaugsson 1993; Tsikalas et al. 1998b; Shuvalov et al. 2002), whereas values for similar-sized craters exhibit a range of 1.4–2.0, with an average of 1.6 (Melosh 1989; Turtle et al. 2005) (see discussion in Chap. 4).

3.2.4 Near-Field Erosional Features

3.2.4.1 Resurge Gullies

Three prominent gullies cutting through the rim faults are identified at Mjølnir (Fig. 3.9). They are located within a 10–15-km-radius outside the crater rim wall, exhibiting meandering and bifurcating patterns. In addition, the location of gullies at Mjølnir exhibit a general concentric form, i.e. their prolongation points out towards the crater center (Fig. 3.9). The spatial distribution of resurge gullies at Mjølnir is in good agreement with that of the few other craters with gullies (4-km-diameter Kärdla crater, Puura and Suuroja 1992; 20-km-diameter Kamensk crater, Movshovich and Milyavsky 1990; and 14-km-diameter Lockne crater, von Dalwigk and Ormö 2001, although the impact origin of gullies in Lockne has been recently disputed by Kenkmann et al. 2007); extending from less than half to a maximum one crater diameter outside the crater rim. The gully on the northwest side of Mjølnir (Fig. 3.9) reaches 60–70 m in depth just outside the crater rim where it is ~5 km wide, and has a length of ~25 km at its deepest parts. The one on the southeast side reaches 30–40 m depth and ~5 km width in the vicinity of the rim, being ~10 km long. Finally, the one in the south is 30–40 m deep, ~1 km wide and 5–10 km long (Fig. 3.9).

Figure 3.15 shows a seismic profile across part of the gully at the southeast side of the crater periphery. The gully shown here has a concave form at its top at the level of impact horizon (reflector TD), where several small to minor erosional undulations locally cut through reflections of the underlying ejecta layer (Fig. 3.15). The undulations may indicate final sedimentary reworking by the resurge water flow. The sediments within the gully exhibit dispersed and scattered seismic reflectivity patterns (Fig. 3.15). They are confined within undisturbed sub-horizontal reflections of pre-impact platform sediments below, and immediately-after-impact layers above that imprint the impact-generated relief. There is no direct geological information from the gullies at Mjølnir, but they display a dispersed seismic reflectivity

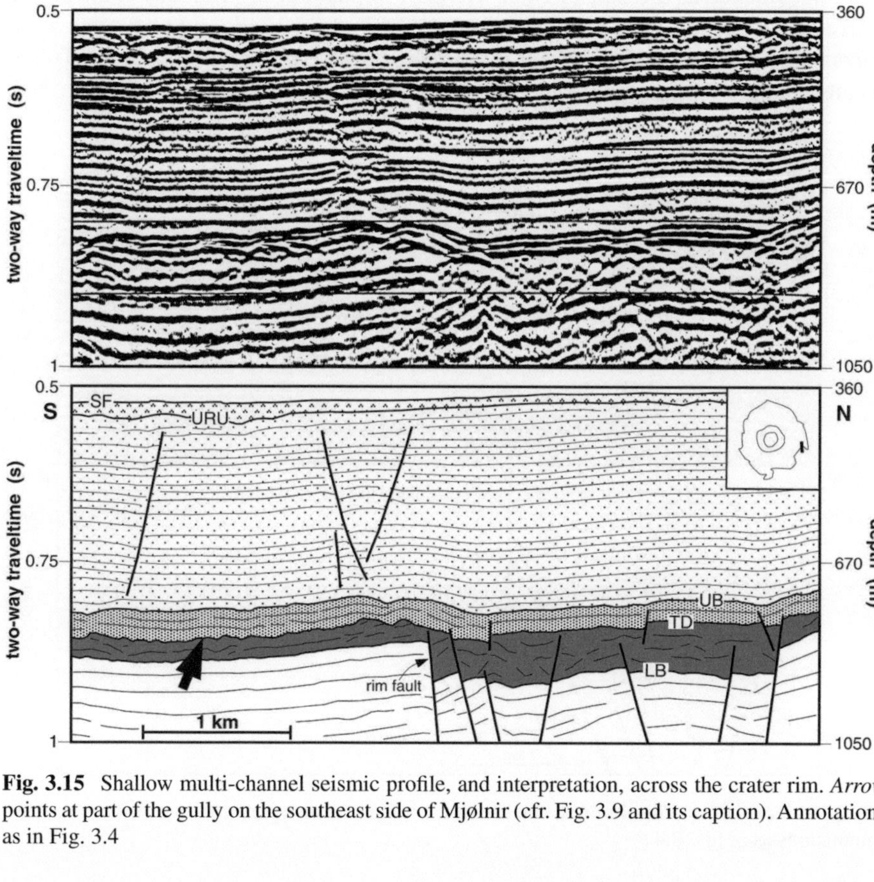

Fig. 3.15 Shallow multi-channel seismic profile, and interpretation, across the crater rim. *Arrow* points at part of the gully on the southeast side of Mjølnir (cfr. Fig. 3.9 and its caption). Annotations as in Fig. 3.4

character and similarities to other craters with gullies. Due to this appearance they are most likely filled with resurge sediments from the surrounding platform mixed with near-field excavated/ejected target material and the denser/heavier portions of the fall-back ejecta.

3.2.4.2 Crater Rim

The presence of a raised crater rim is inferred from the study of terrestrial and planetary impact craters (Melosh 1989; Spudis 1993; Grieve and Pesonen 1996; Turtle et al. 2005). The volume-balance model indicates that a crater with the structural/morphological parameters and extensive collapse as Mjølnir should have a rim height of 40–100 m (see Chap. 4). However, Mjølnir not only lacks a raised crater rim but it locally exhibits a small, but distinct, inward bending of strata on the hanging wall crest of the prominent rim faults (Figs. 3.15 and 3.16). The tilting is in the order of 7–10°, 15–20 m relative to the surrounding platform level and it is present both at the ejecta layer (LB-TD) and the immediately-after-impact deposited layer TD-UB. Although the Mjølnir crater experienced structural reactivation and differential subsidence as a result of extensive post-impact burial (Tsikalas et al. 1998a, 2002a; Tsikalas and Faleide 2007), Figures 3.15 and 3.16 clearly show that post-impact reactivation of the rim fault is not related to the observed rim tilting. This is because the pre- and post-impact strata outside the crater appear almost horizontal,

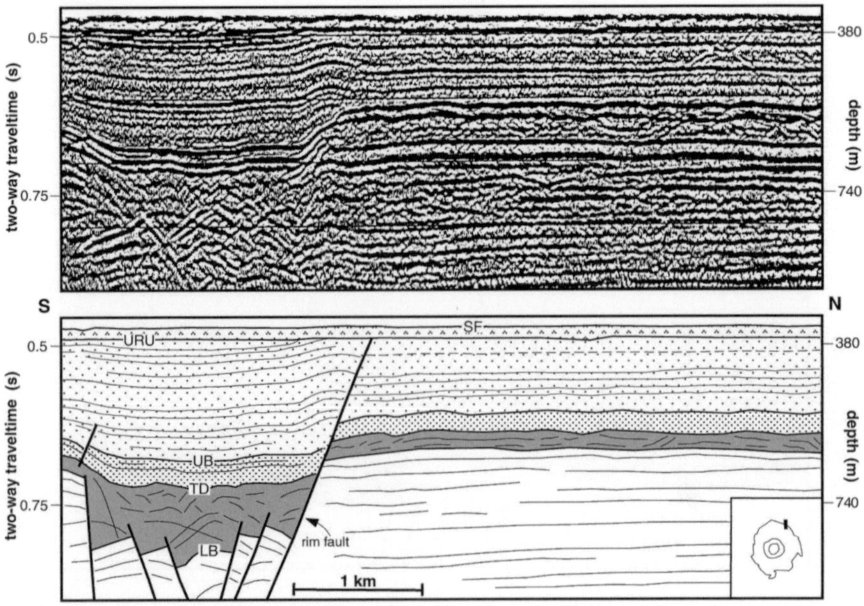

Fig. 3.16 Detail of shallow multi-channel seismic profile, and interpretation, across the crater rim. Annotations as in Fig. 3.4

and because the post-impact structural reactivation and sediment thickness variations took place solely within the crater boundaries. Thus, the observed tilting of the ejecta layer is primarily an impact-related feature. Figure 3.16 also shows several small-scale undulations of possible erosional origin on top of the ejecta layer outside the crater, indicating final reworking of material by the resurge water flow.

3.3 Deep Structure

3.3.1 Impact-Induced Disturbance

Five interpreted seismic sections, crossing the Mjølnir crater at an increasing distance away from the center, are presented in Fig. 3.17. The profiles were depth converted using smoothed seismic velocity functions (Fig. 3.18), obtained from 361 stacking velocity analyses along conventional multichannel profiles.

3.3.1.1 Seismic Reflectivity Patterns

The seismic profiles confirm the presence of a large volume of disturbed seismic reflectivity patterns. The pervasive disturbance is sharply bounded above by reflector TD, which corresponds to the crater surface after the impact (Figs. 3.3, 3.4 and 3.17). The disturbance has a gradational lower boundary and contrasts strongly with the regularly stratified sedimentary platform. There is a systematic loss of reflection coherency from the deeper and outer parts of the disturbance to its upper and central parts, caused by a progression of seismic facies from disrupted layering and diffractions to chaotic and reflection-free zones. The profiles illustrate the consistency of the above phenomena (Fig. 3.17). We divide the seismic disturbance into an intensely disturbed zone and a less disturbed transitional zone (Fig. 3.17). The main criterion used to delimit the intensely disturbed zone is the complete loss of correlatable platform stratification. This largely corresponds to the location where downward bending Triassic strata encounter steeply upward bending strata beneath the annular basin and the central high. In the shallow periphery of the structure, the disturbance is floored by uniformly layered platform strata (Figs. 3.6 and 3.17). Within the intensely disturbed zone the strata appear to have been involved in considerable transport of material, similar to the megabreccia unit at the Slate Islands impact crater (Dressler and Sharpton 1997). The zone contains only few reflector segments that remain subhorizontal and encloses the reflection-free seismic facies, probably related to the destruction of large-scale coherent features by the cratering process. In contrast to the intensely disturbed zone, elements of correlatable platform stratification can be recognized within the transitional zone. There only few reflector segments bend slightly downward or upward, indicating lesser displacement. The intensely disturbed zone fades out just above the top Permian reflector, while the surrounding transitional zone reaches below the reflector (Fig. 3.17). Although the quality and resolution of the seismic data gradually decrease with depth, it is important to determine the depth extent of the Mjølnir crater, as this will

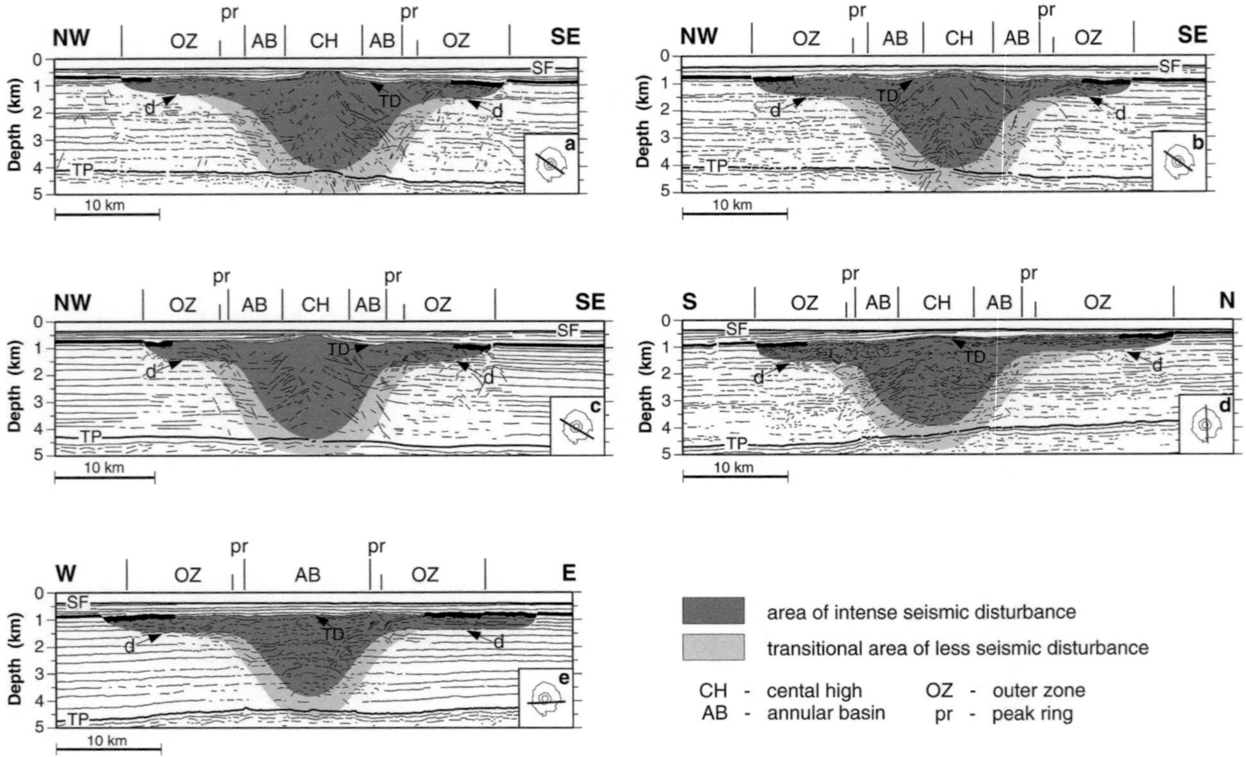

Fig. 3.17 Five interpreted multichannel profiles crossing the central Mjølnir crater. Profiles *a*, *b* and *c* are unmigrated, while profiles *d* and *e* have been migrated by the finite difference algorithm. The stratigraphic interval confining the time of impact is shown in black and is bounded above and below by reflectors UB (upper boundary), lower Barremian, and LB (lower boundary), upper Callovian/middle Oxfordian, respectively. D is low-angle décollement. Other annotations as in Fig. 3.4

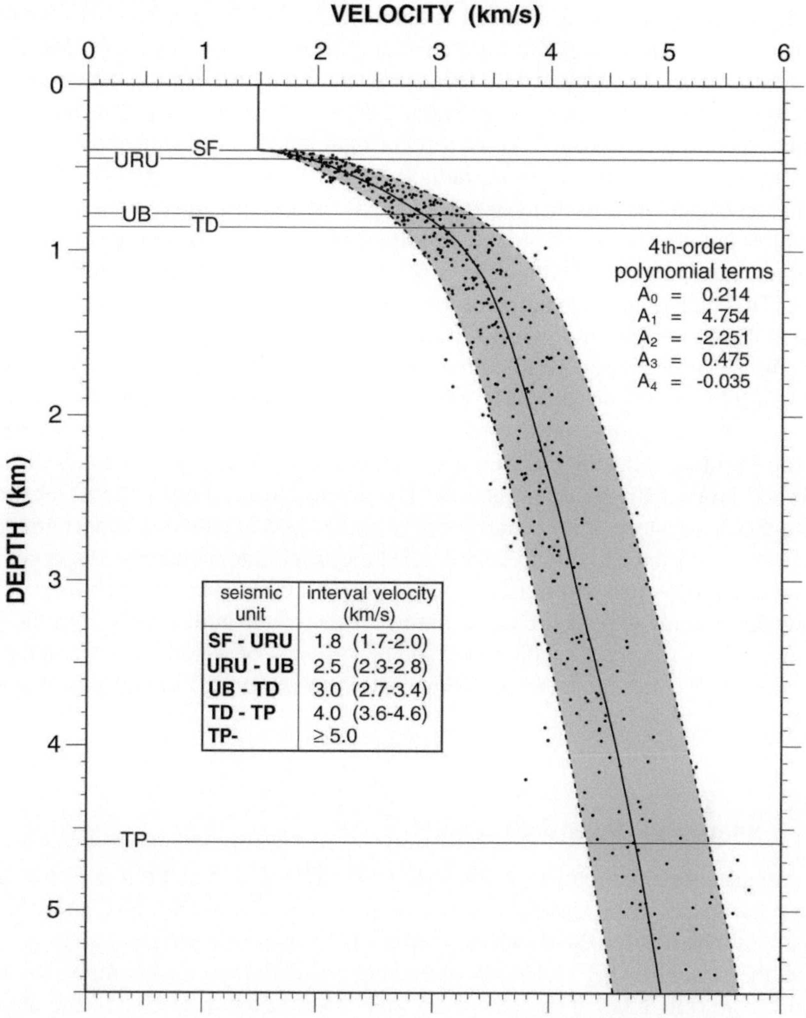

Fig. 3.18 Typical seismic interval velocity function for the Mjølnir crater derived from stacking velocity analyses along conventional multichannel profiles. The cross-plot is based on data from velocity stations at the transition between the annular basin and the outer zone. The *shaded-area* bounded between the *dashed-lines* provides the velocity range with depth, while the *thick solid line* approximates a weighted, fourth-order polynomial curve that was fitted to the data. Annotations as in Fig. 3.4

be used to estimate key structural, geometric and volumetric parameters of the structure. The prominent top Permian reflector defines a smooth and continuous plane, gently tilted toward the south.

Although small deviations on the reflector are present beneath the central part of the structure and two of the seismic sections show evidence of missing reflector segments (Fig. 3.17, profiles b and c), this need not imply that the reflector was deformed by the impact, as minor faults and small irregularities are common regional features of the top Permian reflector (Gérard and Buhrig 1990). Moreover, the missing reflector segments are located beneath the central high where imaging problems due to distortion of wave paths are to be expected. Detailed mapping of seismic travel time anomalies on the top Permian reflector has shown that they correlate spatially with the overlying radially zoned impact structure (Tsikalas et al. 1998b, c). Integrated geophysical modeling of the seismic travel time and gravity anomalies has also shown that the travel time anomalies can be explained as a pull-up effect caused by radial variations in seismic velocity within the disturbed region above. This makes an explanation of the travel time anomalies in terms of structural uplift or pre-impact relief unlikely (Tsikalas et al. 1998c, 2002b). The persistence of the undulations on the top Permian reflector after depth conversion (Fig. 3.17) probably results from the insensitivity of the smoothed velocity functions (Fig. 3.18) to small, yet significant, velocity anomalies within the seismic disturbance. Thus the Top Permian reflector places a lower boundary on the volume of significantly deformed strata beneath the structure. The seismic reflectivity patterns observed within the disturbance at Mjølnir are easily understood in terms of impact cratering processes, as the intensity of brecciation and the amplitude of material displacement are expected to decrease from the center of the structure toward the periphery. This, combined with the expected chaotic nature of the crater infill makes both the systematic variation in seismic coherency and progression of seismic facies compatible with an impact crater model (e.g., Grieve and Pesonen 1992; Pilkington and Grieve 1992).

3.3.1.2 Shape and Dimensions

The extensive seismic reflection database of Mjølnir (Fig. 3.2; Table 3.1) was used to map the shape of the intensely disturbed zone. The disturbance has the form of a deep, parabolic bowl at the center and turns into a shallow broad brim at the periphery of the structure (Fig. 3.19). In particular, the disturbed rock volume thickens inward from 0 to 1 km at the periphery to 1–2 km at the transition to the annular basin and reaches a maximum of ~3.6 km beneath the central high (Fig. 3.19). Based on the contour map in Fig. 3.19a, the estimated volume of the intensely disturbed zone is 850 km^3, whereas the volume of the entire disturbance, including the transitional zone, is ~1,400 km^3.

The characteristic shape of the impact-induced disturbance is an unusual feature of the Mjølnir crater. The literature on large-scale features of complex, terrestrial impact craters is dominated by structural interpretations exhibiting only a simple bowl-shaped disturbance beneath the crater, without any broad and shallow brim at the periphery (e.g., von Engelhardt 1990; Milkereit et al. 1994; Anderson et al. 1996; Grieve and Pesonen 1996). Probable exceptions are the Flynn Creek structure (Roddy 1977), the Chesapeake Bay crater (Poag 1996; Poag et al. 2004), and possibly the Avak structure (Kirschner et al. 1992). Mjølnir is most similar to the Flynn

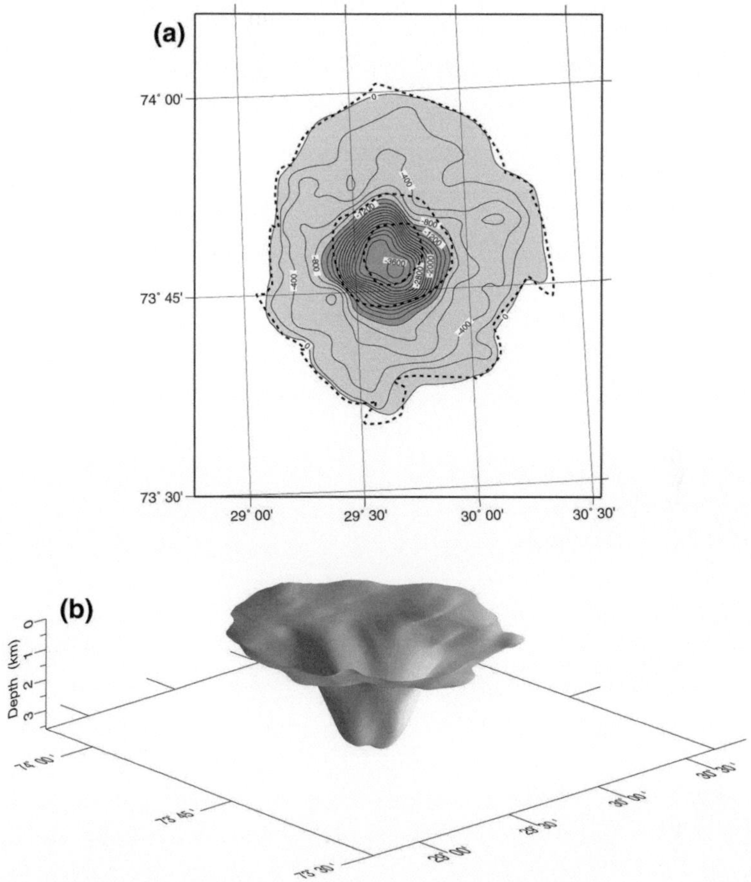

Fig. 3.19 Morphological characteristics of the intensely disturbed zone derived through mapping of the outer limit of the zone and subsequent application of a second-degree polynomial surface in order to remove the overall south-southwestern tilting of the involved strata. Automatic gridding into 0.5×0.5 km cells and subsequent contouring were carried out using the Generic Mapping Tools (GMT) software (Wessel and Smith 1998). **a** Contour plot of residual thickness. Contour interval 200 m. *Dashed lines* outline the structure's radial zonation boundaries. **b** Illuminated perspective diagram of the surface in Fig. 3.19a. The view is from 20° above the horizon looking northeast (azimuth 40°); light sources are located at azimuths 30°, 290°, and 340°. Vertical scale exaggeration ~5.5×

Creek structure, where a shallow excavated shelf surrounds the central region of deep excavation, giving the disturbance a shape similar to the form of an overturned sombrero with a deep parabolic crown and a shallow broad brim (Roddy 1977).

Chapter 4
Impact Geophysics and Modelling

Filippos Tsikalas, Jan Inge Faleide, Stephanie C. Werner, Trond Torsvik, Steinar Thor Gudlaugsson, and Olav Eldholm

4.1 Features Related to the Cratering Process

4.1.1 Excavated Crater and Breccia

During impact, the passage of the shock wave results in extensive in situ fracturing and autochthonous target rock brecciation. Target material is excavated and ejected in ballistic trajectories upward and outward from the impact site. As excavation of the brecciated volume advances, the excavated crater is formed. It delimits the provenance of material expelled from the crater and provides the void space for subsequent infilling of allogenic material and breccia (e.g., Melosh 1989; Turtle et al. 2005). Therefore, the disturbance beneath Mjølnir is expected to be associated with two types of breccia: allochtonous and autochthonous (Tsikalas et al. 1998b, 1999).

The observed seismic disturbance shows variations in the pattern of seismic reflectivity that allow us to identify and map two distinct seismic units. The upper unit exhibits minor reflectivity in places but is mostly characterized by reflection-free zones (Figs. 4.1, 4.2, 4.3, and 4.4). The lower unit comprises tilted fault blocks at the periphery of the structure (Figs. 3.6 and 4.4). The fault blocks show internal stratification, correlatable to the surrounding undisturbed platform (Figs. 3.6 and 4.4). Farther inward, clear definition of the fault blocks is lost and the unit becomes progressively dominated by disrupted and chaotic reflections, diffractions, and reflection-free zones (Figs. 3.14, 4.2, 4.3, and 4.4). This makes it difficult, in most cases, to identify exactly the boundary between the two units as the central part of the disturbance is approached. However, on some profiles the boundary is marked by a characteristic, near-continuous, low-frequency reflector (Fig. 4.4). The upper unit is interpreted to correspond to the allochtonous breccia, which is expected to consist of crushed fall-back and backwash material, possibly containing melt fragments. Similarly, the lower unit probably corresponds to the extensively

F. Tsikalas (✉)
Eni Norge AS, NO-4064 Stavanger, Norway; Department of Geosciences, University of Oslo, Oslo, Norway
e-mail: filippos.tsikalas@eninorge.com

H. Dypvik et al. (eds.), *The Mjølnir Impact Event and its Consequences*,
Impact Studies, DOI 10.1007/978-3-540-88260-2_4,

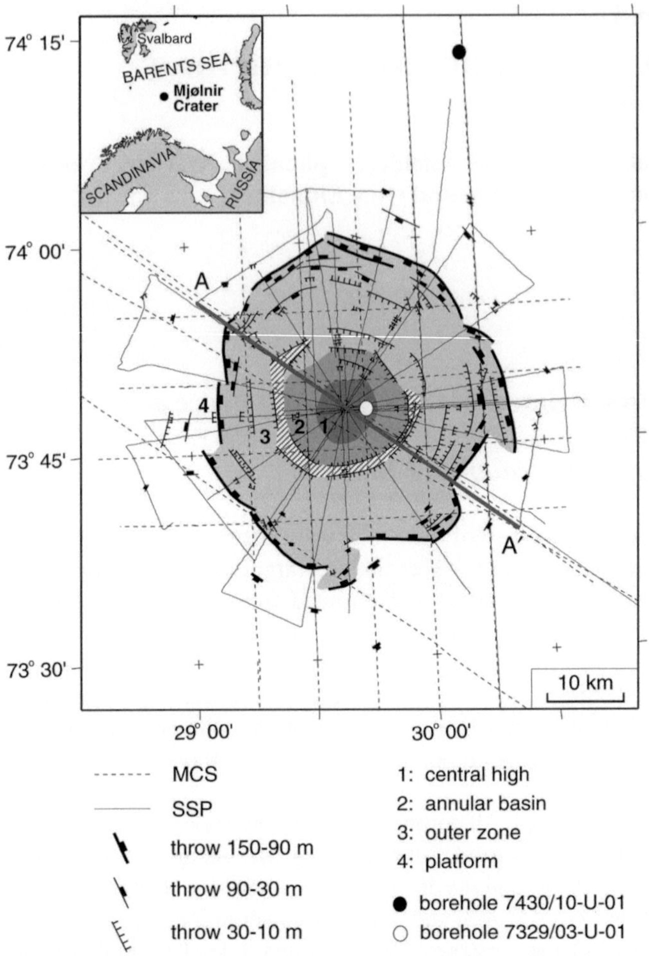

Fig. 4.1 Seismic reflection profiles superimposed on the Mjølnir structure defined by its radial zonation boundaries and the impact-induced structural elements. MCS, conventional multichannel profiles; SSP, shallow multichannel and shallow high-resolution single-channel profiles. Hatched-raster denotes clear definition of the raised peak ring

Fig. 4.2 Five interpreted multichannel profiles crossing the central Mjølnir Structure. Profiles **a**, **b** and **c** are unmigrated, while profiles **d** and **e** have been migrated by the finite difference algorithm. The stratigraphic interval confining the time of impact is shown in *black* and is bounded above and below by reflectors UB (*upper boundary*), lower Barremian, and LB (*lower boundary*), upper Callovian/middle Oxfordian, respectively. TD (top seismic disturbance, impact horizon), the first continuous reflector above the disturbed seismic reflections; DZ, area of intensely disturbed seismic reflections; TZ, transitional area of less disturbance. SF, sea floor; URU, late Cenozoic upper regional unconformity; TP, top Permian; **d**, low-angle décollement

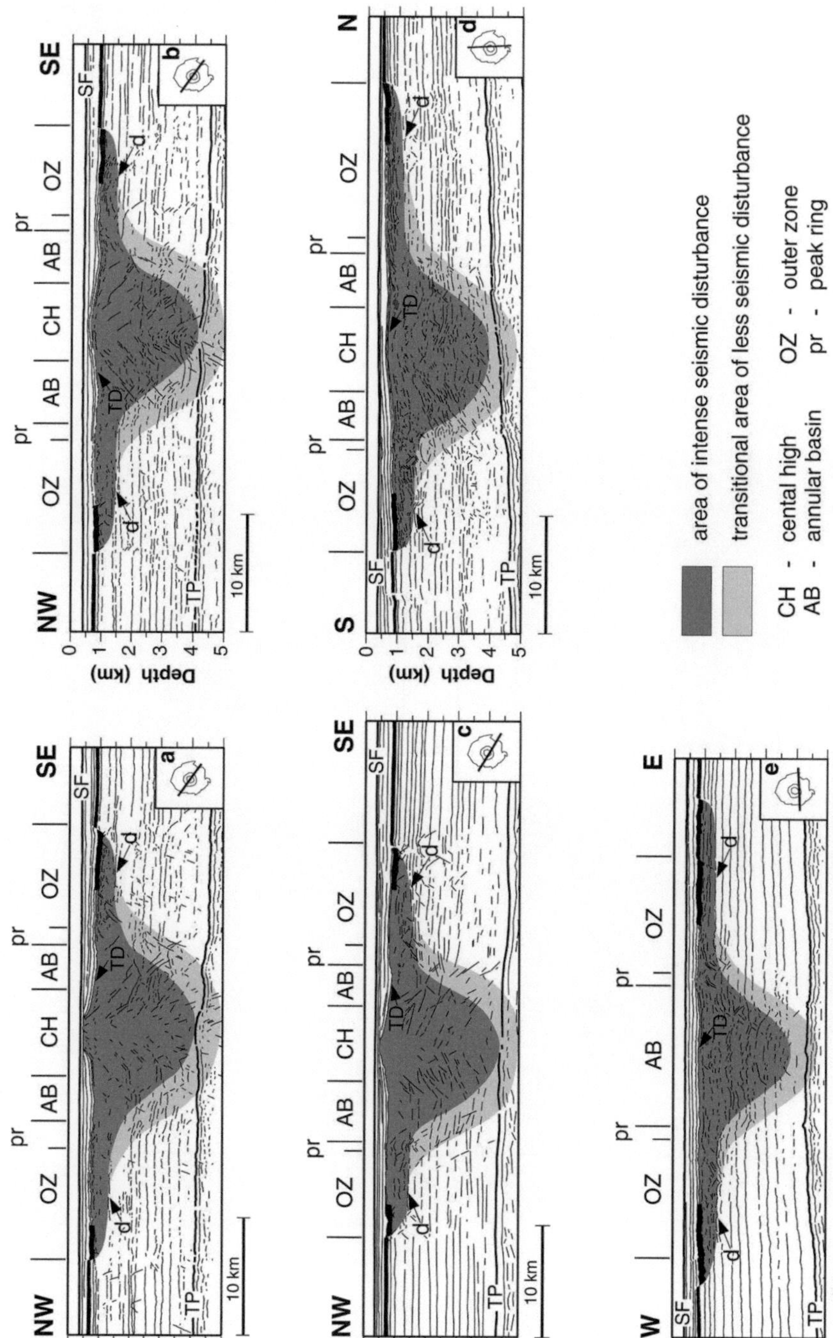

Fig. 4.2 (continued)

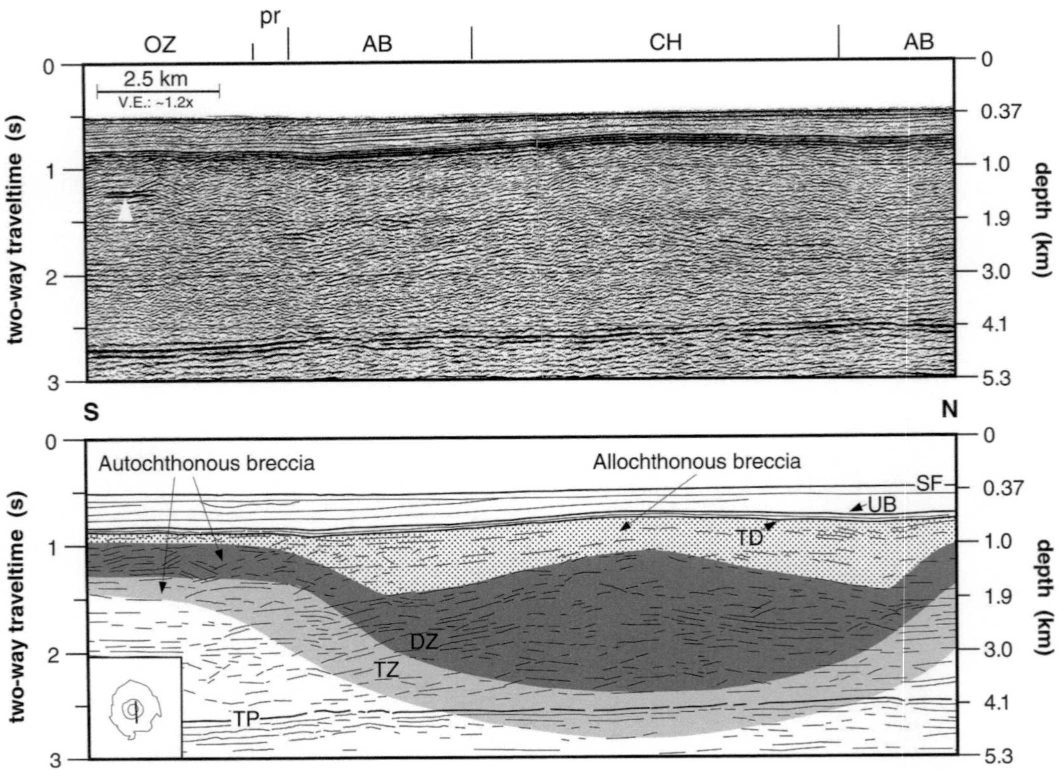

Fig. 4.3 Conventional multichannel seismic profile, and interpretation, across the central part of the structure (Fig. 4.2, part of profile **d**). The allogenic breccia is depicted by the coarse raster, while the autochthonous breccia is defined by both the area of intense disturbance, DZ, and the transitional area of less disturbance, TZ. White arrow (*upper left* in *upper profile*) marks the position of a single high-amplitude reflector, described in the text in connection with the search of identifiable melt bodies. Other annotations as in Fig. 4.2

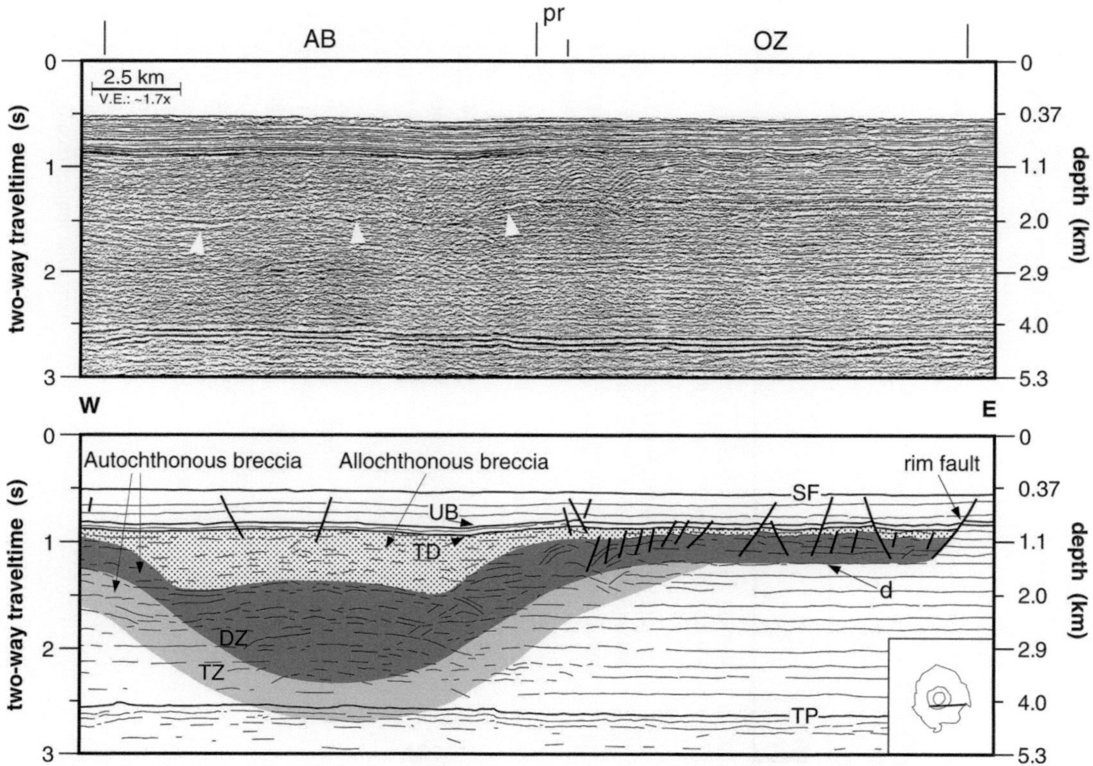

Fig. 4.4 Conventional multichannel seismic profile, and interpretation, across the crater rim (Fig. 4.2, part of profile **e**). The allogenic breccia is depicted by the coarse raster, while the autochthonous breccia is defined by both the area of intense disturbance, DZ, and the transitional area of less disturbance, TZ. *White arrows* mark the position of the near-continuous, low-frequency reflector beneath the annular basin (AB); **d**, low-angle décollement. Other annotations as in Fig. 4.2

Table 4.1 Mjølnir crater seismic features and dimensions

Morphological features	Dimensions	Character
Central high	8 km diameter	– Rises ~250 m above lower Barremian platform level – Top truncated by glacial erosion yielding a circular subcrop surface
Annular basin	4 km width	70 m depression
Outer zone	12 km width	Complex, composed of (width, km): – Peak ring (1–3) – Intermediate zone (3–6.5) – Marginal fault zone (2.5–7)
Cratering process features		
Excavated crater and breccias	True crater floor consists of: (1) a 4- to 6.5-km-wide trough with 1–1.3 km depth beneath annular basin; (2) a 0.05-to 0.2-km-thick unit in crater periphery filling in underlying faults blocks	– Two types of breccia: *allochthonous* and *autochthonous* *Allochthonous breccia*: some reflectivity locally, but mostly characterized by reflection-free zones; its top corresponds to apparent crater floor; its base defines true crater floor *Autochthonous breccia*: tilted fault blocks at periphery; turns into disrupted and chaotic refractions, diffractions and reflection-free zones towards crater center
Impact melts	Estimate: ~10–30 km^3 volume	Most probably dispersed melts in crater periphery; not yet cored
Gravitational collapse	16 km diameter and 4.5 km deep transient cavity; collapse factor, 2.5	Mjølnir exhibits much larger collapse than typical terrestrial craters (average collapse factor, 1.6)
Structural uplift	1.5–2 km	Within range (although on the lower limit) of predicted structural uplift based on empirical relations from the terrestrial impact record

in situ fractured autochthonous breccia (Grieve and Pesonen 1992). The top of the allochtonous breccia corresponds to the apparent crater floor, whereas its base defines the true crater floor (Grieve 1991). The cross-sectional shape of the allogenic breccia (Figs. 4.3 and 4.4; Table 4.1) can be described as consisting of two elements:

1. a~4- to 6.5-km-wide annular trough beneath the present annular basin with fairly steep and well-defined flanks and a maximum depth of ~1.0–1.3 km
2. a thinner unit that fills in the underlying block-faulted relief beneath the outer zone, ranging in thickness from 0.05 to 0.2 km (Figs. 3.6 , 3.14 , and 4.4).

4.1.2 Impact Melts

The energy release during a large meteorite impact is sufficient to shock heat target rock material to the point of melting. In addition, considerable melting occurs during the decompression and pressure release stages (e.g., Melosh 1989). A portion of the melts generated typically remains within the final crater in the form of compact melt bodies or sheets within the allogenic breccia or as melt dykes in the autochthonous breccia beneath the true crater floor (e.g., Melosh 1989). Such melt bodies are expected to give rise to irregular, high-amplitude reflectors. Careful search through the entire data set at Mjølnir (Fig. 3.2) showed almost total absence of significant high-amplitude reflectors within the seismically disturbed volume (Tsikalas et al. 1998a–c). Only one single high-amplitude reflector that might be associated with a melt body was found (Fig. 4.3). Within the relevant travel time range, the vertical and horizontal resolution of the deep-penetrating multichannel profiles (Fig. 4.2) is 20–50 and 300–570 m, respectively. Similarly, the values for the shallow multichannel profiles and the shallow single-channel profiles are 10–20 and 80–160, 5–10 and 40–80 m, respectively (Figs. 3.6 and 3.14). The presence of significant amounts of melt as compact, macroscopic melt bodies with dimensions comparable or larger than these dimensions can, therefore, probably be excluded.

It has been suggested that the amount of melt generated during impact is similar in sedimentary and crystalline targets (Kieffer and Simonds 1980; Cintala and Grieve 1994). However, due to high porosities and extensive expansion of volatiles on their release from shock compression, the resulting melts exhibit a more dispersed and scattered, rather than solid, character in the case of water-covered and water saturated sedimentary targets (Kieffer and Simonds 1980). Instead of distinct melt bodies, a polymictic allogenic breccia deposit, suevitic breccia (Stöffler et al. 1977), containing highly shocked melted clasts and glass, is present. The dispersed character of such features makes their identification in seismic profiles difficult. According to the empirical relationship of Cintala and Grieve (1994), an impact leading to a structure of Mjølnir's size is expected to have generated a melt volume of ~10–30 km^3. Individual magnetic anomalies are localized in the outer zone, close to the transition to the annular basin, and reveal no systematic distribution of amplitude or signs of melt (Tsikalas et al. 1998b, 1999). Due to lack of identifiable melt-derived reflectors at Mjølnir, Tsikalas et al. (1998b) interpreted the low-amplitude magnetic anomalies over the structure in terms of dislocation of weakly magnetized platform strata, possibly associated with local concentrations of dispersed melts or minor melt dykes in the peripheral region (Table 4.1).

4.1.3 Gravitational Collapse

The amount of crater collapse can be determined by comparing the dimensions of the transient cavity with the final, so-called apparent, crater size (Turtle et al. 2005). During the compression stage of the impact process, target rocks are displaced laterally and downward. The resulting parabolic-shaped cavity, referred to as the transient cavity, is about 3 times deeper than the excavated crater (Melosh 1989; Turtle et al. 2005). Because of gravitational collapse of the crater walls and rebound of the crater floor during decompression and modification of the crater, the transient cavity is a short-lived feature. In complex craters, collapse of the transient crater results in an increased crater diameter and the development of fault blocks at the periphery, an elevated crater floor, and a centrally located high (Melosh 1989; Grieve and Pesonen 1992; Pilkington and Grieve 1992).

The extent of the transient cavity at Mjølnir was estimated from the shape of the intensely disturbed zone imaged in the seismic reflection data. When extrapolated to the level of the impact horizon, reflector TD, the bowl-shaped disturbance has a diameter of 16 km, coinciding with the outer perimeter of the annular basin (Fig. 4.2). The maximum thickness of the intense disturbance is ~3.6 km and it just reaches down to the top Permian reflector. Because Mjølnir has compacted significantly under the load of a maximum postimpact overburden of 2–2.5 km (Tsikalas et al. 1998a; Tsikalas and Faleide 2007), this thickness is an underestimate. Decompaction of the Mjølnir crater and the surrounding sedimentary platform, utilizing a porosity-depth function based on well logs from the southwestern Barents Sea (Tsikalas 1992; Tsikalas et al. 1998a), results in a thickness of 4.5–5 km for the intensely disturbed zone. In addition, the seismically defined transient cavity diameter of 16 km, in combination with an empirically determined average ratio between the transient crater depth and diameter of 0.28 (Melosh 1989; Turtle et al. 2005), gives a 4.5 km depth for the transient cavity. This estimate agrees well with the decompacted thickness of the intense disturbance. Thus a paraboloid reaching down to the decompacted top Permian level and extending to the boundaries of the present annular basin approximates the transient crater.

The ~1.0–1.3 km maximum thickness of the allochthonous breccia is a minimum estimate of the depth of the excavated crater. Decompaction, utilizing the same porosity-depth function as above, results in an allochthonous breccia thickness of ~1.4–1.8 km. This value agrees very well with a depth of ~1.5–1.6 km expected on the basis of the empirically determined ratios of 1:3 and 1:10 between the excavated crater depth and transient crater depth and diameter, respectively (Melosh 1989). The volumes of the excavated and transient cavities were estimated by fitting two paraboloids of revolution to the outer boundary of the 16-km-diameter transient crater and to the decompacted base of the ~1.4–1.8-km-thick allochthonous breccia and the 4.5-km decompacted depth to the top Permian reflector, respectively. The volumes were found to be 140–180 km^3 for the excavated crater and 450 km^3 for the transient crater (Table 4.1).

The ratio of the final, so-called apparent, crater diameter to the transient cavity diameter is referred to as the collapse factor. The value of this parameter is known to lie in the range of 1.4–2.0, with an average value of 1.6 (Melosh 1989). Assuming the average value, an impact structure with a transient cavity diameter of 16 km would therefore be expected to have a final diameter of 26 km. By contrast, Mjølnir's 40-km final diameter corresponds to a collapse factor of 2.5. Hence the Mjølnir crater expanded to an unusually large degree during gravitational collapse, following a trend that is typical for many marine impact craters (e.g., Turtle et al. 2005).

Detailed mapping and analysis of the Mjølnir crater has revealed an unusually shallow crater depth and absence of a prominently raised crater rim (Figs. 3.10 and 4.2) (Tsikalas et al. 1998a). These unusual features can be explained in terms of the large degree of collapse and extensive infilling coeval with the collapse, comparable to the Flynn Creek structure (Roddy 1977). The large degree of collapse is attributed to the low-strength siliciclastic sedimentary target, typical in marine impacts. The clear expression of concentric graben-like faults at the periphery of the structure resembles detached sedimentary blocks floored by low-angle décollement surfaces (Figs. 3.6 , 4.2, and 4.4). Outward widening of the crater walls by inward collapse on tilted fault blocks is also clearly imaged in the shallow high-resolution seismic profiles (Figs. 3.6 and 3.14). The impact in a marine environment created a water cavity (Fig. 4.5), resulting, among others, in turbulent back-rush water flow that probably transported large amounts of material back to the crater, accounting for both the unusually extensive infilling and the absence of a raised crater rim (Tsikalas et al. 1998a). Although the extensive collapse experienced by the Mjølnir crater is unusual in comparison with typical terrestrial craters, there is clear empirical and experimental evidence that impact craters in fluid-rich sediments collapse to a larger degree than those in crystalline targets (Roddy 1977; Melosh 1989; Dypvik and Jansa 2003; Turtle et al. 2005). In addition, the shallow expression of the final crater is consistent with laboratory experiments showing that meteorite impact in unconsolidated, water-covered sedimentary targets may result in subdued crater morphology (Gault and Sonett 1982; McKinnon 1982).

4.1.4 Structural Uplift

Numerical modeling of cratering mechanics indicates that the central high in complex craters originates within the transient cavity by a process referred to as structural uplift, involving rebound of the crater floor and upward elevation of deep, denser strata relative to the surroundings (Melosh 1989; Morgan et al. 2000). The structural uplift in the center and the collapse of the periphery are probably intimately linked. The mechanical basis for the process that allows the crater to slump and yet preserve a stable high in the middle is not well understood but probably related to the rheology created by the impact. It has been suggested that the crater collapses in a regime of acoustically fluidized target rocks and as the acoustic energy is dispersed the fluidized rocks in the central high simply locks in (Melosh 1989; Grieve 1991).

Integrated geophysical modeling of potential field and seismic travel time anomalies associated with the Mjølnir crater has demonstrated the presence of rocks with

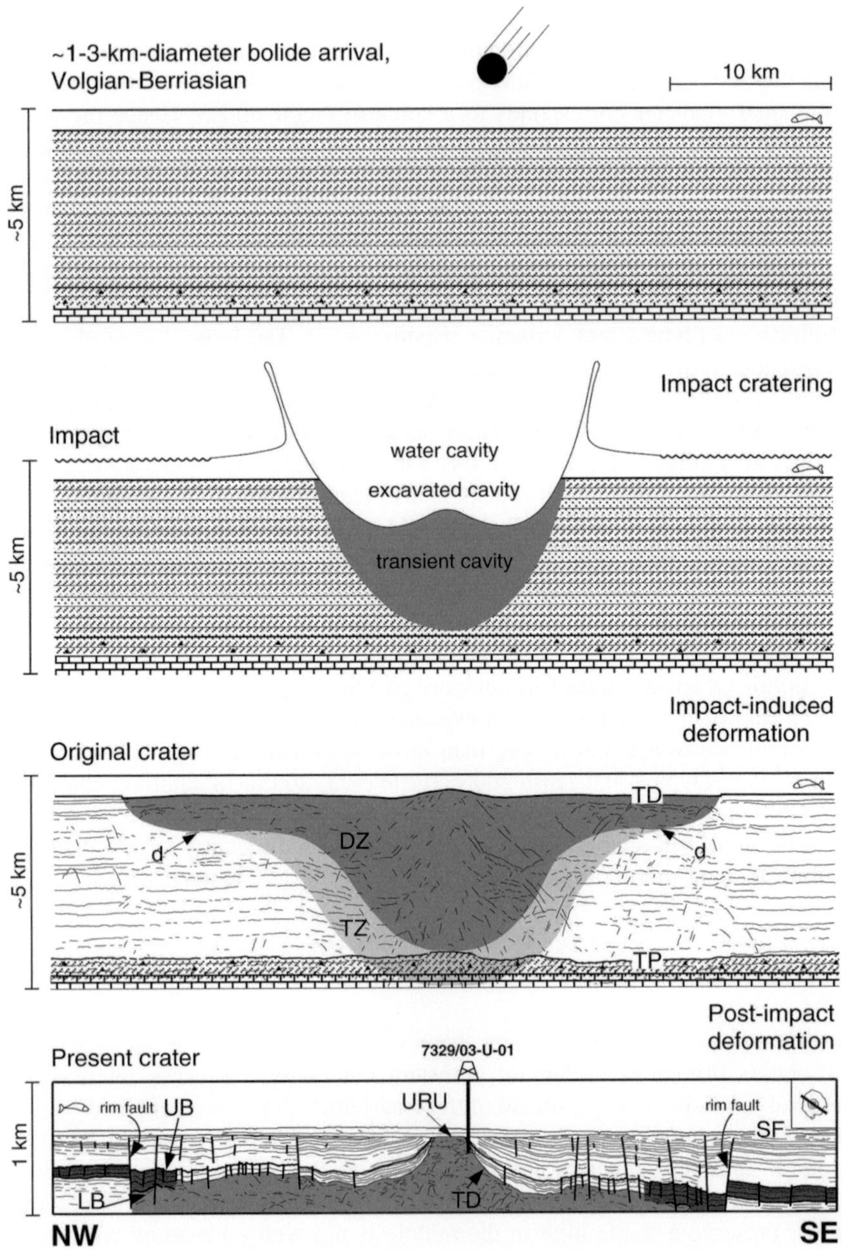

Fig. 4.5 Schematic cross sections showing the Mjølnir physical impact and deformation types. DZ, area of intense disturbance; TZ, transitional area of less disturbance; TP, top Permian; d, low-angle décollement. Although the impact disturbance gradationally fades out at depth, probably extending below TP, this reflector is not influenced by the impact induced structural uplift and thus constrains the volume of appreciably deformed strata. Other annotations as in Fig. 4.2

significantly higher densities and velocities beneath the central high, than the undisturbed platform outside the structure. These physical property anomalies have been attributed to the prevalence of processes related to crater floor uplift at the central high over impact-induced brecciation and gravitational collapse (Tsikalas et al. 1998b, c, 1999, 2002b).

Several of the seismic profiles crossing the Mjølnir crater exhibit features that can be attributed to structural uplift. In particular, there is evidence of upward bending reflector segments beneath the central high and the annular basin, indicating elevation of deep strata to shallower levels (Fig. 4.2). The upwards bending of the strata persists after migration (Fig. 4.2, profiles d and e) and can be used to estimate the amount of structural uplift. However, due to the chaotic and incoherent character of the seismic disturbance at Mjølnir, it is only possible to provide a rough, minimum estimate of the amount of uplift from the seismic profiles. The faulting during collapse is predominantly of normal, steep-angle type (Melosh 1989), thus subhorizontal structures may be used as marker horizons to estimate the amount of uplift (Juhlin and Pedersen 1987). Such estimates are made by measuring the difference in depth between the extrapolated top of selected upward bending reflector segments beneath the central high and their most likely equivalent subhorizontal interfaces. In a few favorable cases (Fig. 4.2, profiles a, b, and d) this method yields direct estimates for structural uplift that approximate 1.0–1.5 km and when decompacted ~1.5–2.0 km (Table 4.1). These values fit with the predicted structural uplift based on the dimensions of the Mjølnir structure and the empirical relations of Pilkington and Grieve (1992) and Cintala and Grieve (1994) that have a best estimate of 2.5 and 2 km, with a possible range of 1–6 and 1–5.5 km, respectively.

4.2 Impact into a Marine Sedimentary Basin

Although the principles of cratering mechanics in subaerial, non-aqueous, targets have been largely established (e.g., Melosh 1989), the nature of the cratering processes will vary, depending on whether the target is crystalline or sedimentary, and on the presence of water (e.g., McKinnon 1982; Ahrens and O'Keefe 1983; Sonett et al. 1991). Subaquatic/submarine target impacts are still poorly understood (Dypvik and Jansa 2003; Dypvik et al. 2004a) and the global cratering record introduces a bias towards crystalline, water poor targets in models of typical impact structures. Nonetheless, integrated studies of craters believed to have formed in a marine environment, attribute several features associated both with the final crater and the related deposits to the presence of water (e.g., Jansa 1993; Poag 1996). Similarly, there is both empirical and experimental evidence that impact craters in fluid-rich sediments collapse to a larger degree than those in crystalline targets (Roddy 1977; Melosh 1989; Dypvik and Jansa 2003). In particular, laboratory experiments have shown that meteorite impacts in unconsolidated, water-covered sedimentary targets may result in more modulated crater topography than statistically inferred from known terrestrial impact craters (Gault and Sonett 1982). In addition, marine impacts have greater chances to be preserved because they are immediately covered by post-impact sediments, which in some cases can reach considerable thicknesses. Finally, sediment loading above the primary impact relief may result in substantial post-impact deformation and structural modification. However, the quantification of such post-impact effects is almost entirely absent from the terrestrial impact record.

In this context, the seismic mapping of the Mjølnir crater profiles has revealed some unusual features (Tsikalas et al. 1998a, b, 2002a). In particular, the expected step-like terraces at the periphery are replaced by prominent fault-blocks floored by apparent low-angle décollement surfaces (Fig. 4.2). Continuous reflectors beneath the fault blocks are best imaged by the shallow multichannel seismic profiles (Fig. 3.6). In addition, the seismic profiles image extensive post-impact deformation expressed by structural reactivation and differential subsidence (Fig. 4.5). It is postulated that the shallow expression of the initial crater is related to a large degree of collapse and of coeval extensive marine infilling. The outward structural expansion resulted from the inward collapse of the initial crater rim along faults floored by apparent low-angle décollement surfaces at the periphery (Tsikalas et al. 1998b, 2002a). Towards the centre, the collapse is characterized by debris/mass flows. The massive collapse was probably caused by the low strength of the siliciclastic sedimentary target. Tsikalas et al. (1998b) have suggested that the collapse of the impact-induced water cavity (Fig. 4.5) and the subsequent rapid surge of seawater into the excavated crater transported large amounts of ejecta and crater wall material back into the crater, accounting for the extensive infilling. Hydraulic excavation by back-rushing turbulent seawater may explain the lack of a raised crater rim. Subsequently, the extensive post-impact deformation is triggered by prograding post-impact sedimentation (Fig. 4.5) and governed by the instability and radially-varying changes in physical properties within the impact-affected rock volume (Tsikalas et al. 1998c, 2002a; Tsikalas and Faleide 2007). It is also evident that the post-impact deformation has considerably enhanced the structural expression of the original, subtle crater (Fig. 4.5; see also Chap. 9). Thus, the present distinct expression of Mjølnir is largely a post-impact burial phenomenon.

The Mjølnir crater is, in terms of its stratigraphy, morphology and structure, similar to the majority of large complex craters (Grieve 1991; Melosh 1989; Pilkington and Grieve 1992; Grieve and Pesonen 1996; Turtle et al. 2005). The central high, annular basin and outer zone are common features of impact craters of this size on Earth. Key features of Mjølnir compatible with its impact-origin include:

(a) the distinct rim faults that form a circular pattern and separate highly deformed strata within the crater from undisturbed platform strata;
(b) the prominent tilted fault-blocks within the marginal fault zone;
(c) the variably thick breccia unit with incoherent seismic reflectivity caused by the impact.

Thus, these seismically-mapped features, together with the discovery of impact signatures (Dypvik et al. 1996), make a strong case for an impact origin of the

Mjølnir crater. However, a few of the observations are not typical of impact craters. These include the unusually shallow relief both of the original and the present crater, and the extensive post-impact deformation over the structure.

Utilizing the well-established empirical relationships of Schmidt and Holsapple (1982) and Melosh (1989) supplemented by seismic observations, it was estimated that a crater volume of ~180 km^3 was displaced from the crater (Tsikalas et al. 1998b). Recently, numerical simulations increased the displaced volume to ~230 km^3 and showed that the layer of primary ejecta is expected to fall off rapidly with distance from the crater center; more than 60% of the ejecta volume will be deposited within the crater's final diameter (Shuvalov et al. 2002; see also Chap. 10). Moreover, the first effort to compare the theoretically predicted and the reconstructed crater reliefs across Mjølnir at the time of impact was conducted by Tsikalas et al. (1998a). In the light of the numerical simulation results (Shuvalov et al. 2002), we now provide updated and better constrained estimates for the predicted Mjølnir crater depth. Here, crater depth refers to the apparent crater depth, i.e., the depth relative to the pre-impact surface to reflector TD, and not to the true crater depth corresponding to its base. We also estimate the volume of excess infilling experienced by Mjølnir as a result of the impact-generated water-cavity collapse and formation of resurge gullies that channellized material flow back to the crater site.

We estimate the theoretically predicted relief across the Mjølnir crater (Fig. 4.6) using the volume balance method of Croft (1985) and Melosh (1989). The method is based on a geometrical model that equates the volume of the parabolic-shaped transient cavity to the volume of the flat-floored final crater, assuming mass conservation under collapse of the material surrounding the transient crater. On the basis of numerical simulations integrated with seismic observations a transient cavity of 16–20 km in diameter was determined, translating to a collapse factor of 2–2.5 for the 40-km-final-diameter Mjølnir crater (Gudlaugsson 1993; Tsikalas et al. 1998b; Shuvalov et al. 2002). Using these values, an average empirical 0.28 ratio between the transient crater depth and diameter (Melosh 1989) implies depths for the transient cavity in the range of 4.5–6 km (Fig. 4.6). On the basis of this range of transient crater dimensions and assuming balanced volumes, we calculate a theoretically expected crater depth for Mjølnir of 250 $\pm$ 100 m (Fig. 4.6). This value differs considerably from the average depth of ~30–40 m and a maximum depth of ~70 m in the annular basin obtained through reconstruction of the original crater relief (Tsikalas et al. 1998a; Tsikalas and Faleide 2004). We also calculate the volume of the excess resurge infilling affecting Mjølnir, i.e., the volume difference between the predicted and the reconstructed crater surfaces to be ~50 km^3, which approximates one-third to one-fifth of the total excavated/ejected volume (Fig. 4.6).

Siliciclastic deposits dominate the post-Permian sedimentary succession of the southwestern Barents Sea (Worsley et al. 1988; Gabrielsen et al. 1990). The Mjølnir crater was formed by an impact into a shallow-marine, Volgian-Berriasian, sedimentary basin (Smelror et al. 2001a; Dypvik et al. 2004b, c). The impact resulted in an unusually shallow structure without a raised crater rim. We postulate that the difference in predicted and observed crater depths is related to the large degree of

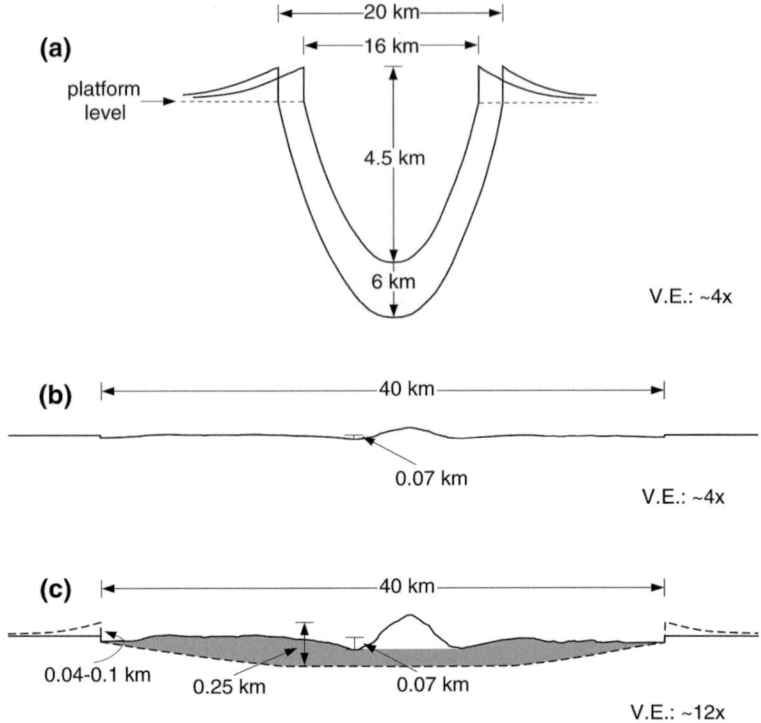

Fig. 4.6 Schematic diagram of crater collapse and infilling at the Mjølnir crater. (**a**) Estimated range of transient cavity dimensions. (**b**) Reconstructed relief from Tsikalas et al. (1998a). (**c**) Theoretically predicted relief utilizing the volume balance method of Croft (1985) and Melosh (1989) for a collapse factor of 2–2.5 (*dashed line*). Reconstructed relief from (**b**) (*solid line*) is shown for comparison. We neglected the volume of the central high because it is considerably less than the final crater volume, and thus does not affect the calculations. Shading denotes the additional infilling affecting Mjølnir (modified extensively from Tsikalas et al. 1998a). v.e. = vertical exaggeration

collapse (Gudlaugsson 1993; Tsikalas et al. 1998b) and unusually extensive infilling coeval with the collapse (Fig. 4.6). Outward widening of the structure resulted from the inward collapse of the initial crater rim on listric faults thought to be floored by a low-angle décollement (Figs. 4.2 and 4.5). The large degree of collapse was probably caused by the low strength of the siliciclastic sedimentary target. However, the volumetric calculations (Fig. 4.6) demonstrate that the shallow crater depth cannot be explained only in terms of collapse. The collapse of the impact-induced water cavity and the subsequent rapid surge of seawater into the excavated crater probably transported large amounts of ejecta and crater wall material back into the crater, accounting for the extensive infilling. Similarly, hydraulic excavation by back-rushing/resurging turbulent seawater resulted in erosion of an uplifted rim and explains the lack of a raised crater rim (Figs. 3.6 , 4.2, and 4.6c) (Poag 1996). These processes left a subtle crater relief (Fig. 4.5), forming the depositional surface for subsequent sedimentation.

4.3 Impact Crater Modelling

4.3.1 Potential Field Data

Bathymetric measurements along the tracks in Fig. 3.2 show a present, smooth, gently undulating sea floor surface at an average depth of 0.36 km. We observe no bathymetric expression of the underlying, buried Mjølnir crater (Fig. 4.7).

The gravity field was measured along the tracks in Fig. 3.2 at 1 min, 0.2 km, sampling intervals and reduced to free-air anomalies. The data are of high quality

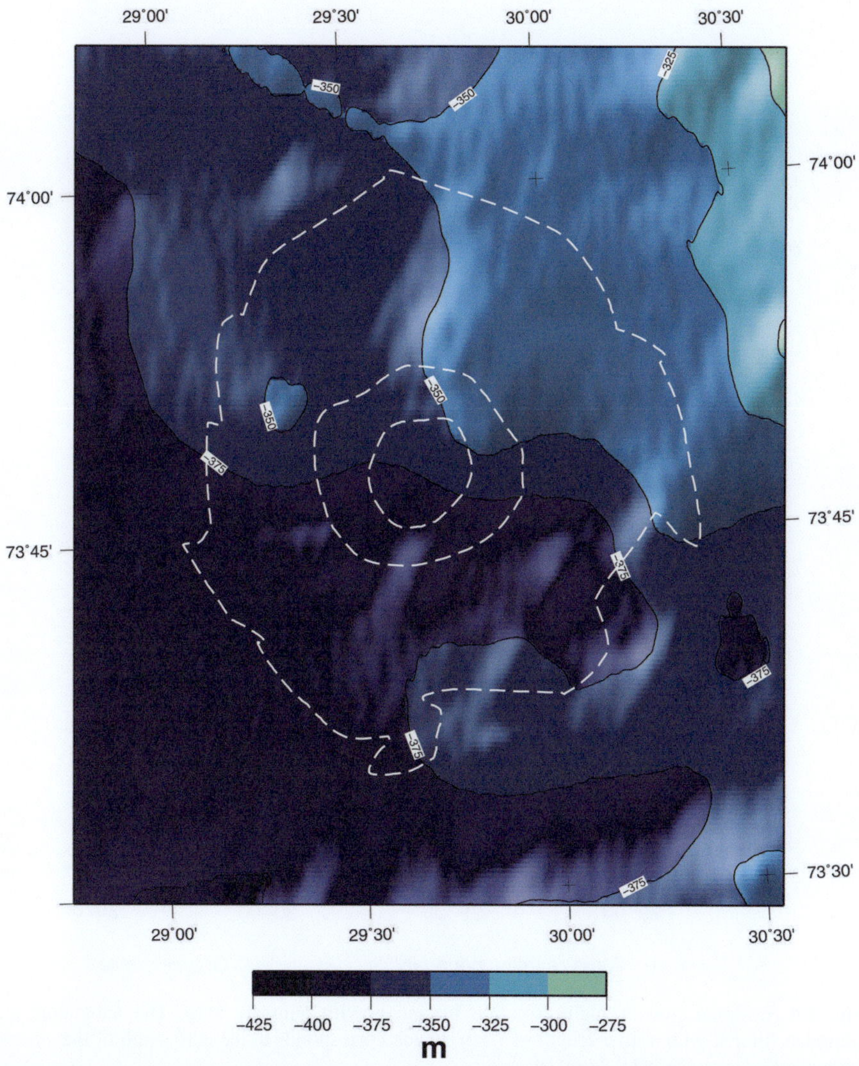

Fig. 4.7 Bathymetric relief and Mjølnir crater zonation boundaries. Contour interval 25 m

and analysis of cross-over errors indicates minimal deviations. The gravity field on the Bjarmeland Platform is dominated by a north-northeast striking regional trend (Breivik et al. 1995), while Mjølnir exhibits a distinct, radial gravity signature comprising a conspicuous gravity high superimposed on a wider low. The observed data were Gaussian-filtered, with a filter length of 7 km. In order to isolate the Mjølnir free-air anomaly, the regional trend was removed by applying a third-order polynomial curve to the filtered data along individual profiles using the Generic Mapping Tools (GMT) software package (Wessel and Smith 1998). The polynomial was chosen to obtain a mean residual anomaly that is approximately zero along each profile. Then, the residual free-air field was automatically gridded into 0.5 × 0.5 km cells and contoured using GMT (Figs. 4.8 and 4.9).

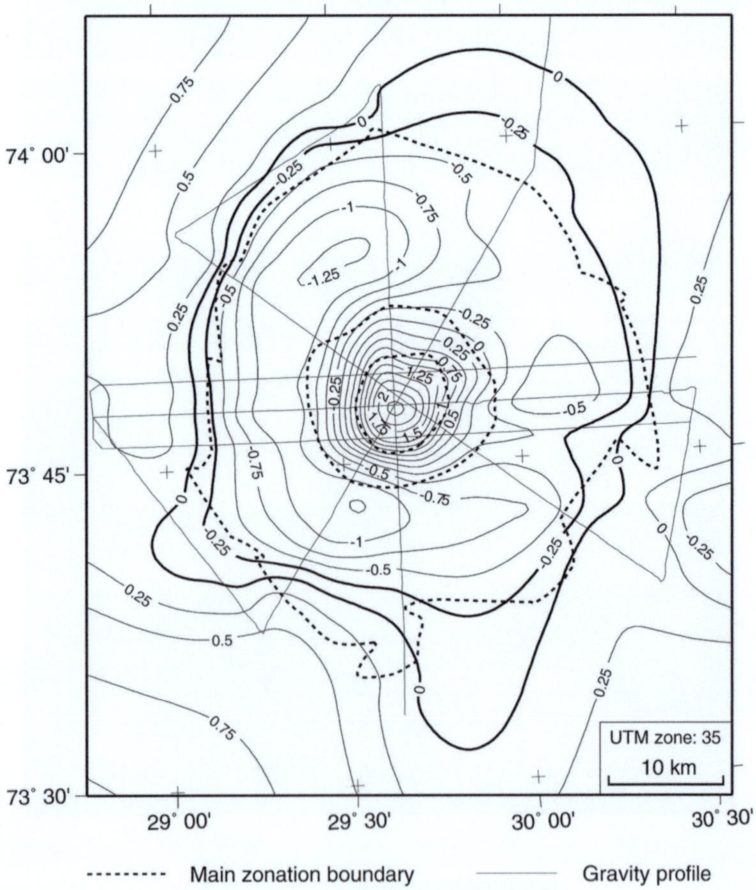

Fig. 4.8 Residual free-air high-resolution marine gravity anomaly map. The anomalies are Gaussian-filtered with a filter length of 7 km, which corresponds to the half-width of the central anomaly. Contour interval 0.25 mGal

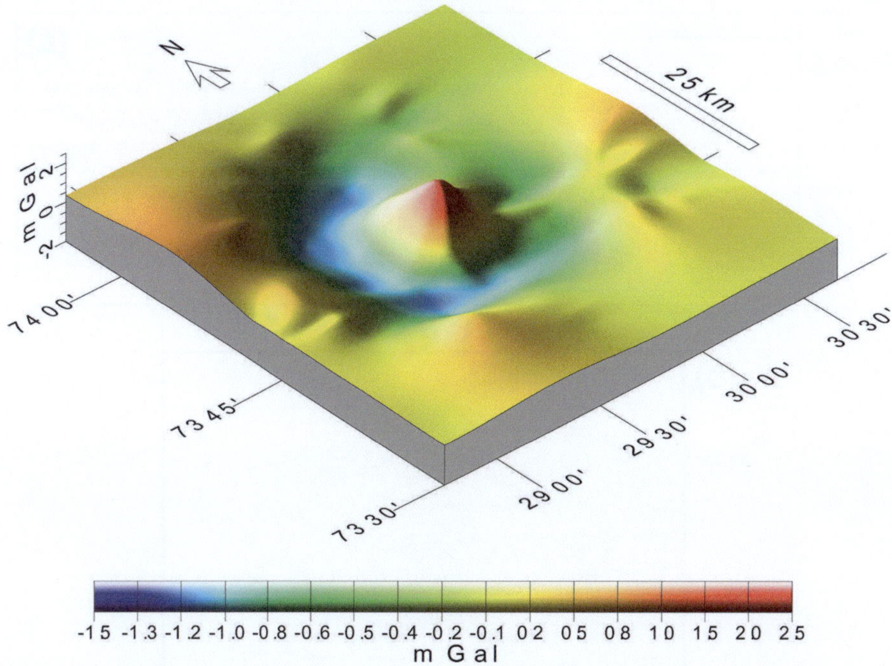

Fig. 4.9 Illuminated perspective image of the surface in Fig. 4.8 (residual free-air high-resolution marine gravity anomaly map). The view is from 30° above the horizon looking northeast (azimuth 40°); light source at azimuth 290°

The total magnetic field was recorded along track (Fig. 3.2) at 7.5 s, 0.025 km, sampling intervals. The total field values over the structure correlate well with coeval values from magnetic base stations at Bjørnøya, Hopen, and Tromsø, located 300-600 km from the Mjølnir crater (Fig. 4.10). Due to the local character of the survey, residual magnetic anomalies were obtained by adjusting the base station and profile records to the same level, and correcting the observed field for temporal variations, by using the base station records as reference. The profiles were corrected for cross-over errors and the data were Gaussian-filtered, applying a filter length of 10 km. The same procedure as for gravity was applied in removal of the regional field gridding, and contouring. The residual anomaly map reveals several local, low-amplitude anomalies within the ±100 nT range (Fig. 4.10).

4.3.2 Marine Gravity Anomalies and Modelling

The residual free-air gravity field shows a circularly symmetric anomaly over Mjølnir (Figs. 4.8, 4.9 and 4.11). The anomaly is composed of an annular low with an outer diameter of 45 km, attaining minimum values of –1.5 mGal over the outer disturbed zone, and a 14-km - wide central gravity high, with a maximum value of

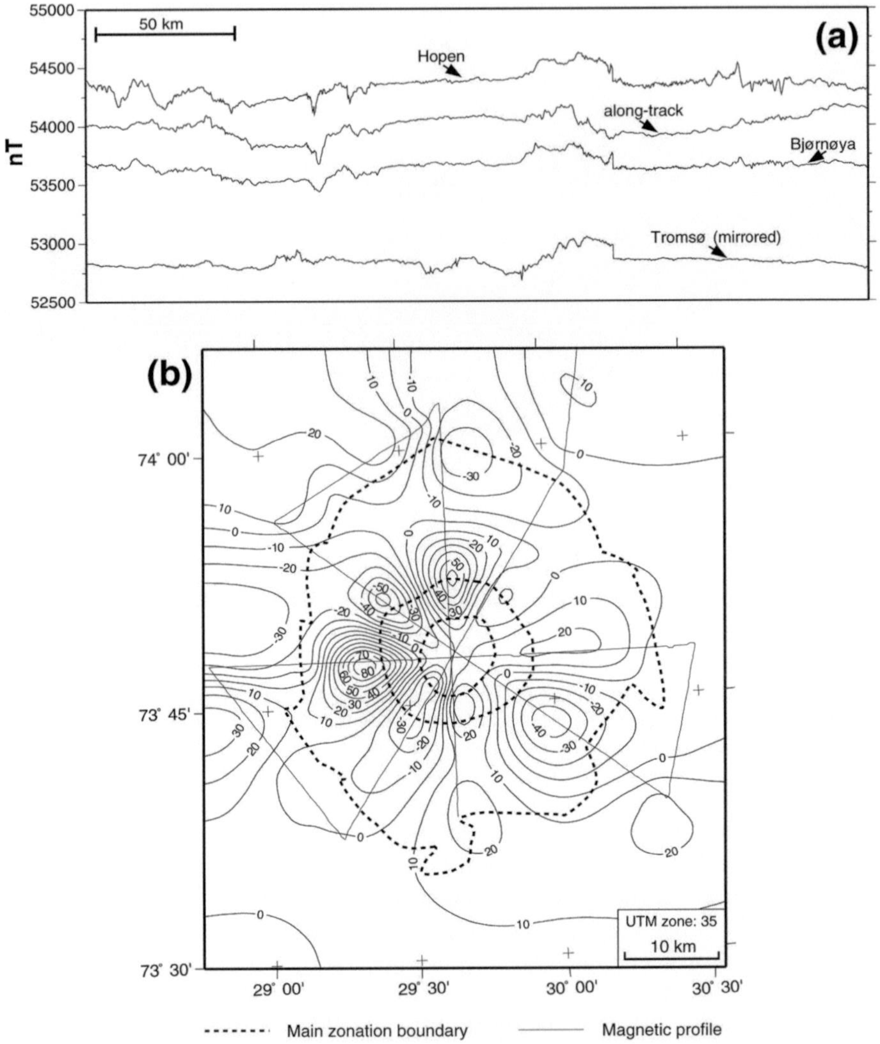

Fig. 4.10 Magnetic data: (**a**) recorded total field values over Mjølnir in correlation with coeval values from magnetic base stations at Bjørnøya, Hopen, and Tromsø (located 300-600 km from the Mjølnir crater), (**b**) residual magnetic anomaly map. The anomalies are Gaussian-filtered, with a filter length of 10 km. Contour interval 10 nT

+2.5 mGal. Due to the close spatial correspondence between the impact structure and the residual field (Figs. 4.8, 4.9 and 4.11), we believe that the anomaly is real and closely related to the radially zoned structure.

In the subsequent modelling, sediment densities were determined by converting seismic interval velocities into densities using the Nafe and Drake curve (Ludwig

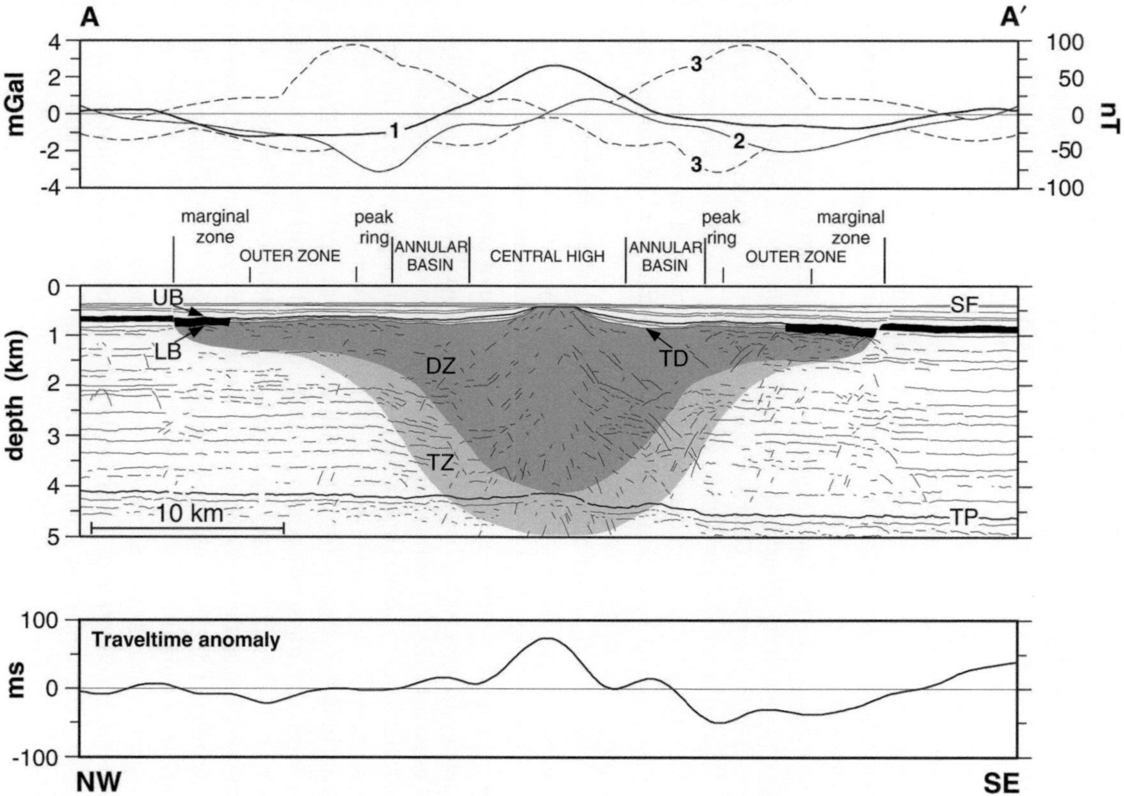

Fig. 4.11 Geophysical type-section of the Mjølnir crater (profile AA′, Fig. 4.1). The observed gravity, magnetic and traveltime data were Gaussian-filtered with filter lengths of 7, 10 and 10 km, respectively. *Top*: 1, free-air gravity anomaly; 2, residual magnetic anomaly; 3, "envelope"-magnetic anomaly. *Middle*: Interpreted seismic section based on both a high-resolution single-channel profile and a multi-channel profile recorded along the same line. Annotations as in Fig. 4.2. The profile was depth converted using smoothed interval velocities derived from stacking velocities (Tsikalas et al. 1998c). *Bottom*: two-way traveltime anomaly with reference to a planar top Permian reflector

et al. 1970; Barton, 1986). A well-log velocity-density relationship has previously been developed for the Hammerfest Basin and Loppa High (confer Fig. 1.7) (Tsikalas 1992). This relationship yields slightly higher values than the Nafe and Drake empirical curve. This difference, however, is not considered important and we have chosen the Nafe and Drake curve because of its common use.

A two-dimensional geometric model was constructed along profile AA′ from the depth-converted seismic interpretation (Figs. 4.1 and 4.11). Mjølnir is embedded in a platform sequence tilted slightly towards south. The removal of the regional effects allows us to consider the geophysical anomalies at Mjølnir as residual anomalies (confer Figs. 4.12 and 4.13). Thus, only the disturbed region is modelled, and the physical properties determined are all relative to the adjacent platform. In constructing the geometric structural model, we have also removed the regional tilt from the seismic profiles. This geometry (Fig. 4.11) provides the initial configuration for the gravity, magnetic and traveltime modelling.

To better simulate the gravity (and magnetic) effects of circular bodies, a pseudo-three-dimensional modelling procedure was applied using the 2.5-dimensional algorithm developed by Northwest Geophysical Associates (1991) based on Talwani et al. (1959), Talwani and Heirtzler (1964), and Won and Bevis (1987). Our model consists of a conical core, 2.2 km in diameter at the top, followed by symmetrical, 0.3-km-thick plates with cross-sectional lengths commensurate with the dimensions of the inverted-sombrero-shaped disturbed zone.

Initially, the entire seismic disturbance was modelled with a uniform density contrast. This resulted in a partial fit with the observed gravity anomaly, either of the central gravity high or of the peripheral low. This approach documents density differentiation within the disturbed zone. A main consequence of an impact event is a brecciated volume of target rocks that exhibits impact-induced increase in porosity (Grieve and Pesonen 1992). In contrast large complex craters, however, are characterized by the formation of a central uplift due to the upward displacement of deeper and sometimes denser strata to shallower levels during rebound of the crater floor and gravitational collapse (Grieve 1991; Melosh 1989; Turtle et al. 2005). The former originates within the transient cavity beneath the central part of the structure and is referred to as structural uplift. Applying the empirical relations of Pilkington and Grieve (1992), we estimate the expected structural uplift for a structure of Mjølnir′s size to be in the order of 3.5 km, with a possible range of 1.5–6.0 km (Tsikalas et al. 1998b). Therefore, a central core with a positive density contrast, with respect to the surroundings, was included in the model. On the other hand, fracturing, brecciation, and gravitational collapse will lower the densities of both the uplifted and the peripheral region, i.e., annular basin and outer zone. The flanks of the central high are determined from the seismic character of the disturbance, separating upward bending core strata from downward displaced strata beneath the central high and annular basin (Fig. 4.12). Such density distribution, i.e., central uplift vs. periphery, is typical of craters with structural uplift (Sharpton and Grieve 1990; Pilkington et al. 1994; Plescia et al. 1994; Espindola et al. 1995; Plescia 1996).

Interactive gravity modelling, applying a range of density contrasts, showed that a satisfactory fit to the observed values was best obtained in a simple way by assigning a single uniform density to the area of intense disturbance and the less

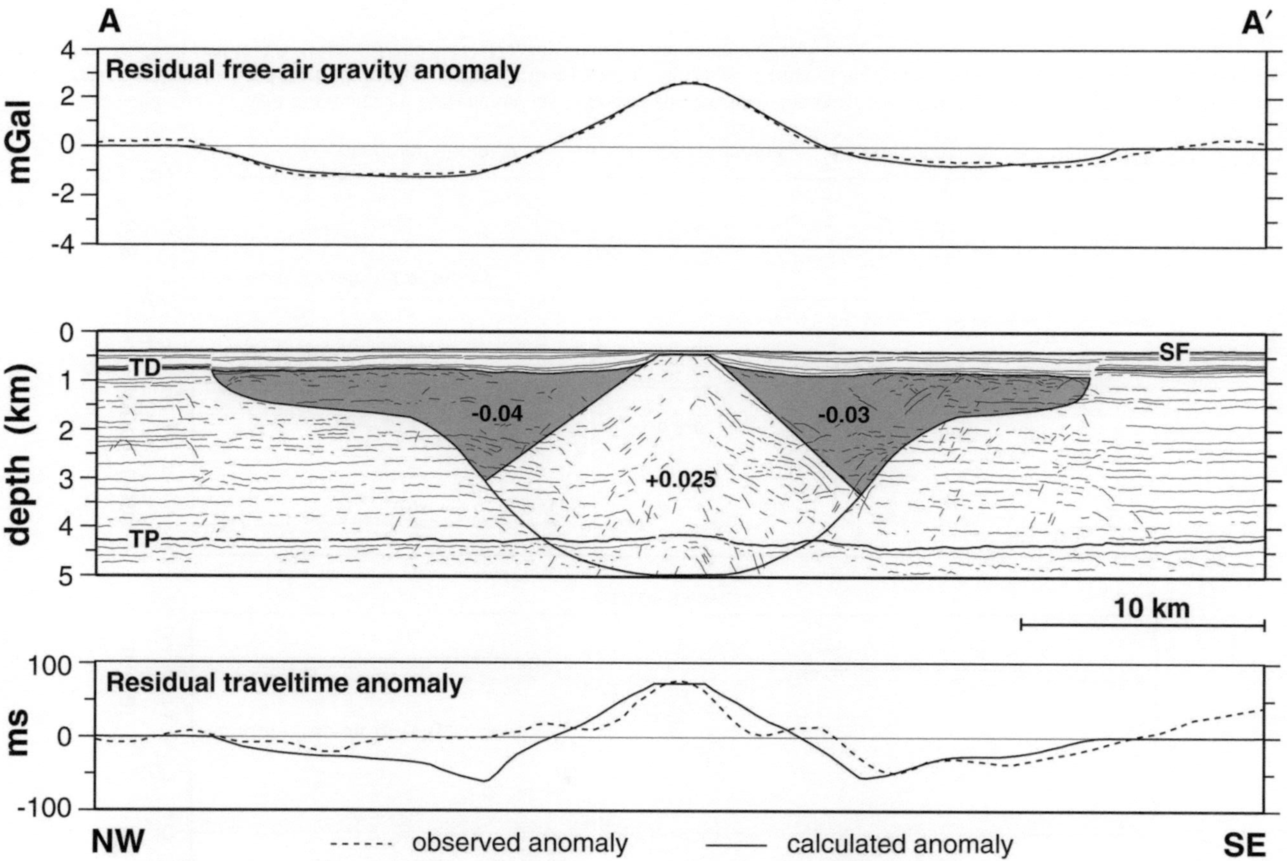

Fig. 4.12 Initial impact structure model and observed and calculated free-air gravity and traveltime anomalies. The model geometry is superimposed on the seismic interpretation in Fig. 4.11, corrected for regional tilt. Density contrasts are given in g/cm^3

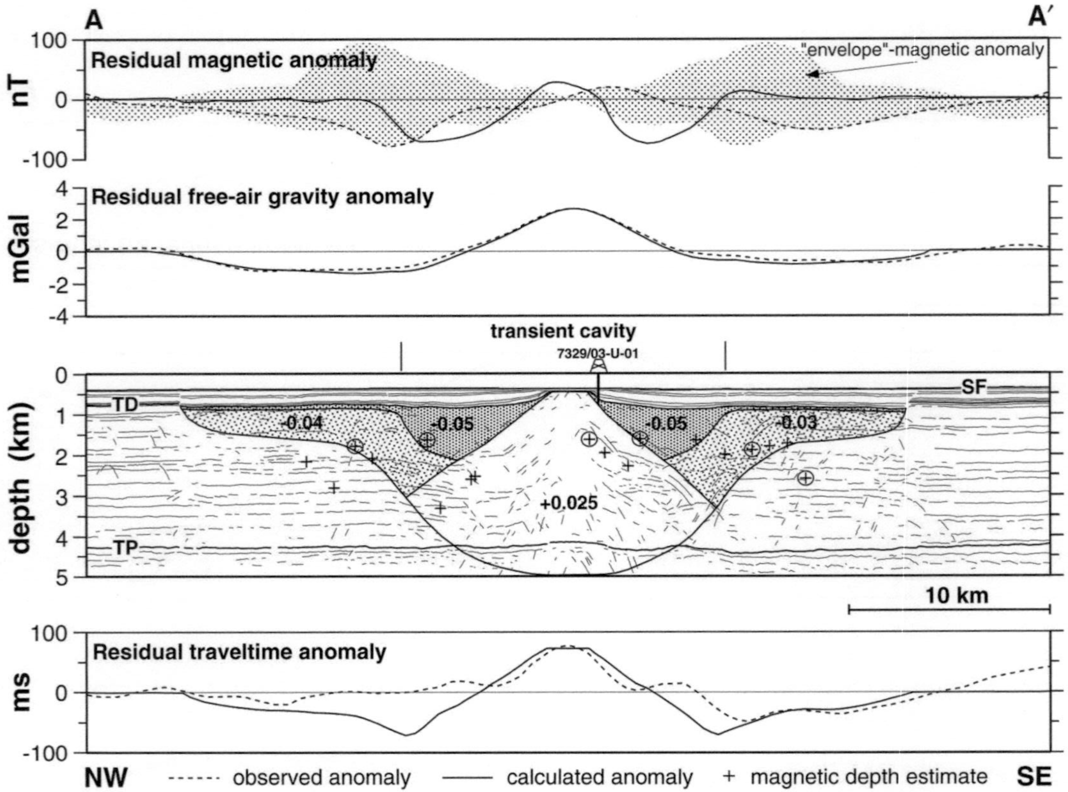

Fig. 4.13 Impact structure model with differentiated brecciation and observed and calculated gravity, magnetic and traveltime anomalies. Magnetic source depth estimates refer to all profiles in Fig. 4.10. In view of the circular structure, the depth estimates have been plotted according to distance from the centre. Circled depth estimates from profile AA′ (Figs. 4.1 and 4.11). Density contrasts are given in g/cm^3

disturbed transitional area (Fig. 4.12). The model yields a positive density contrast of 0.025 g/cm^3 for the central high, whereas the peripheral region requires a negative density contrast of -0.03 to -0.04 g/cm^3 reflecting the circular gravity low. In addition, the disturbance beneath the outer zone had to increase in thickness by about 0.2 km, an increase within the resolution of the interpreted seismic data (Fig. 4.12). The modelled and observed gravity show a close fit, and the root-mean-square deviation is only ±0.2 mGal. Consequently, the gravity field is consistent with the geometry and lateral density distribution of an impact feature.

4.3.3 Marine Magnetic Anomalies and Modelling

The magnetic anomaly along the type-section (Fig. 4.11) exhibits a broad, low-amplitude field ranging from –75 to 20 nT. The 20 nT central anomaly, offset about 1.8 km with respect to the central topographic high, is surrounded by well defined lows corresponding to the outer zone. The magnetic character along the type-section, combined with its northwest-southeast azimuth, indicates a potential remnant magnetization acquired during a period of a reversed magnetic field. Figure 4.11, however, reveals no systematic distribution of the amplitudes or signs of the individual anomalies in the outer zone, close to the transition to the annular basin. For example, the northwestern low along the type-section (Fig. 4.11) is surrounded by two distinct positive anomalies. Nonetheless, to evaluate the origin of the anomalies, we have qualitatively investigated the spatial correlation of anomalies and seismically interpreted structural features. Figure 4.10 shows that local anomalies tend to cluster at the outer zone-annular basin region. To further illustrate this, regardless of track direction, we have projected all profiles crossing through the centre onto profile AA′ and constructed the corresponding minimum and maximum magnetic "envelope" anomaly curves (Fig. 4.11). Similar to gravity modelling, magnetic modelling was performed along the type geophysical section (Fig. 4.13).

It is well established that impacts result in the formation of polymictic breccia deposits, so-called suevitic breccias (Stöffler et al. 1977), containing fragments of highly shocked clasts and dispersed melt glass. These breccias occupy a stratigraphic position in marine target areas similar to the allochthonous breccia and the melt bodies in crystalline targets (Grieve and Pesonen 1992), overlying extensively fractured autochthonous breccias. The amount of melt generated by impacts into sedimentary targets is expected to be similar to that generated in crystalline target-rocks. Being much more dispersed and scattered (Kieffer and Simonds 1980), it is more difficult to identify than compact melt bodies or sheets. This may explain the absence of high amplitude, melt-derived reflections (Figs. 4.11 and 4.13). Hence, we have not modelled the impact-induced melts as distinct and compact bodies. However, even dispersed melt clasts may be magnetized during cooling, and have the potential to produce local magnetic anomalies.

Consequently, we developed a more detailed model where the annular basin and the outer zone are divided into an upper suevite and a lower autochthonous body. The seismically-defined diameter of the bowl-shaped transient cavity approximates the extent of the present annular basin, i.e., 16 km. In addition, the thickness of the allochthonous breccia has been determined to ~1.0–1.3 km, based both on seismic observations and morphometric calculations (Tsikalas et al. 1998b, 1999). Therefore, the main allochthonous, suevitic, breccia body was modelled as a lens within the annular basin (Fig. 4.13). Gravity modelling produces a satisfactory fit to the observations, if the density contrast of the suevitic breccia is from 0.05 to –0.04 g/cm^3, while the autochthonous breccia body and the central uplift retain the contrasts of the earlier two-density contrast gravity model (Fig. 4.12). By introducing the predicted differentiation into allochthonous/autochthonous brecciation based on seismic profiles and morphometric relationships, we conclude that any density contrast between the two bodies is probably very small.

The siliciclastic sedimentary sequences on the Bjarmeland Platform are considered to be largely non-magnetic. Furthermore, the depth to magnetic basement is estimated to at least 6–7 km (Johansen et al. 1993), and individual magnetic-depth estimates to the top of the crystalline basement approximate 10 km (Skilbrei 1993a). Furthermore, slope and half-slope estimates of magnetic source depths for the local anomalies in Fig. 4.10 (Peters 1949; Grant and West 1965; Nettleton 1976) yield a depth range of 1.6-3.4 km with an average value of 2.1 km. The above methods might over- or under-estimate the magnetic source depth depending on the width-to-depth ratio and extent of the magnetic body (Skilbrei 1993b). Our estimates clearly indicate shallow sources within the seismically disturbed region, rather than deep basement sources. We have therefore modelled the magnetic anomaly sources as part of a polymictic, weakly magnetized suevite unit with susceptibility of 7×10^{-3} SI and total magnetization contrast of 1 A/m (Fig. 4.13). The magnetization is constrained by the anomaly magnitudes and compares favorably with the measured remanent magnetization of 2.3 A/m for suevite from the Ries crater (Pohl et al. 1977). The model (Fig. 4.13) accounts for the suppressed magnetic signature at the centre and generates magnetic anomalies over the annular basin that is within the observed ± 100 nT range, although exhibiting somewhat steeper gradients (Fig. 4.13). The modelling includes a paleopole at 59° N, 171° E, based on paleomagnetic measurements in Bathonian-Hauterivian dolerite intrusions in Svalbard (Halvorsen et al. 1996). The introduction of a reversed magnetic field during the time of impact is consistent with the Late Jurassic-Early Cretaceous frequent M-series reversals (Gradstein et al. 2004), but does not constrain the time of magnetization.

The magnetic anomalies at Mjølnir are within the expected range for a sedimentary-type crater (Pilkington and Grieve 1992), i.e., a non-magnetic central uplift. We do not observe a broad magnetic low which is frequently predicted for impact structures of Mjølnir's size in crystalline targets (Grieve and Pesonen 1992; Pilkington and Grieve 1992). Instead, we have several low amplitude anomalies, both positive and negative, concentrated in the outer zone, near the transition to the annular basin (Fig. 4.10). The confinement of the dispersed-character melts to the allochthonous unit will only yield anomaly sources within the annular basin proper. However, the slope and half-slope source depth estimates indicate that possible magnetic bodies may be located within the autochthonous rather than in the allochthonous unit (Fig. 4.13). It is unlikely, though, to have significant amounts of injected melts into the autochthonous breccia body as careful search through the entire seismic dataset at Mjølnir showed almost total absence of significant high-amplitude reflectors within the seismically disturbed volume (Tsikalas et al. 1998a–c, 1999). This indicates that the methods may overestimate the depth to the magnetic source bodies or that other source than melt bodies may contribute to the anomalous field, such as dislocation of weakly magnetized platform strata.

4.3.4 Traveltime/Velocity Anomalies and Modelling

Stacking velocities from multichannel seismic reflection profiles and interpretation of shallow refracted arrivals recorded by sonobuoys have not revealed any statistically significant lateral changes in seismic velocities between the Bjarmeland Platform and the Mjølnir crater (Tsikalas et al. 1998c). Nevertheless, small but significant lateral velocity anomalies may exist, without being resolved in these data. Real velocity anomalies would cause pull-up or pull-down effects on continuous reflectors below the structure. If we assume that such reflectors are approximately planar, even small velocity anomalies may be detected by local reflector relief because traveltime anomalies measured in this way are more easily detected than the corresponding stacking velocity anomalies.

Regionally, the top Permian reflector marks the upper boundary of thick evaporites, carbonates and silicified clastic deposits in the Barents Sea (Gabrielsen et al. 1990; Gérard and Buhrig 1990; Nøttvedt et al. 1993; Gudlaugsson et al. 1998). This distinct, high-amplitude, south-southwest dipping reflector dominates in the pre-Jurassic succession of the Mjølnir region. Therefore, lateral changes in seismic velocity will introduce undulations of the top Permian horizon. In fact, we observe small deviations in the top Permian reflector beneath the entire structure.

We mapped the top Permian reflector along the regional grid of multi-channel seismic reflection profiles and Gaussian-filtered the data with a filter length of 5 km. We then performed cross-over corrections, fitted a third-order polynomial surface to the data and determined residual traveltime anomalies. The subsequent gridding and contouring bring out several top Permian traveltime anomalies. These appear mainly beneath the central high and the annular basin forming a near-circular, positive (pull-up) traveltime anomaly (Fig. 4.14). The anomaly is 16 km in diameter and rises approximately to +80 ms beneath the central high, decreasing towards the annular basin (Fig. 4.14). In the outer zone this circular anomaly is surrounded by diffuse lows and highs that probably reflect smaller-scale pre-impact relief, such as carbonate buildups, along the top Permian reflector. The close correspondence of the positive traveltime anomaly with the central part of the Mjølnir crater (Figs. 4.11 and 4.14) strongly suggests it is real and caused by the structure. In the modelling, we assume that the anomaly is a pull-up effect caused by lateral variations in seismic velocity within the disturbed region and we try to model both the gravity and traveltime anomalies simultaneously.

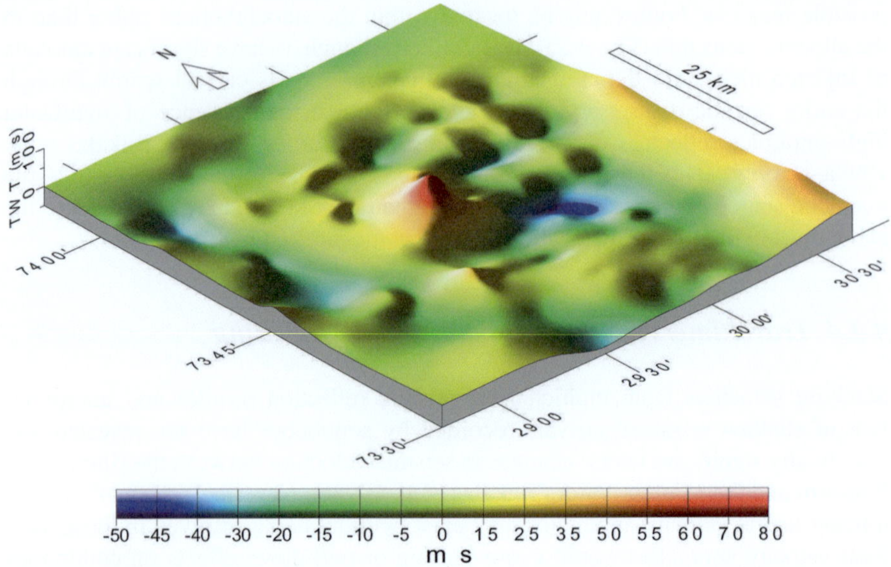

Fig. 4.14 Illuminated perspective image of the residual traveltime anomaly map surface at the top Permian reflector. The view is from 30° above the horizon looking northeast (azimuth 40°); light source at azimuth 290°

The effect of the density contrasts in Fig. 4.12 on the seismic traveltime was tested with reference to the top Permian reflector assuming that densities and seismic velocities are related by the Nafe and Drake empirical relationship (Ludwig et al. 1970; Barton 1986). The average velocity between the top of the structure and the top Permian level is estimated to 4.0 km/s and corresponds to a density of 2.39 g/cm^3. Similarly, the modelled density anomalies correspond to velocity anomalies from +175 to -200 m/s. The calculated traveltime anomaly approximates the observed relief of the top Permian reflector, especially in the centre. This shows that the traveltime anomaly is a simple consequence of the lateral density distribution obtained by the gravity modelling (Fig. 4.12). The traveltime signature may be explained along the same lines as the gravity signature, i.e., the traveltime anomaly beneath the central uplift results from a positive velocity anomaly beneath the central high. The small deviations of the observed and the calculated anomalies in the peripheral region probably reflect a pre-impact relief of the top Permian reflector.

With reference to the undisturbed Bjarmeland Platform, the modelled integrated mass anomalies in Fig. 4.13 correspond to velocity anomalies from +175 to -250 m/s. The calculated traveltime anomaly (Fig. 4.13) provides a slightly closer approximation to the observed central anomaly than that of the initial model (Fig. 4.12). In particular, the introduction of allochthonous breccia lenses with lower velocity beneath the annular basin reproduces slightly better the steep gradient (Fig. 4.13) of the observed traveltime anomaly in the center.

4.4 Modelled Porosity Anomalies

The porosity-depth relationship for a compacted sedimentary section is best approximated by the exponential function (Fig. 4.15):

$$\varnothing = \varnothing_0 e^{-cz} \tag{4.1}$$

where $\varnothing_0$ is the initial porosity, c is the compaction constant, and z is the total, present and eroded, overburden (e.g., Wyllie et al. 1956, 1958; Gardner et al. 1974; Sclater and Christie 1980). By introducing an impact-induced porosity change (Fig. 4.15), the new porosity-depth function, $\varnothing'$, which defines the compaction in the crater, becomes:

$$\varnothing' = \varnothing_0 e^{-cz} + \Delta\varnothing \tag{4.2}$$

where $\Delta\varnothing$ is the porosity anomaly induced by the impact. If we consider a constant impact-induced porosity anomaly during burial (Fig. 4.15):

$$\varnothing' = \varnothing + \Delta\varnothing \tag{4.3}$$

thus,

$$\Delta\varnothing = \varnothing' - \varnothing \tag{4.4}$$

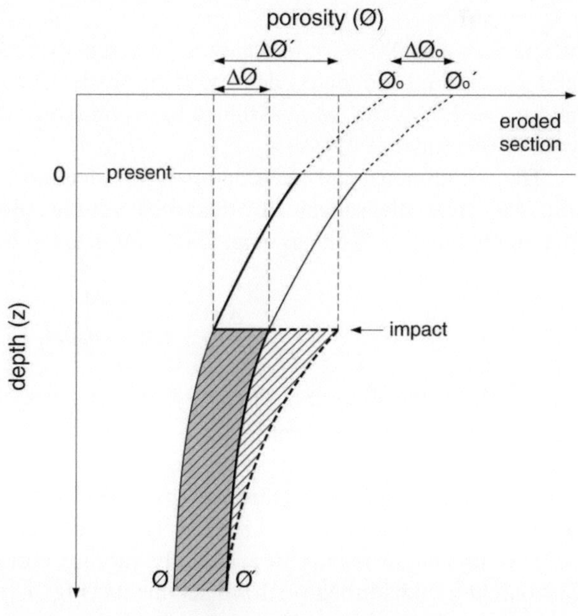

Fig. 4.15 Generalized porosity-depth functions for the Mjølnir impact crater. *Shading* denotes a constant porosity anomaly function, whereas *hatching* denotes an exponentially decreasing function. The various porosities, Ø, are described in the text

We calculate, at 0.1 km sampling intervals, separate porosity anomalies from the density distribution, seismic traveltime distribution, and postimpact sediment deformation calculated by Tsikalas et al. (1998a, c). The lateral porosity distribution is determined as the average of the three approaches along a typical depth converted cross section (Fig. 4.12). Although the assumption of a constant porosity anomaly may not be realistic, it is introduced because, in a simplistic manner, it quantifies the contrast between the crater and the undisturbed sediments.

We also decompact the present lateral porosity distribution to the crater configuration immediately after impact (Fig. 4.15). The decompacted porosity anomaly, $\Delta\emptyset'$, provides a more appropriate estimate for comparison with other craters and for assessing petroleum potential. It can be defined as (Fig. 4.15):

$$\emptyset'\emptyset_0 \mathrm{e}^{-\mathrm{cz}} + \Delta\emptyset' \mathrm{e}^{-\mathrm{cz}} = \emptyset + \Delta\emptyset' \mathrm{e}^{-\mathrm{cz}} \tag{4.5}$$

thus combining this equation (see Eq. 4.5) with Eq. (4.4),

$$\Delta\emptyset' = \frac{\Delta\emptyset}{e^{-cz}} \tag{4.6}$$

4.4.1 Density-Derived Porosity Anomaly

It can be shown (see Tsikalas et al. 2002b for detailed calculations) that the relationship between density and porosity anomalies is expressed by:

$$\Delta\rho_b = (\rho_w - \rho_m)\Delta\emptyset \tag{4.7}$$

where $\Delta\rho_b$ and $\Delta\emptyset = \emptyset' - \emptyset$ are the bulk density and porosity anomalies, and ρ_w and ρ_m are fluid and matrix density, respectively. Density values of $\rho_w = 1.0$ g/cm^3 and $\rho_m = 2.65$ g/cm^3 were assigned based on typical rock properties (e.g., Telford et al. 1990; Rider 1991).

The two-dimensional mass anomaly is calculated by integrating equation (see Eq. 4.7) over the seismically disturbed volume, thus the porosity anomaly is determined by (see Tsikalas et al. 2002b for detailed calculations):

$$\Delta\emptyset = \frac{\Delta M}{(\rho_w - \rho_m)H'_f} \tag{4.8}$$

where H'_f is the thickness of the seismic disturbance at a given location; and $\Delta M = \Delta\rho_b H'_f$ the corresponding two-dimensional mass anomaly determined from the gravity model (Fig. 4.12). Note that the gravity anomaly along the cross section is extracted from the residual free-air gravity field that shows a circularly symmetric anomaly over Mjølnir (Figs. 4.8 and 4.9) (Tsikalas et al. 1998a).

This procedure results in a laterally-varying porosity anomaly relative to the surrounding undisturbed platform sediments (Fig. 4.16). The porosity is increased

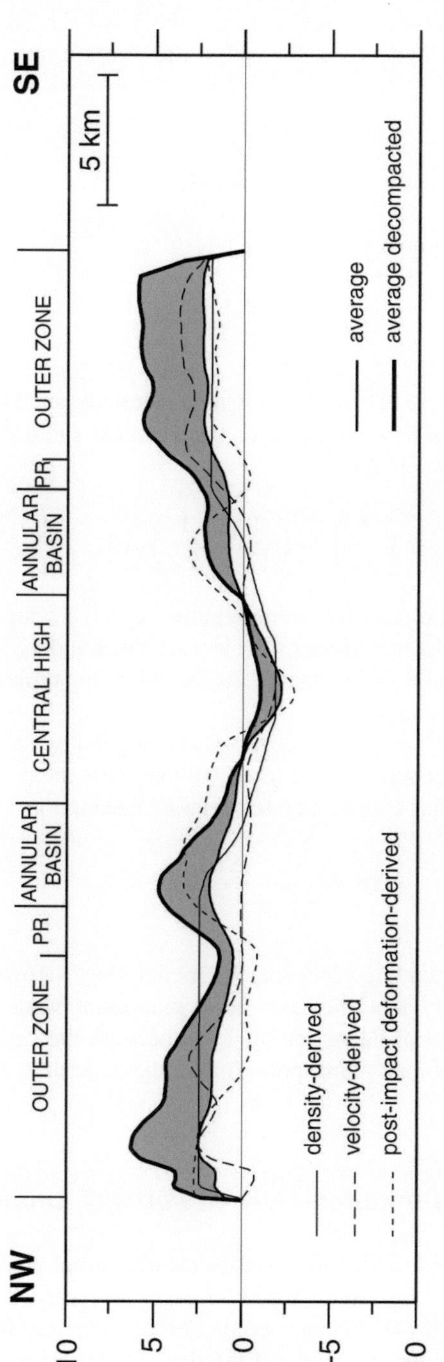

Fig. 4.16 Relative porosity anomaly models along the type cross section (AA′, Figs. 4.1, 4.11 and 4.12). Shading denotes the change between the present and the decompacted porosity anomaly distributions. PR = peak ring.

by 1.8–2.4% at the peripheral flanks, whereas the bowl-shaped central porosity anomaly decreases by as much as 1.8% within the ~16-km-outer-diameter of the annular basin (Fig. 4.16).

4.4.2 Velocity-Derived Porosity Anomaly

The traveltime anomaly (Fig. 4.12) is defined by:

$$\Delta T = 2\Delta\left(\frac{1}{V_b}\right)H'_f \tag{4.9}$$

where $\Delta(1/V_b)$ is the inverse velocity, or slowness, anomaly, and H'_f the thickness of the seismic disturbance. In terms of porosity, the slowness anomaly is (see Tsikalas et al. 2002b for detailed calculations):

$$\Delta\left(\frac{1}{V_b}\right) = \left(\frac{1}{V_w} - \frac{1}{V_m}\right)\Delta\emptyset \tag{4.10}$$

where $\Delta\emptyset = \emptyset' - \emptyset$ is the porosity anomaly; and V_w and V_m are fluid and matrix velocity, respectively. Velocity values of $V_w = 1.45$ km/s and $V_m = 5.5$ km/s were assigned based on regional well-log data (Tsikalas 1992) and typical rock properties (e.g., Telford et al. 1990; Rider 1991).

The traveltime anomaly at a given location is calculated by integrating equation (see Eq. 4.10) over the seismically disturbed volume, thus the porosity anomaly is determined by (see Tsikalas et al. 2002b for detailed calculations):

$$\Delta\emptyset = \frac{\Delta T}{2\left(\frac{1}{V_w} - \frac{1}{V_m}\right)H'_f} \tag{4.11}$$

Compared with the surrounding platform, the model (Fig. 4.16) yields a radially-varying porosity anomaly that increases to a maximum value of 3.6% in the peripheral region, whereas it decreases by 1.9% beneath the central high. Again, the extent of the central porosity low approximates the outer diameter of the annular basin (Fig. 4.16).

4.4.3 Postimpact Deformation-Derived Porosity Anomaly

The high-resolution seismic reflection profiles clearly reveal postimpact sedimentary thickness variations and lateral facies changes governed by the underlying crater relief (confer Fig. 3.14) (Tsikalas et al. 1998c). After the impact depression (Fig. 4.5) was filled by sediments (Fig. 4.17a), deposition continued, and the crater became buried by sediments 1.5–2 km thicker than at present (e.g., Nardin and Røssland 1992; Nyland et al. 1992). The progressive loading triggered structural

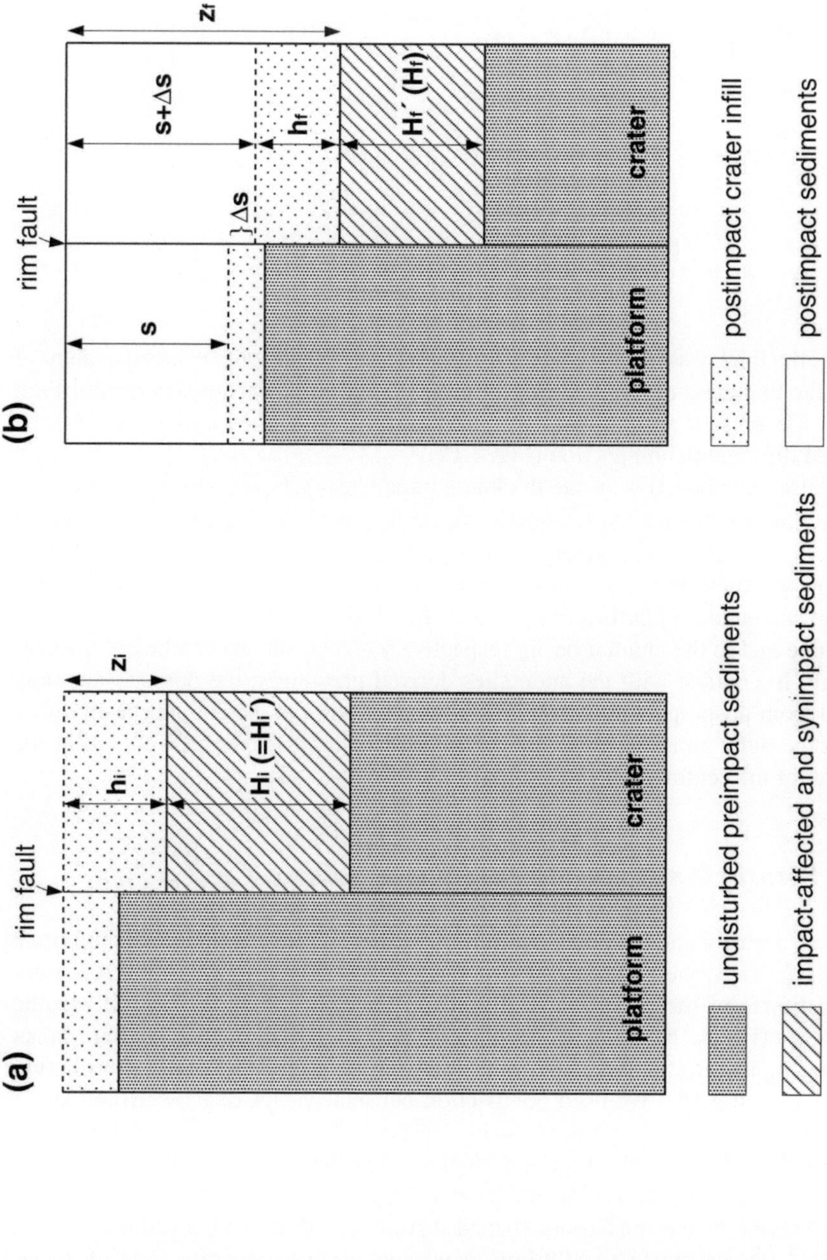

Fig. 4.17 Schematic diagram of postimpact differential compaction across the crater rim: **(a)** immediately after the postimpact infilling of the crater, and **(b)** at present. Parameters are described in the text

reactivation and differential compaction (Figs. 3.14, 4.5, and 4.17b) (Tsikalas et al. 1998c).

We have shown (see Tsikalas et al. 2002b for detailed calculations) that the porosity anomaly is related to the differential compaction by (Fig. 4.17):

$$\Delta \emptyset = \frac{1}{\mathrm{H}_\iota' - H_f'} \times \left[\Delta s - \frac{\emptyset_0}{c} e^{-c(s+h_f+H_f')} e^{-c\Delta s} (1 - e^{-c\Delta s}) \right] \quad (4.12)$$

where:

$$H_i' + \left(\frac{\emptyset_0}{c(1-\Delta\emptyset)} e^{-ch_i} \right) e^{-cH_i'} = H_f' + \frac{\emptyset_0}{c(1-\Delta\emptyset)} \times \left[e^{-ch_i} - e^{-c(s+\Delta s+h_f)} (1 - e^{-cH_f'}) \right] \quad (4.13)$$

$$h_i + \left(\frac{\emptyset_0}{c} e^{-ch_i} \right) = h_f + \frac{\emptyset_0}{c} \left[1 - e^{-c(s+\Delta s)} (1 - e^{-ch_f}) \right] \quad (4.14)$$

H_f' is the thickness of the seismic disturbance; H_i' the decompacted equivalent of H_f'; h_f the thickness of the post-impact crater infill; h_i the decompacted equivalent of h_f; s the present and eroded overburden above the infilled crater; and Δs the observed differential compaction (Fig. 4.17).

Because decompaction of the thickness parameters (H_f', h_i) requires knowledge of the porosity anomaly ($\Delta\emptyset$), equations (see Eqs. 4.12, 4.13, and 4.14) are solved iteratively. The result is a porosity-anomaly curve that reflects the postimpact lateral sediment thickness variations and the radial crater zonation (Fig. 4.16). Compared with the surrounding platform, maxima of 1.5–3 and 3–3.5% are calculated in the outer zone and in the annular basin, respectively; decreasing by 3% below the central high. In contrast with the anomalies derived previously, the porosity anomaly derived from postimpact deformation has its maximum in the annular basin rather than in the outer zone, whereas the central porosity low is restricted to within the diameter of the central high (Fig. 4.16).

4.4.4 Porosity Anomaly and Pore Space Volume

The three porosity anomaly distributions are grossly similar in trend and amplitude (Fig. 4.16). Thus, we first assume that the average anomaly provides a representative estimate of the present lateral porosity changes within the intense seismic disturbance (Fig. 4.16). Then, we reconstruct the situation immediately after impact by using equation (see Eq. 4.6) to decompact the present porosity distribution (Figs. 4.15 and 4.16). We used compaction constant values of $c = \pm 0.42\ \mathrm{km}^{-1}$ based on regional well-log data (Tsikalas 1992) for the positive and negative porosity anomalies, respectively. Figure 4.16 show that the maximum porosity reaches 6.3% at the periphery, and -1% at the central high.

The volume of the mushroom-shaped disturbance (Fig. 4.5) is estimated to be ~850 km^3, increasing to ~1,400 km^3 if we include the transition zone (Tsikalas et al. 1998b). Of these volumes 450–600 and 400–800 km^3 belong to the peripheral region and central core, respectively. The volume estimates and the modelled porosity distribution (Fig. 4.16) imply an average impact-induced excess pore space of 13–33 km^3 in the periphery, and a 13–6 km^3 reduction of pore space in the central crater, for the present and immediately-after-impact settings, respectively.

4.4.5 Porosity Anomaly and Hydrocarbon Potential

Many impact structures are associated with hydrocarbons (Table 4.2) (Donofrio 1981, 1998; Grieve and Masaitis 1994). In fact, the impact cratering process may result in structural and stratigraphic re-organization that may be conducive to hydrocarbon accumulation. In brittle crystalline target rocks, impact-induced fracturing and brecciation may significantly increase porosity and permeability. The effect of fracturing and brecciation may be less important in enhancing reservoir potential in sedimentary targets, but impact may lead to structural traps by creating listric-type faults, mainly along the crater rim. Nonetheless, impacts in marine sedimentary basins, which theoretically provide the greatest hydrocarbon potential, are presently poorly understood (Grieve and Pesonen 1992, 1996; Pilkington and Grieve 1992). In particular, key questions relate to how much fracture volume will be generated or whether this volume will be maintained against porosity-reducing processes such as postimpact mechanical compaction and diagenesis.

The Barents Sea has experienced extensive hydrocarbon exploration. The Bjarmeland Platform is adjacent to the Nordkapp, Maud, Ottar, and Hammerfest basins (confer Fig. 1.7), which are considered to have produced hydrocarbons at some stage of their geological history (Fjæran and Spencer 1991; Larsen et al. 1992). The estimated 13–33 km^3 excess pore space volume at Mjølnir is several times larger than the reserves of the hydrocarbon-associated impact structures in Table 4.2. Hence, the impact-induced fracture volume along the periphery (Fig. 4.12) may be considered a potential reservoir unit.

High-resolution single-channel seismic profiles reveal several intra-sedimentary features, within and, in most cases, above the structure, commonly considered as potential gas indicators (e.g., Hovland and Jodd 1988). These include: local amplitude increase and smearing of individual reflections; columnar disturbances; and acoustic blanketing appearing as loss of seismic coherence (Fig. 4.18). In contrast, no such features are recorded beyond the crater boundary. On the other hand, deep multichannel profiles do not show discernible anomalous amplitude signatures, probably due to their lower frequencies, which inhibit resolution at shallow depth. Amplitude anomalies of this kind may be indicative of gas occurrence and vertical gas seepage within and above the structure. The amplitude anomalies along the periphery are underlain by tilted fault blocks and brecciated strata (Figs. 3.6 and 4.18). Thus, gas might migrate from deeper stratigraphic levels. Furthermore, the potential seismic hydrocarbon indicators correspond spatially to the 400–650 km^3 volume of impact-deformed platform strata of increased porosity

Table 4.2 Commercial hydrocarbon accumulations associated with impact structures. The Red Wing Creek structure is the most prolific impact-related oil field in the USA (Donofrio 1998)

Structure	Diameter (km)	Age	Hydrocarbons	Reserves/ (Production)
Ames Oklahoma, USA (Carpenter and Carlson 1997)	13	Early Ordovician	Oil and gas production from dolomite on crater rim and from brecciated granite and dolomite on crater floor	25 MMbbl oil, 0.42×10^9 m^3 gas (2,600 bbl oil/day, 0.85×10^5 m^3 gas/day)
Red Wing Creek North Dakota, USA (Gerhard et al. 1982)	9	Triassic/ Jurassic	Oil and gas production from carbonate breccia in central uplift; ca. 820 m oil column (ca. 490 m net pay)	20 MMbbl oil, 0.71×10^9 m^3 gas (1,000 bbl oil/day, 0.57×10^5 m^3 gas/day)
Viewfield Saskatchewan, Canada (Sawatzky 1977)	3.2	Triassic/ Jurassic	Oil production from carbonate breccia and from raised rim. Porosity, 14%; permeability, 400 md; net pay, 4-50 m	11 MMbbl oil, 0.14×10^9 m^3 gas (600 bbl oil/day, 0.085 $\times 10^5$ m^3 gas/day)
Avak Alaska, USA (Kirschner et al. 1992)	12	Early/Late Cretaceous	Gas production from listric fault traps along crater rim caused by impact-triggered submarine landslides	1.1×10^9 m^{3}gas (0.37×10^5 ft^3 gas/day)
Steen River Alberta, Canada (Winzer 1972)	25	Pre-late Cretaceous	Precambrian basement complex uplifted 760 m	3–5 MMbbl oil (550 bbl oil/day)
Newporte North Dakota, USA (Donofrio 1981)	3.2	End-Cambrian	Oil from sands draped over the raised rim, and some production from highly fractured Precambrian gneiss-schist	15 MMbbl oil (280 bbl oil/day)

bbl = barrel; MMbbl = million barrels = 0.16×10^6 m^3.

along the periphery (Figs. 4.12 and 4.16) (Tsikalas et al. 2002b). We rule out that the seismic-amplitude anomalies are "tuning" effects resulting from sediment thickness variations, because they are only found at the crater periphery where porosity is enhanced.

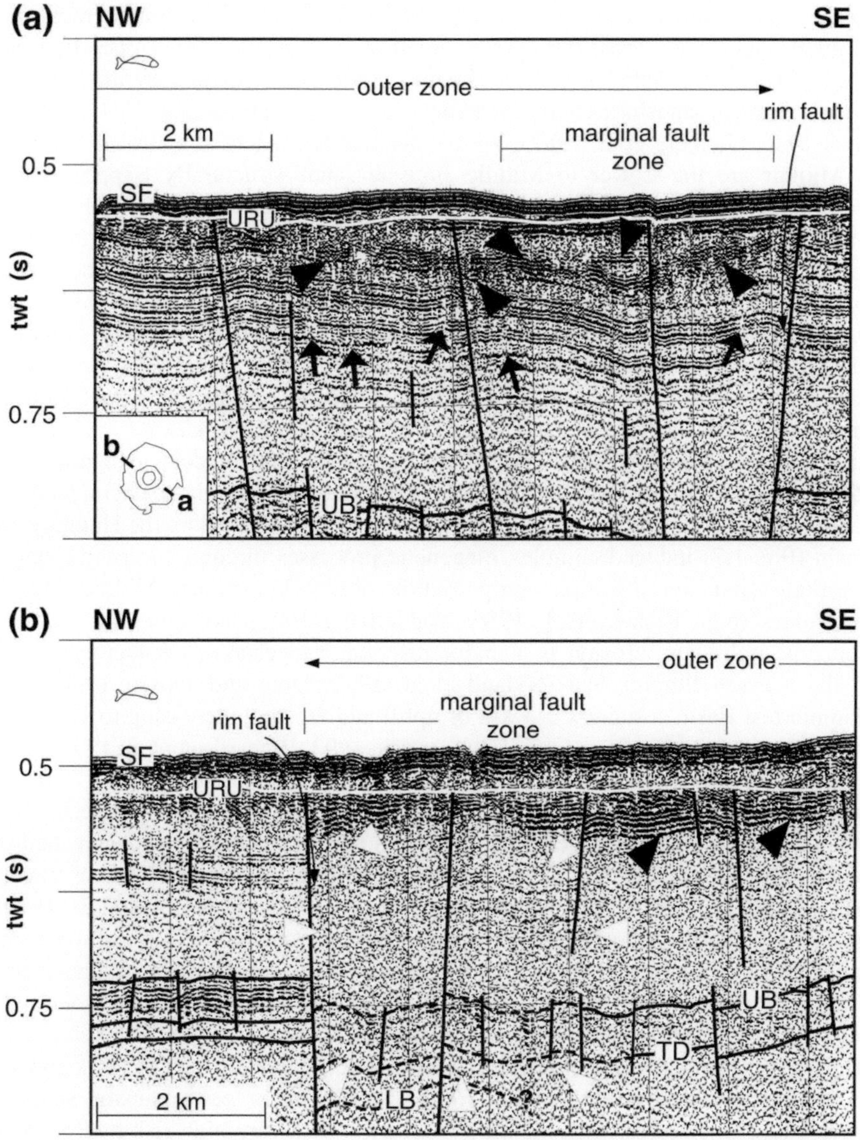

Fig. 4.18 Examples of high-resolution single channel seismic profiles across the crater rim exhibiting possible gas-related seismic-amplitude anomalies. *Black triangles* = enhanced reflections and smearing; *white triangles* = acoustic blanketing; *black arrows* = restricted columnar disturbances ("chimneys")

Most Barents Sea exploration wells have been drilled in the Hammerfest Basin (Fig. 1.7) where gas accumulations and small, non-commercial, oil finds have been encountered (Grung-Olsen and Hanssen 1987; Linjordet and Grung-Olsen

1992). Lower Jurassic and Upper Jurassic-Lower Cretaceous source rocks are of good quality, and most discoveries are structural traps within Upper Jurassic-Lower Cretaceous shales (Fig. 3.7). The most important reservoirs are the Lower to Middle Jurassic sandstones of the Nordmela and Stø formations (Fig. 3.7) (Berglund et al. 1986; Gjelberg et al. 1987). We believe that the potential hydrocarbon plays at Mjølnir are the Lower to Middle Jurassic sands structurally trapped within the impact-generated tilted fault blocks, and the synimpact, extensively fractured, allochthonous breccia (Figs. 3.6 and 3.7). At the periphery of the crater, the faults appear to diminish within a low seismic-coherence zone above prominent continuous reflectors (Fig. 4.5), which suggests inward displacement of the blocks during gravitational collapse over an apparent low-angle décollement (Tsikalas et al. 1998a–c, 1999; Tsikalas and Faleide 2004; Tsikalas 2005). In fact, a similar structural-trap configuration of listric-type faults caused by impact-induced submarine slides, characterizes the huge Avak gas field (Table 4.2).

The porosity and permeability of the sandstone and breccia at Mjølnir are controlled by the degree and interplay of postimpact mechanical and chemical compaction. It has been shown that the extensive burial experienced by the Hammerfest Basin (Fig. 1.7) induced complex diagenetic processes through successive stages of silicification, stylolitization, and partial dissolution in Lower to Middle Jurassic sandstones (e.g., Riches et al. 1986; Walderhaug 1992). Nonetheless, the later Cenozoic uplift and erosion is a major reservoir and charge risk factor in most of the western Barents Sea (Nyland et al. 1992; Doré and Jensen 1996). The Hammerfest Basin sustained 0–1 km of uplift and erosion, increasing to 1.5–2 km on the Bjarmeland Platform (e.g., Vågnes et al. 1992; Richardsen et al. 1993). This event may explain the present shallow exposure of the crater (e.g., Fig. 3.14). It may also have prevented potential hydrocarbons in the Lower to Middle Jurassic sands and the synimpact allochthonous breccia plays to be retained at such shallow depths. Moreover, many rim faults may not be efficiently sealed because of truncation by the Late Cenozoic erosional unconformity (URU, Figs. 3.6 and 3.14). On the other hand, the entire Bjarmeland Platform is regionally tilted to the south due to extensive Late Cretaceous-Tertiary uplift and to isostatic response to glacial erosion (Gabrielsen et al. 1990; Faleide et al. 1993). Therefore, the south-southwest part of the structure presently lies more than 1 km below the seafloor, being least affected by uplift and erosion. Here, the postimpact sediments may be thick enough to retain an adequate reservoir pressure, preventing major gas expansion and seal breaching. Nonetheless, we consider the crater to be a very high-risk hydrocarbon prospect.

4.5 Potential Non-impact Origin

The structural models in Figs. 4.11, 4.12, and 4.13 are compatible with the observed gravity, magnetic and traveltime anomalies of typical complex impact craters in general (Grieve and Pesonen 1992; Pilkington and Grieve 1992). Nonetheless, one has to evaluate whether other processes than impact could have formed the structure. Gudlaugsson (1993) qualitatively discussed and evaluated alternative interpretations, such as salt and clay diapirs, and igneous intrusions, whereas Tsikalas et al. (1998c) quantitatively investigated their geophysical signature by modelling a range of geologically reasonable configurations. Here, we briefly discuss the alternative non-impact interpretations that provide the most realistic fit to the observed anomalies. Although geological data have confirmed the impact origin of Mjølnir, we offer this discussion for completeness and for revealing the Mjølnir studies as paradigms in the investigation of potentially new impact structures.

4.5.1 Clay Diapir

The presence of a central positive gravity anomaly and the apparent lack of a negative velocity anomaly with a corresponding pull-down effect beneath the structure are incompatible with an active clay diapir (Fig. 4.12). On the other hand, a non-active clay diapir may reproduce the observed anomalies, although a feature of this kind is not supported from regional considerations. Clay diapirs usually form in geodynamic environments characterized by overpressures, commonly induced either by rapid deposition of fine-grained siliciclastic sediments or by compression. The regional depositional environments and continuity in lithological composition and burial history of the Bjarmeland Platform strata, however, argues against such conditions (Gudlaugsson 1993). Even when such conditions arise, clay diapirs almost always form in clusters. Thus, it is difficult to find a plausible mechanism for clay diapirism or any structural or depositional anomaly that might have initiated the formation of an isolated shale diapir (Fig. 4.10). We also note that the extensive deformation above the structure has been attributed to differential compaction of the embedded impact crater during post-impact burial (Tsikalas and Faleide 2007).

4.5.2 Salt Diapir

The Mjølnir crater is located about 70 km north of the Nordkapp Basin, a major salt diapir province (confer Figs. 1.7 and 3.1). Numerous diapirs reach the seafloor, and individual structures have diameters up to 20 km (Bergendahl 1989). Lower Permian-Upper Carboniferous evaporites were mobilized during rapid basin subsidence culminating with diapirism in Early to Middle Triassic time. Renewed salt movements took place due to tectonic episodes in Late Jurassic-Early Cretaceous and Early Tertiary times (Gabrielsen et al. 1992b; Jensen and Sørensen 1992).

Through intensive gravity- and traveltime modelling of a possible diapiric structure, it was shown that salt diapirs with a significant stem would produce gravity and traveltime anomalies that differ significantly from the amplitudes and shape of the observed anomalies both over the diapir proper and in the periphery (Fig. 4.19) (Tsikalas et al. 1998c). A reasonable fit to the observed anomalies requires a relatively shallow salt body within the circumference of the Mjølnir crater. The presence of a shallow, thin salt layer should be clearly imaged by seismic reflection profiles; however, no evidence for shallow salt is present (Figs. 3.14 and 4.2). In

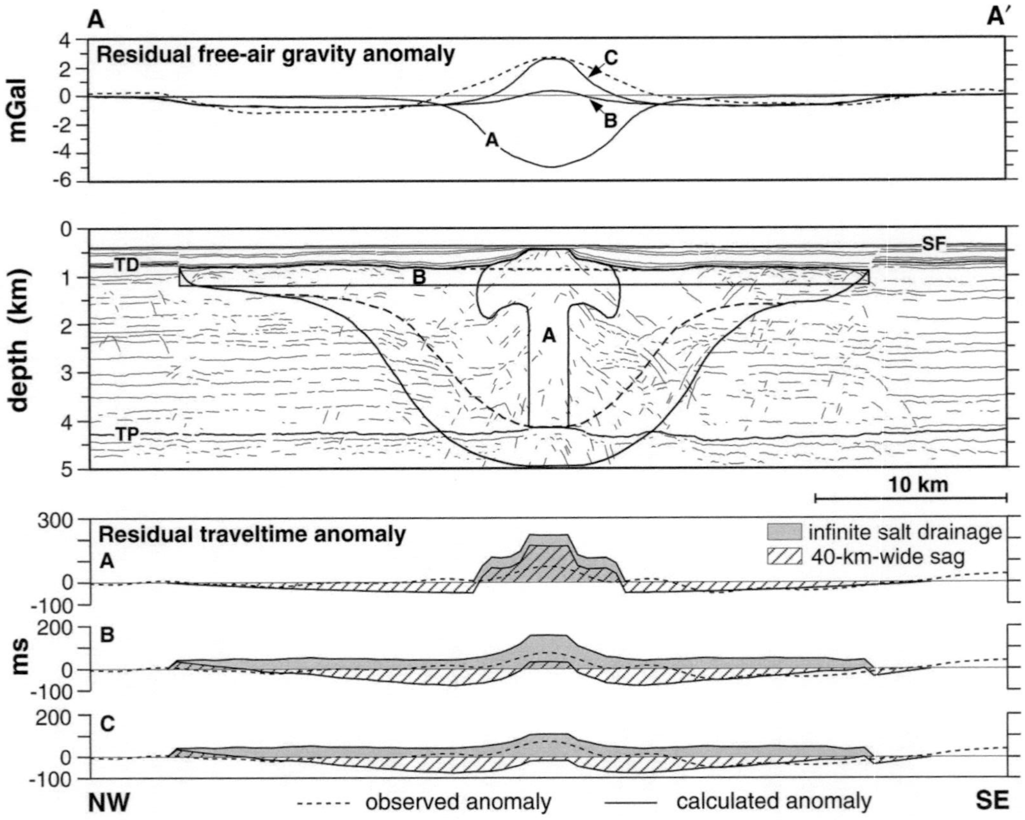

Fig. 4.19 Free-air gravity and traveltime anomalies from salt diapir models. *Dashed line* outlines the area of intense seismic disturbance. (**a**) Simple mushroom shaped diapir; (**b**) Salt plate model without caprock; (**c**) Model (**b**) with caprock. Model densities are discussed in the text

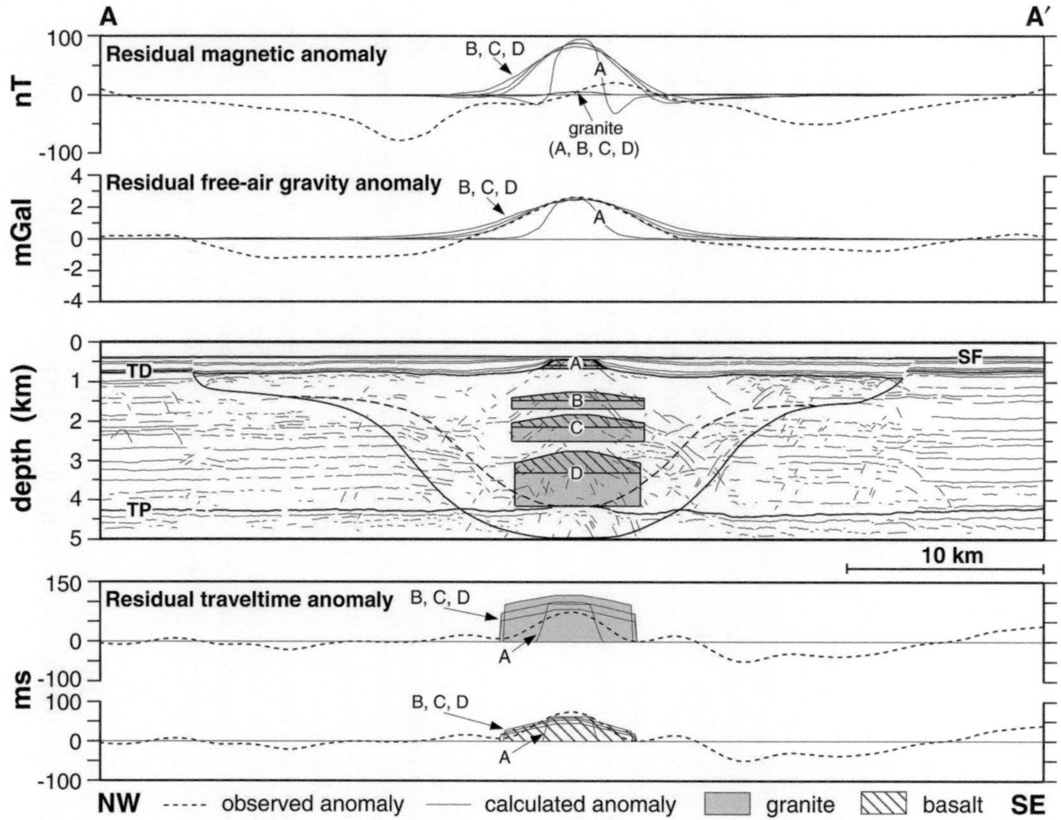

Fig. 4.20 Geophysical anomalies from igneous intrusion models. *Dashed line* outlines the area of intense seismic disturbance. Note that all four granite intrusion models only yield a minimal, ~5 nT, magnetic perturbation. The intrusions were modelled using typical rock properties from Telford et al. (1990); density of 2.90 and 2.60 g/cm^3, susceptibility of 70 and 2.5×10^{-3} SI, and velocity of 5.5 and 5.5 km/s, for the basaltic and granitic intrusions, respectively

addition, a seismic unit similar to the disturbed region at Mjølnir is not observed in other typical rim-synclines in the Barents Sea (Gudlaugsson 1993). Although some of the diapirs in the Nordkapp Basin may have developed distinct overhangs (Koyi et al. 1993), there is no evidence for the presence of extensive salt canopies. Moreover, there is no evidence of major salt-withdrawal sag at the Mjølnir crater. Therefore, it can be inferred that salt diapirism is not a viable explanation (Tsikalas et al. 1998c).

4.5.3 Igneous Feature

The Mjølnir crater is located in a marginal position relative to the Late Jurassic-Early Cretaceous northern Barents Sea igneous province. The province comprises extrusive basalts in Kong Karls Land (Svalbard) and Franz Josef Land (Smith et al. 1976; Heafford and Kelly 1988; Kelly 1988; Johansen et al. 1993) and dolerite intrusions in Svalbard (Halvorsen et al. 1996) and the northern Barents Sea (Faleide et al. 2008). In view of the Volgian – Ryazanian age of the Mjølnir deformation event (Smelror et al. 2001a), we consider the possibility that the Mjølnir structure represents a volcanic or intrusive igneous feature (Fig. 4.20). In fact, the structure has previously been interpreted as a Late Jurassic-Early Cretaceous magmatic intrusion (Brekke et al. 1992).

A dolerite intrusion is considered unlikely because magnetic anomalies over such features in the northern Barents Sea are in the order of ~600–1,200 nT (Åm 1975; Gustavsen 1995), contrasting with the moderate-amplitude magnetic anomalies over Mjølnir (Fig. 4.10). In addition, dolerite intrusion K/Ar ages from Svalbard indicate two distinct magmatic episodes at 110 ± 10 Ma (Aptian-Albian) and 149 ± 17 Ma (Bathonian-Hauterivian) (Halvorsen et al. 1996). The former clearly postdates the formation of the Mjølnir structure, while the latter age range bounds the structure, though it is likely that the emplacement predates the impact horizon. Several alternative models have been constructed to reconcile the observed geophysical data with a volcanic feature, and basaltic and granitic intrusions (Fig. 4.20). None of those models were able to simulate the structural and geophysical observations, and thus any possible igneous feature origin for Mjølnir was rejected (Tsikalas et al. 1998c).

4.6 Alternative Interpretation of Mjølnir Crater Dimensions Based on Regional Gravity and Aero-magnetic Profiles and Modelling

The substantial exploration interest in this petroleum frontier area of southwestern Barents Sea has resulted in a great number of seismic, gravity, and magnetic surveys (Fig. 4.21a). Seismic data originally led to the discovery and definition of the shape of the subsurface structure of Mjølnir crater. Nevertheless, the seismic data only revealed the paleo-relief of the impact structure and the volume influenced by the

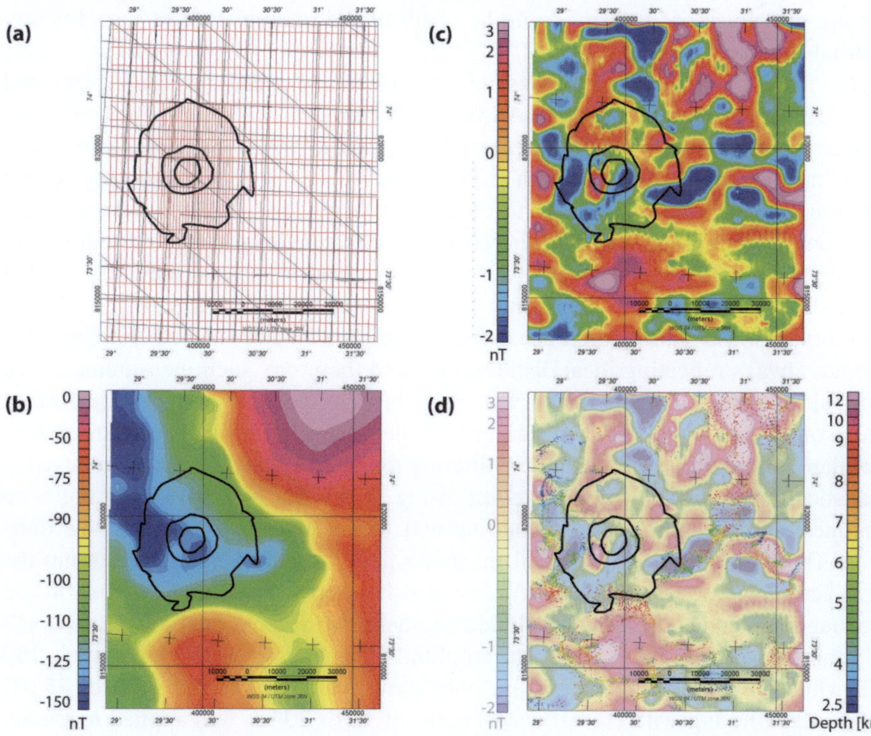

Fig. 4.21 Aeromagnetic anomaly map as surveyed during the BAS06 campaign. (**a**) Aeromagnetic measurements were preformed along flight lines shown in *red*. Ship recordings of gravity and seismics are shown as *black track lines*. (**b**) The same in comparison with the mapped Mjølnir zonation boundaries from seismic reflection data (Tsikalas et al. 1998a–c). (**c**) 20-km Gaussian high-pass filtered magnetic anomaly map; (**d**) Example of the results of an Euler deconvolution with a structural index of SI = 1 and a tolerance of 7% applied to the 20-km Gaussian high-pass filtered magnetic anomaly in comparison with mapped Mjølnir zonation boundaries

impact event. Prominent is the disturbance of the stratification of the sediments due to the impact event while circumjacent stratification is still well observable.

4.6.1 The Mjølnir Aero-magnetic Anomaly

Earlier (ship-borne) magnetic measurements were collected together with gravity and seismic measurements in three profiles crossing the structure diagonally. After processing (e.g., Tsikalas et al. 1998c), the remaining low-amplitude anomalies range between –75 and 20 nT. A magnetic anomaly (20 nT) was interpreted as a central anomaly, which is slightly offset from the topographic center. This relative high is surrounded by relative lows correlating to the outer crater zone, and the anomaly pattern looks like cloverleaves (Fig. 4.21b). Considering the profile set up, the cloverleaf pattern, could be result of the interpolation of insufficiently diurnal-corrected profiles.

During 2006, a new data set was collected of which a small portion is presented here. This aeromagnetic survey (BAS-06) was flown in an in-line-tie-line configuration above the southwestern Norwegian Barents Sea area. The general N-S oriented lines have a spacing of 2 km with E-W oriented tie-lines of 6-km spacing. In the Mjølnir area, the lines were filled to 1 km line spacing (Fig. 4.21a, red track-lines). The sensor, a cesium magnetometer, recorded the magnetic field at an altitude of 230 m with an airplane ground speed of 225 km/h; the resulting spatial sampling interval is 12–14 m. Magnetic measurements, notably in the high Arctic are often affected by diurnals. Base-station reference measurements at distances of 300–600 km away are not always reflecting local disturbances, and therefore, such measurements are difficult to correct. During the survey, diurnals were recorded and roughly 10% of the profiles had to be re-flown due to high noise level. The data were treated by standard processing which included: filtering for noise created by the manoeuvring, statistical-levelling and micro-levelling using "Geosoft OASIS Montaj" software routines, and median-filtering (Mauring and Kihle 2006) to correct for the diurnals. The data portion of the total magnetic field variations measured during the new aeromagnetic survey BAS-06 presented here covers the area of Mjølnir and surroundings and shows low-amplitude anomalies with a regional low close to the Mjølnir impact site (Fig. 4.21b). The amplitude range found in this region is ca. 150 nT, while the area related to Mjølnir only shows variations in the order of 25 nT. High-pass filtering with cut-off wavelengths of 10 or 20 km was applied to extract the shallow sources in the study area. The patchy distribution of impact melt could result on a crustal magnetic anomaly map in higher frequency content than, e.g., visible on gravity maps. On the other hand, a smoothing of magnetic anomalies around an impact site has been observed for structures in crystalline environment (e.g., Suvasvesi N crater, Finland; Pesonen et al. 1996).

A possible correlation between a relative minimum in the magnetic field and the crater outline can be observed (Fig. 4.21b). The dataset suffers from a low signal-to-noise level. No short-wavelength anomalies related to the crater can clearly be separated when a high-pass filter is applied (Fig. 4.21c). The statistical uncertainties are at the level of 1–3 nT for the entire survey, and therefore wavelengths less than 10 km cannot be separated above the noise level (Fig. 4.21c). Using a cut-off wavelength of 20 km reliable anomalies can be separated (range about 6 nT), but they are unlikely to be induced by crater related structures. An example of Euler deconvolution is shown for a 20-km Gaussian high-pass filtered anomaly map (Fig. 4.21d). The structural index is that of sheet-like source distribution, such as expected for melt patches. Euler deconvolution applied to the gridded data only yielded depths outside the expected crater depth-diameter dimensions, independently of which structural index was chosen or if applied to filtered or unfiltered data.

We applied Euler deconvolution (Reid et al. 1990) also to in-line profiles that have a much higher sampling rate and frequency content than the gridded data. The only layer, which was confidentially identified, is the seafloor. Considering the host material of the Bjarmeland platform being largely non-magnetic, anomaly variations of only a few nT are expected. Mostly undisturbed horizontal strata characterize the Bjarmeland Platform, and no significant anomaly generated by the sediments is expected in this area around the impact structure. The sources contributing to the long-wavelength content of the observed crustal field are related to the bedrock horizon, situated much deeper than any volume influenced by the impact crater. From the susceptibility and density measurements and the core interpretation, only detrital magnetic minerals and siderite-cemented beds or nodules aside the bedrock can be considered as the source for magnetisation observed in this study area. Such a result – the macroscopic invisibility of the crater – is expected for sedimentary areas such as in the Barents Sea, unless the projectile "delivered" magnetic material.

None of the features found in the new data set (Fig. 4.21b) resemble the magnetic anomaly pattern of the interpolated shipborne data map by Tsikalas et al. (1998c). Actually, a few of the features are inverted, although this could be related to an artefact of overcompensation due to trend correction on the earlier data set. However, a common feature of both the shipborne- and aero-magnetic surveys (and corresponding data processing) is the low amplitude of any observed magnetic anomaly within Mjølnir, and thus absence of shallow impact-related macroscopic melt bodies, in agreement with available seismic reflection dataset.

4.6.2 The Mjølnir Regional Free-Air Gravity Anomaly

The Barents Sea area is substantially covered by seismic lines and gravity profiles (Fig. 4.20a, black track-lines). The crater appears well preserved under layers of unconsolidated and consolidated sediments about 500 m thick, and overlain by a shallow water column (about 350 m). Tsikalas et al. (1998a–c) utilized seismic reflection profiles and well data to study in detail post-impact sedimentary, as well as the deeper structural extent (Fig. 4.22).

The most prominent feature of the crater is the central uplift, which is well preserved under the sediments, and more or less the only feature recognizable in the gravity map. Tsikalas et al. (1998b, c) described the extent of uplift with a diameter of about 8 km, surrounded by a trough of about 4 km in width and further a 12-km wide outer zone. We observe a relative gravity low of about 3–4 mGal coinciding with the extent of the "seismic disturbance" and a relative high (2 mGal) along with the uplifted structure observed in the seismic profiles. The full extent of the crater is unclear (Fig. 4.22). Similarly, the anomaly was described by Tsikalas et al. (1998c), but a detailed comparison is impossible, because their data were filtered and trend-corrected. Subtractions of regional fields as well as band-pass filtering usually result in distortion of the residual anomaly and of the shape of the interpreted subsurface structure. Stripping methods, which use detailed geological information to calculate and subtract the known contributions from the observations (Hammer 1963), lead to a better understanding of the residual anomaly. The data are presented here with and without further trend correction, but no filtering.

This results, as pointed out above, in a shift of the relative maximum of the gravity anomaly above the structural central peak and a slight distortion of the anomaly

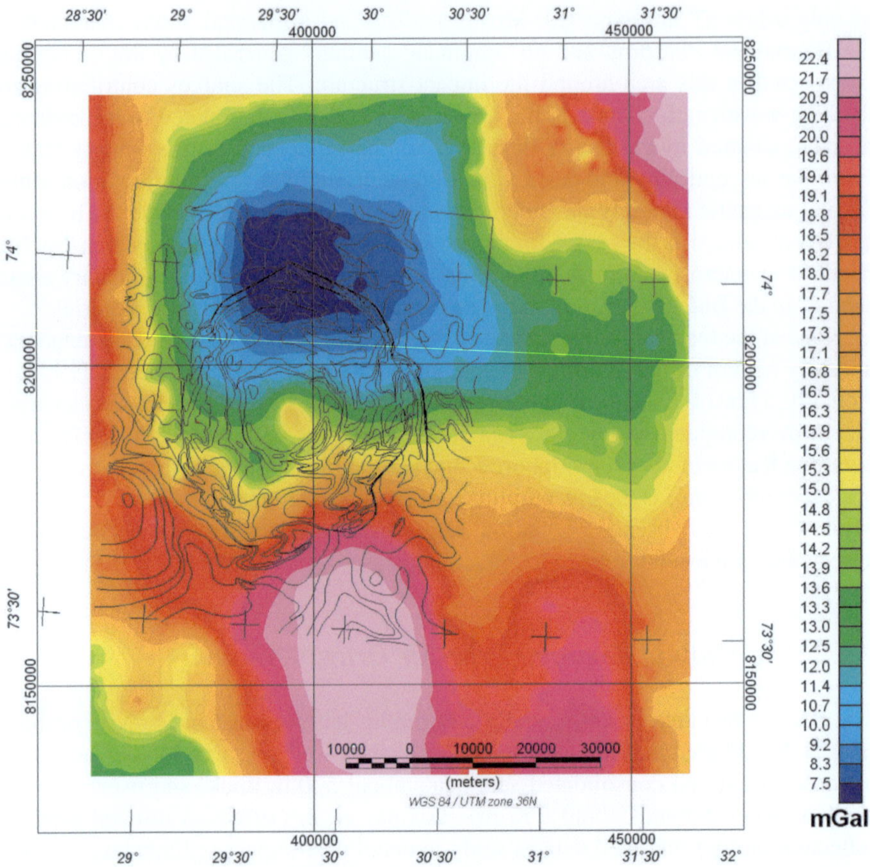

Fig. 4.22 Regional free-air marine gravity anomaly map in comparison with the mapped Mjølnir zonation boundaries from seismic reflection data (Tsikalas et al. 1998a–c)

representing the crater. Such an offset due to data correction leads to an interpretation of an asymmetric shape. If considering a more regional view, no simple regional trend can be defined (Fig. 4.22).

4.6.3 Alternative Interpretation of Mjølnir Crater Dimensions

We modelled the sub-surface structure using a forward 2D modelling method (originally developed by Talwani et al. 1959). We decided to model along a profile in NW-SE orientation (Fig. 4.22) that coincides with seismic sections based on high-resolution single-channel and multi-channel recordings (Gudlaugsson 1993; Tsikalas et al. 1998a–c). From the aeromagnetic data, no signal can be clearly related to a source related to the impact structure, therefore we focused on the

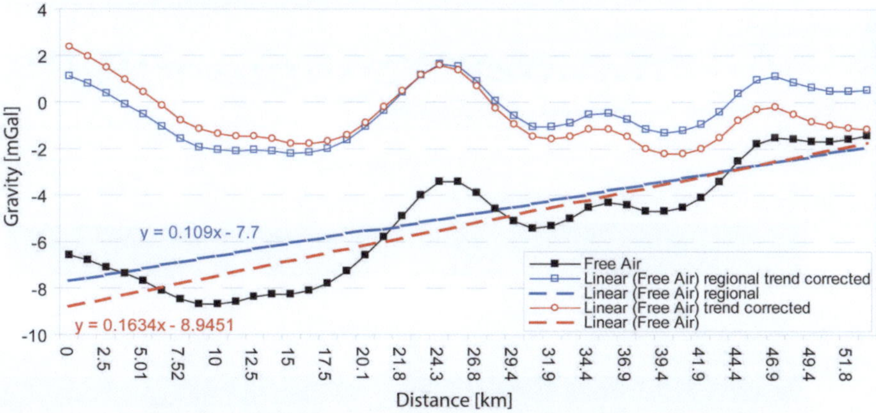

Fig. 4.23 Profile plot derived from the regional free-air gravity anomaly map in Fig. 4.22 compared with the derived linear trends, as well as the trend corrected version of the anomaly

gravity data. For the gravity field modelling, simple structural assumptions have been made, making use of the densities derived from the core (see Fig. 5.16), typical morphologies of a complex crater (Melosh 1989), and relevant seismic lines (Gudlaugsson 1993; Tsikalas et al. 1998a–c).

We used the gravity anomaly as observed, with and without correction for a regional linear trend. Figure 4.23 shows the original and trend-corrected profile. Such linear trends often represent long wavelength contributions from the basement or even deeper anomaly sources. We modelled the impact crater for both uncorrected and trend-corrected anomaly residuals to further constrain the basement interface, which is not resolved in the available seismic sections.

The resulting but preliminary model is shown in Fig. 4.24 for the trend-corrected case, together with two seismic profiles along which we modelled. Typically for complex craters, two major characteristics contribute to the gravity anomaly:

1. the cavity including possibly reduced density due to shattering of the surroundings
2. the uplift of deeper and denser layers in the centre

The former features a broad relative low and the latter adds a central relative high to the gravity anomaly. The model accounts for this and consists of eight model bodies:

1. Water layer
2. Unconsolidated sediments 1,
3. Unconsolidated sediments 2,
4. Consolidated layered sediments,
5. Breccia,

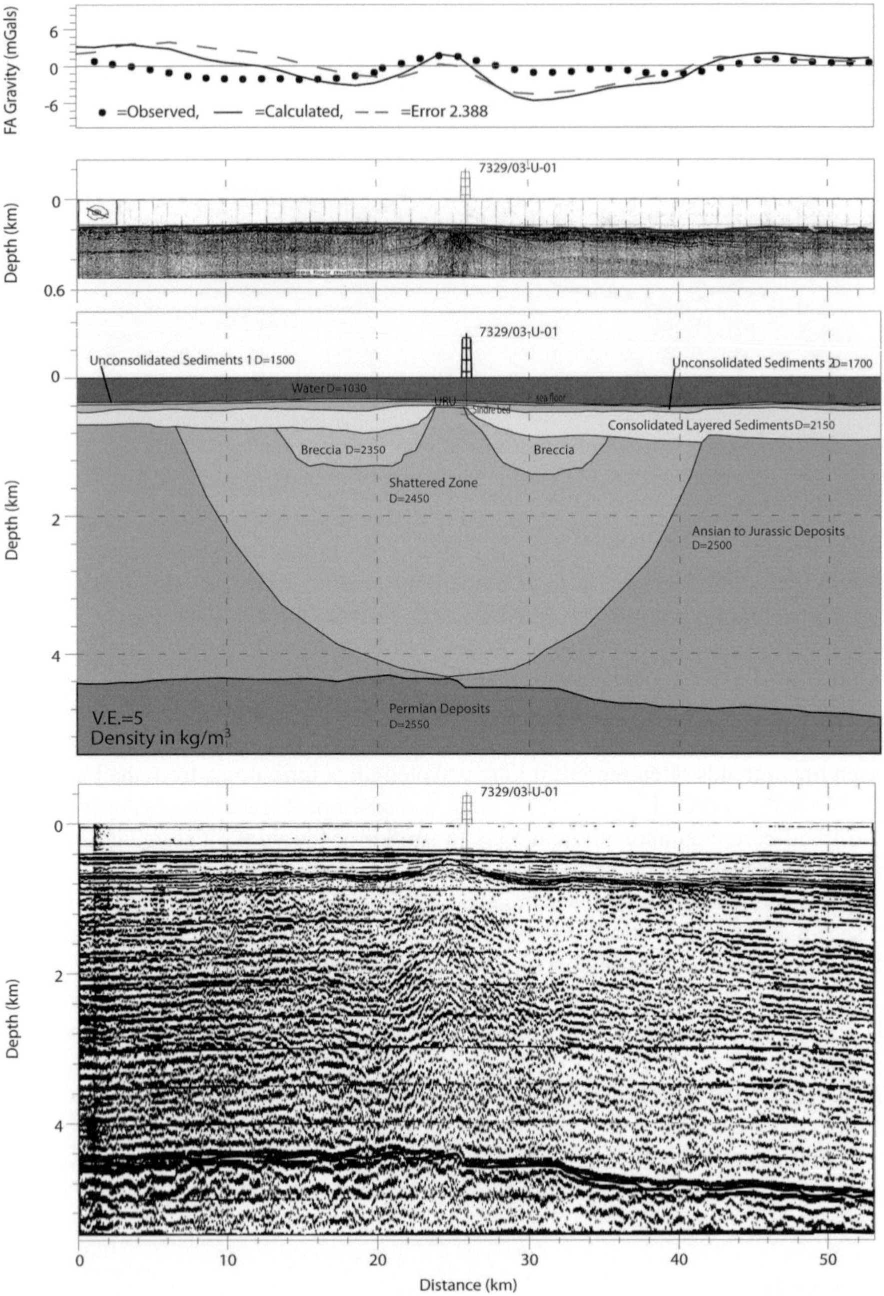

Fig. 4.24 The modelling results for a trend-corrected anomaly, together with the corresponding seismic reflection profiles and the density model

Table 4.3 Model bodies, model densities, body characteristics and boundary description referring to Figs. 4.24 and 4.25

Model body	Density (kg/m^3)	Body characteristics boundary (seismic horizon)
Water	1,030	Seafloor
Unconsolidated sediments 1	1,500	Stratification
Unconsolidated sediments 2	1,700	Seismic reflector Stratification
Consolidated layered sediments	2,150	Late Cenozoic upper regional unconformity Stratification
Allochtonous and autochtonous breccia	2,400	Base Cretaceous (Reflector) Disturbed stratification (void)
Shattered zone	2,450	Disturbed stratification
Anisian to jurassic deposits	2,500	Undisturbed stratification
Permian deposits	2,550	Top Permian (reflector) Top basement (no seismic record)
Crust	2,670	Interpreted

6. Shattered zone,
7. Anisian to Jurassic deposits,
8. Permian deposits.

The characteristic densities and the unit description and boundary justifications are given in Table 4.3.

The observed free-air gravity anomalies and the calculated gravity anomalies are given in Fig. 4.24. Densities assigned to the uppermost model bodies are chosen similar to the densities measured for the core samples (see Fig. 5.16), although increased compaction could very well have taken place at greater depth. The major uncertainties for this model are the densities of the uppermost layers representing unconsolidated sediments for which the first 70 m of the core were lost. Naturally, the uppermost layers influence the model response the most, especially if the density contrast is high, and do not leave much interpretational space for the deeper structure, here made up of bodies with very low density contrast. The first layers are unconsolidated Quaternary sediments, which may have densities less than 1,700 kg/m^3 and as low as 1,300 kg/m^3. In our model, horizons were picked from the high-resolution seismic profile, but the densities (1,500 and 1,700 kg/m^3) could be overestimated, and thus result in shallower anomalies.

The dominant feature is the structural uplift of higher density material compared to the surrounding compacted or unconsolidated sedimentary layers. The upper boundary is well defined through the seismic profiles available, as opposed

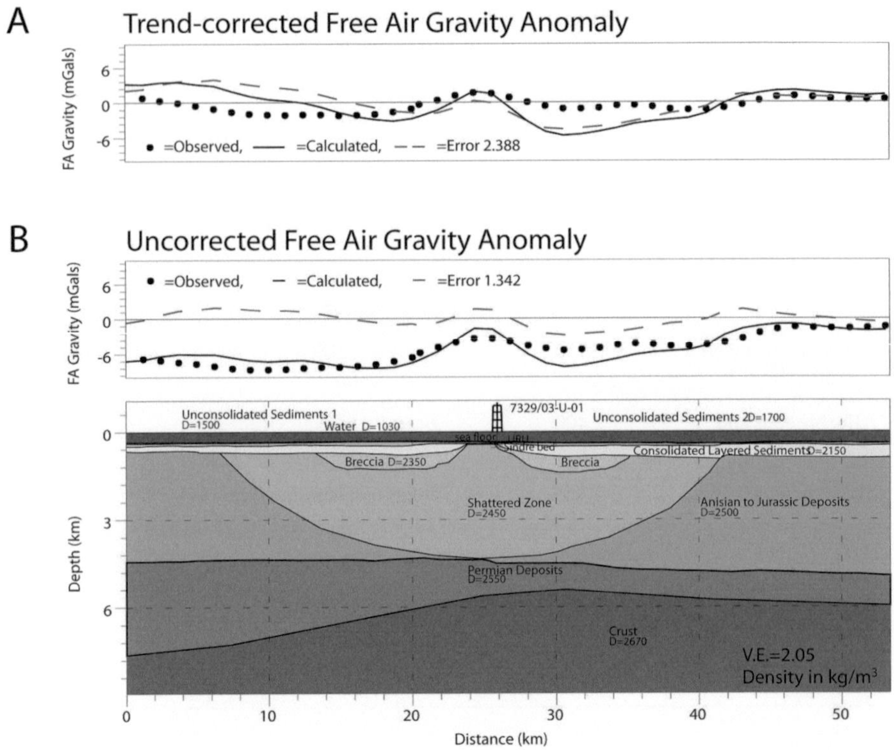

Fig. 4.25 A comparison between the modelling results for the trend-corrected model as shown in Fig. 4.24 and the uncorrected one. The only difference is an additional interface (*top basement*), which separates the sedimentary body from bedrock. This interface is not constrained by seismic reflection data

to the densities related to the layers. We observe that the amplitude of the observed gravity anomaly is somewhat smoother and shallower than the modelled values. One possible reason for this is that the modelling approach using a 2-dimensional cross-section, which in the case of a circular structure as craters results in an overestimation of the amplitude of relative lows and highs.

Assuming a symmetric distribution of the model bodies an asymmetric distribution of the calculated gravity anomaly is observed, when compared to the trend corrected observation. Therefore, we calculate the anomaly for the same model configuration, but with an additional layer, which represents the underlying bedrock and the top basement interface and compared to the uncorrected gravity signal (Fig. 4.25). The resulting modelled anomaly shows a better fit than for the corrected on, which is mainly due to the fact that the bedrock interface is not represented by a simple slope but more complex shape. For the trend correction, only a dipping layer was assumed when as simple slope was subtracted. From these models, the Mjølnir crater diameter itself is only about 20 km, and much smaller compared to earlier estimates (Tsikalas et al. 1998a–c). Note that this value corresponds to the maximum transient cavity diameter (16–20 km, Gudlaugsson 1993; Tsikalas et al. 1998b; Shuvalov et al. 2002) where the greatest part/volume of impact-related allochthonous breccia (Ragnarok Formation) resides (Figs. 1.11, 4.13, and 4.15). The area influenced by the impact event, due to shattering or disturbing the subsurface in the vicinity of the crater is larger. Also the intense and wide-ranging gravitational collapse as modelled by Shuvalov et al. (2002) might affect the morphology of that site, but cannot be resolved by gravity measurements.

4.7 Impact-Induced Changes in Physical Properties

Terrestrial craters have geophysical characteristics that are largely associated with the passage of a shock wave and the initiation of subsequent crater forming processes. The most conspicuous geophysical signature is a residual negative gravity anomaly (Pilkington and Grieve 1992) caused by low-density material resulting from both lithological and physical changes associated with the cratering process (Grieve and Pesonen 1992). The gravity low commonly extends to or slightly beyond the crater rim. In addition, there is a tendency for the gravity signature of >30-km-diameter complex craters to exhibit a central positive high, ascribed to central structural uplift (e.g., Pilkington et al. 1994; Plescia 1996). However, it is the counteracting processes of structural uplift and brecciation that determine the final density distribution within the impact-affected rock volume.

The Mjølnir gravity data are of high quality and analysis of cross-over errors indicates minimal deviations (Tsikalas et al. 1998c). The residual free-air gravity field exhibits a circular symmetric anomaly over the structure. The anomaly is divided into an annular low with an outer diameter of 45 km attaining minimum values of –1.5 mGal over the periphery, and a central 14-km-wide gravity high with a maximum value of +2.5 mGal (Figs. 4.8 and 4.9) (Tsikalas et al. 1998c). The spatial correspondence of the seismically defined structure with the central positive gravity high and the surrounding annular low suggests a causal relationship (Figs. 4.11, 4.12, and 4.13). According to the compilation of gravity anomalies of impact craters in sedimentary targets by Pilkington and Grieve (1992), a structure of the size of Mjølnir should produce an annular low of about –7 mGal. Nonetheless, the observed value of –1.5 mGal is still within the –1 to –11 mGal annular anomaly range determined by a number of craters. The moderate Mjølnir gravity signature is ascribed to several combined effects. The primary cause of the annular gravity low is the impact-induced porosity increase from extensive fracturing and brecciation. However, less pervasive brecciation in the relatively soft sediments may contribute to a smaller amplitude gravity low (Pilkington et al. 1995). In addition, strong modification of the density field takes place as a result of mass-transport during gravitational collapse and rebound of the crater floor displacing denser strata to shallower depths beneath the centre of the structure. We infer that the lowering of density by impact-induced porosity, which is expected to be at maximum in the central part of the crater, does not overcome the effect of the elevated high density rocks. Therefore, the lateral density distribution associated with the seismically defined disturbance at Mjølnir is characterized by a central high with relatively elevate density contrast, whereas the peripheral region has negative contrast. Subsequently, the impact cratering processes are followed by lesser, long-term alteration due to differential compaction of post-impact sediments (Tsikalas et al. 2002b; Tsikalas and Faleide 2007), which may reduce the density contrast between the crater periphery and platform, thus decreasing the amplitude of the annular gravity low. It is important to realise however that ~1.5–2 km of the post-impact sediments have been eroded later (e.g., Nyland et al. 1992).

The magnetic signature of impact craters is variable. The most common impact effect is a broad magnetic low, primarily observed in crystalline target impacts, resulting from disturbance of the regional magnetic trends.

In addition, the presence of local magnetic anomalies can be ascribed to uplift of magnetized rocks or impact-generated melt bodies (Pilkington and Grieve 1992). The amount of melt generated by impacts into sedimentary targets is expected to be comparable to that generated in crystalline target rocks, but with a more dispersed and scattered rather than coherent character (Kieffer and Simonds 1980; Cintala and Grieve 1994). A common feature of both the shipborne- and aero-magnetic surveys (and corresponding data processing) is the low amplitude of any observed magnetic anomaly within Mjølnir, and thus absence of shallow impact-related macroscopic melt bodies, as also evidenced by the available seismic reflection dataset (Figs. 4.10, 4.11, 4.12, 4.13, and 4.21). Magnetic modelling does not conclusively identify any magnetized melt sources and the majority of the anomalies may be mainly attributed to dislocation of weakly magnetized platform sediments. Although the magnetic signature above impact craters is more variable than the gravity signature, the magnetic response of Mjølnir is also within the range expected for impacts in sedimentary targets (Pilkington and Grieve 1992). Finally, the robust traveltime anomaly at the top Permian reflector (Figs. 4.12, 4.13, and 4.14) is ascribed to the positive velocity anomaly beneath the central high.

Fracturing and brecciation is expected to induce changes in the seismic velocity (Pilkington and Grieve 1992). The lateral changes may, in turn, produce pull-up or pull-down effects on continuous reflectors below the impact structure. Extensive analysis of stacking velocities and sonobuoy profiles (Fig. 3.2 and Table 3.1) has not resolved major lateral changes with reference to the Bjarmeland Platform strata. However, even small lateral velocity anomalies may induce robust traveltime anomalies in planar reflectors. Indeed, the seismic profiles reveal a small pull-up of the high-amplitude, originally planar top Permian reflector beneath the structure. The traveltime anomaly is 16 km in diameter and rises to +80 ms beneath the central crater (Figs. 4.11 and 4.14) (Tsikalas et al. 1998c, 2002b).

Integrated geophysical modelling results support the lateral differentiation of the Mjølnir seismic disturbance into a central high and a peripheral region (Fig. 4.12). A qualitative model to explain the modelled distribution of physical properties postulates interaction of several cratering processes including formation of transient cavity, brecciation, gravitational collapse, and structural uplift (Tsikalas et al. 1998c, 1999). The primary effect of the impact event is an impact-induced porosity-increase due to extensive fracturing and brecciation. Subsequent modification of the density field takes place as a result of mass-transport during gravitational collapse and structural uplift of the crater floor displacing deep, denser strata to shallower levels beneath the central structure. Furthermore, differential compaction of post-impact sediments results in long-term alteration of the impact-induced density field. Absence of high-amplitude seismic reflectivity (Tsikalas et al. 1998a, b) precludes the presence of impact-generated compact melt bodies or sheets of reasonable dimensions. In fact, the modelling (Tsikalas et al. 1998c) demonstrates that the observed low-amplitude magnetic anomalies can be interpreted in terms of dislocation of weakly magnetized platform strata, perhaps associated with local concentrations of dispersed melts or minor melt dikes in the peripheral region (Figs. 4.11 and 4.13). Moreover, the geophysical data are not compatible with alternative origins for the structure, such as salt or clay diapirs and igneous complex (Tsikalas et al. 1998c). In contrast, the Mjølnir geophysical signatures are consistent with those produced by similar-sized craters elsewhere (Tsikalas et al. 1998c). Thus, the spatial correspondence of structural features and geophysical anomalies substantiate the impact interpretation.

4.8 Mjølnir as an Oblique Impact Event

Numerical simulations of meteorite impacts (e.g., O'Keefe and Ahrens 1999; Pierazzo and Melosh 1999) have shown that the shock wave produced in an oblique impact is roughly hemispherical in spite of the obliquity, although it is weaker than the shock wave produced by a vertical impact at the same velocity. This is in accordance with the cratering record, which reveals that the circularity of craters on different planets is a first-order morphological characteristic (Spudis 1993; Alexopoulos and McKinnon 1994; Grieve and Pesonen 1996). The circularity underscores the first-order symmetry created by large impacts down to very low angles of incidence as a result of near-symmetry in the propagation of the shock wave front and symmetry of gravitational collapse during the modification stage (e.g., Melosh 1989), although the far-field shock stress distribution may retain asymmetry in terms of intensity (Dahl and Schultz 2000, 2001). Recently a number of detailed diagnostic structural and geophysical signatures of oblique terrestrial, and planetary, impacts based on interior crater features have been appreciated. These characteristics include asymmetries on e.g. crater morphology and shape, on central high/peak ring morphology and placement, and on nature and distribution of ejecta (e.g., Manson crater, Schultz and Anderson 1996; Chicxulub crater, Schultz and D'Hondt 1996; Toms Canyon probable impact structure, Poag and Poppe 1998; cratering record on Venus, Schultz 1992). Here, a re-assessment of the established structural and geophysical Mjølnir signatures is carried out searching for evidence that point out to an impact obliquity.

4.8.1 Elongated Crater Diameter

The seismic reflection database has revealed in detail Mjølnir's distinct radial zonation pattern of the impact-induced relief (Figs. 4.1 and 4.2) composed of a

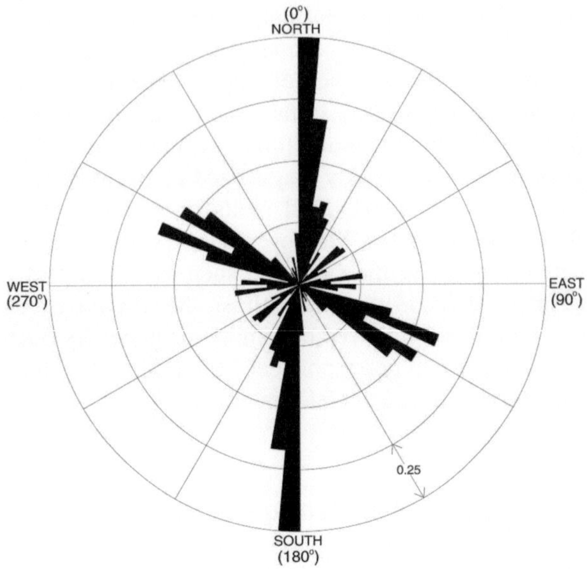

Fig. 4.26 Histogram of the crater diameter vs. azimuth at 1° intervals around the entire Mjølnir periphery normalized to the 40 km average crater diameter. The normalized histogram lengths range 0–1

12-km-wide complex outer zone, including a marginal fault zone and a modestly elevated peak ring, a 4-km-wide annular depression, and an uplifted central high, 8 km in diameter. Sharp boundary faults form a 150–100-m-high, near-circular rim wall and separate highly deformed strata within the crater from intact platform strata (Figs. 4.1 and 4.2). The most representative diameter for the near-circular Mjølnir crater periphery was found to be 40 km (Tsikalas et al. 1998a, b).

The well-established Mjølnir crater structure and morphology (Fig. 4.1) is further used as the basis for detailed measurements on the crater diameter. In particular, the crater diameter is estimated at 1° intervals for a 360° circle at Mjølnir's periphery, and after normalization to the 40 km average crater diameter the residuals are plotted versus the azimuth direction. The resulting histogram (Fig. 4.26) reveals that there is a dominant elongation of the structure in a N-S/NNE-SSW direction, and a second, smaller, elongation trend at a ESE-WNW direction (Tsikalas 2005).

4.8.2 Seismic Disturbance Asymmetry

The regional grid of multichannel seismic profiles combined with the fan-like geometry of the single-channel seismic surveys provides a dense coverage of the structure at all directions (Fig. 4.1). Figure 4.27 shows three deep multichannel seismic profiles stacked with three shallow single-channel profiles. All profiles cross the central high and are aligned along the geometric crater center. Note that the latter is defined as the center position where a 40-km-diameter circle best fits the near-circular Mjølnir crater periphery. For each profile the ratio of the two radii, on the left and right side, between the crater center and the rim faults is estimated (Fig. 4.27;

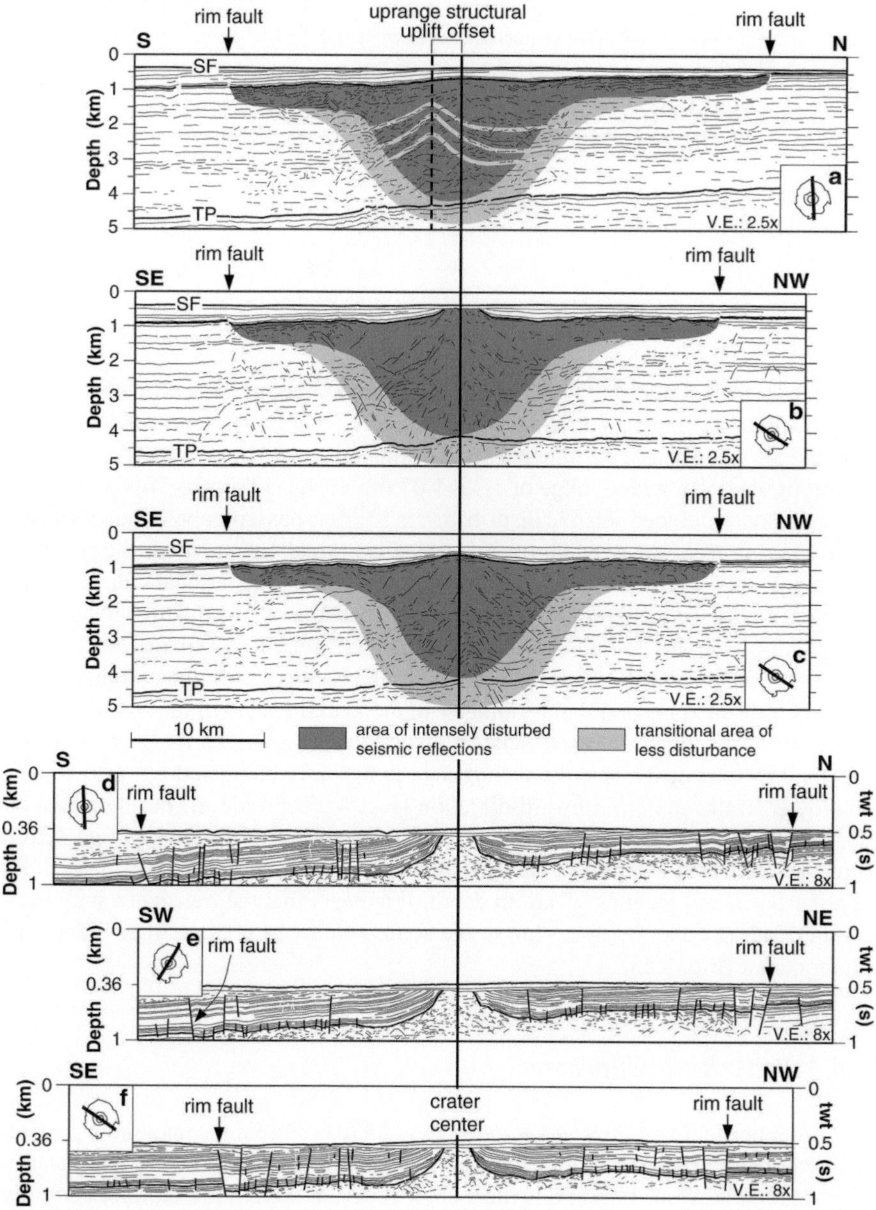

Fig. 4.27 Interpreted multichannel (*top three*) and single-channel (*bottom three*) seismic reflection profiles stacked along the geometric crater center. In profile (**a**) selected reflector segments are connected to visualize the clear up-range structural uplift offset

Table 4.4 Crater radii asymmetry factors (dimensionless) defined as the ratio of the two radii, on the left and right side, between the geometric crater center and the rim faults of each of the profiles (a–f) illustrated on Fig. 4.27

Profile (Fig. 4.27)	Asymmetry factor (direction)
a	1.33 (N)
b	1.12 (NW)
c	1.14 (NW)
d	1.07 (N)
e	1.17 (NE)
f	1.15 (NW)

Table 4.4). If the crater radii on each profile were symmetrical, a ratio of ~1.0 is to be expected. The analysis provides a crater radius asymmetry factor for each profile and reveals a consistent asymmetry towards the northward direction (NW/N/NE) in the order of ~1.16, with a range of 1.33–1.07 (Table 4.4) (Tsikalas 2005).

The impact-induced seismic disturbance at Mjølnir has a parabolic bowl-shape at the center of the structure and turns into a shallow broad-brim towards the periphery (Figs. 4.2 and 4.27). This distinct shape evolved during gravitational collapse of the transient cavity wall and progressive outward expansion of the crater by inwards dipping fault-blocks floored on low-angle décollement surfaces (Tsikalas et al. 1998a, b, 1999; Tsikalas and Faleide 2004). The crater radius asymmetry (Table 4.4) is directly translated to a similar asymmetry in the lateral extend of the shallow broad-brim part of the impact-induced seismic disturbance (Fig. 4.27). It also appears that the shallow part of the seismic disturbance is not only elongated but also slightly shallower in the same northward-direction (Fig. 4.27, profile a). Furthermore, the decompacted-thickness contour map of the intensely disturbed zone in Fig. 4.28 is directly related to the transient cavity bounded approximately at the perimeter of the annular basin and reaches ~5 km in depth. It appears that the transient cavity maximum depth is offset by 2–2.5 km to the south-southwest relative to the geometric crater center (Fig. 4.28).

4.8.3 Peak-Ring Character

The outer perimeter of the annular basin, thought to represent the maximum possible diameter of the short-lived transient cavity during the excavation and modification stages (Tsikalas et al. 1998b; Shuvalov et al. 2002), as well as the subfloor compression crater limit (Schultz 1992; Schultz and Gault 1992), exhibits a slightly raised relief (Figs. 4.1, 4.2, and 4.5). Although irregular in shape, and varying in width from 1 to 3 km, the raised relief gives the impression of a subdued ring structure, similar to those typically found in peak ring craters, and was thus referred to as a peak ring feature by Tsikalas et al. (1998b).

In the greatest detail, the peak ring at Mjølnir is clearly defined as a raised near-arcuate feature delineated by opposite dipping faults with 10–30 m throws (Fig. 4.1).

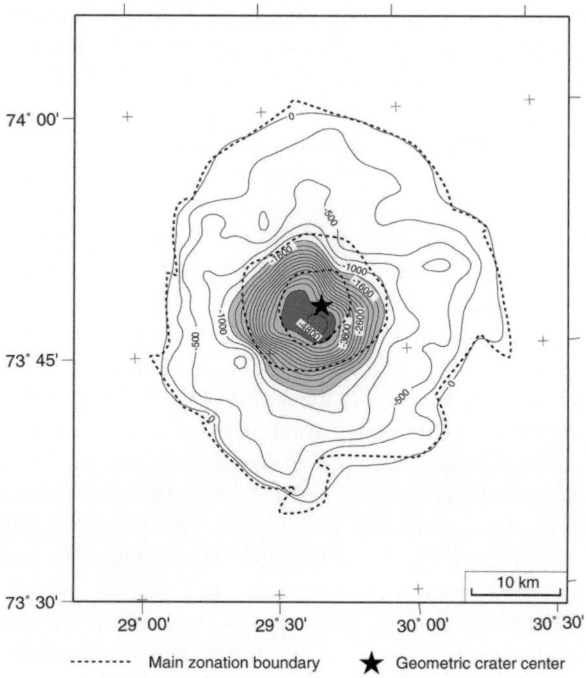

Fig. 4.28 Morphological characteristics of the decompacted impact-induced intensely disturbed zone (confer Figs. 4.11 and 4.27; *dark-grey raster*) derived through mapping of the entire seismic reflection dataset. *Light-shading* denotes the *central bowl-shape* part of the disturbance approximately at the limit of the annular basin; *dark-shading* denotes the transient cavity maximum depth. Contour interval in meters

This characteristic shape becomes less clear in the N- and NE-directions where the raised relief is breached and the peak ring remains open, being replaced by faults facing the crater center (Fig. 4.1).

4.8.4 *Offsets in Brecciation and Structural Uplift*

Impact craters on sedimentary targets have the advantage, in comparison with similar structures on crystalline targets that the regular, pre-impact stratification of these targets provides reference horizons against which the impact-induced structures can be identified and mapped by seismic reflection studies. Such studies have provided an effective means of mapping the large-scale geometrical structure at depth with a high-degree of horizontal and vertical resolution (e.g., Morgan and Warner 1999).

At Mjølnir, seismic mapping and analysis of the deeper structure levels in combination with gravity and seismic-velocity modelling, and with detailed numerical simulations have provided greater insight into several cratering processes, such as brecciation and excavation, gravitational collapse of the transient crater, and structural uplift (Tsikalas et al. 1998a–c; Shuvalov et al. 2002). The integrated analysis supports the lateral differentiation of the Mjølnir seismic disturbance into a central uplift and a peripheral region (Fig. 4.2). The primary effect of the impact event is an impact-induced porosity increase due to extensive fracturing and brecciation (Pilkington and Grieve 1992; Tsikalas et al. 2002b). Subsequent modification of the density field takes place as a result of mass transport during gravitational collapse and structural uplift of the crater floor displacing deep, denser strata to shallower levels beneath the central structure (Fig. 4.2) (Tsikalas et al. 1998a–c; Shuvalov et al. 2002).

The models for Mjølnir are directly supported by the observed free-air gravity and seismic velocity anomalies. In particular, the residual free-air gravity field exhibits a circular anomaly over the structure (Fig. 4.29a). The anomaly is divided into an annular low, with an outer diameter of 45 km, attaining minimum values of –1.5 mGal over the periphery, and a central 14-km-wide gravity high, with a maximum value of +2.5 mGal (Tsikalas et al. 1998c). It appears that the 0-mGal gravity anomaly contour exhibits a distinct elongated-shape in the SW-NE direction (Fig. 4.29a). In addition, the annular gravity low (<–0.5 mGal), which is directly connected with the region of most intense fracturing and brecciation, closely resembles a U-shaped central pit open to the northeast (Fig. 4.29a).

The deep seismic profiles provide evidence of upward bending reflector segments beneath the central high and the annular basin, indicating elevation of deep strata to shallower levels (Fig. 4.27). By measuring the difference in depth between the extrapolated top of selected upward bending reflector segments beneath the central high and their most likely equivalent subhorizontal interfaces the structural uplift was estimated to be 1.0–1.5 km and when decompacted ~1.5–2.0 km, fitting the theoretical expectations for the Mjølnir dimensions (Fig. 4.27) (Tsikalas et al. 1998a). The north-south profile in Fig. 4.27 (profile a) clearly shows a maximum structural uplift lateral offset of 2–2.5 km towards the south from the geometric crater center. Similarly, the gravity central peak that corresponds to the maximum structural uplift (Tsikalas et al. 1998c) is laterally offset by ~1.5–2 km to the southwest from the geometric crater center (Fig. 4.29a). Furthermore, fracturing and brecciation are expected to induce changes in the seismic velocity expressed as pull-up or pull-down effects on continuous reflectors below the impact structure. Indeed, the seismic profiles reveal a small pull-up of the high-amplitude, originally planar top Permian reflector beneath the structure (Fig. 4.2). The mapped traveltime anomaly is 16 km in diameter and rises to +80 ms beneath the central crater (Fig. 4.29b), corresponding to a +175 m/s seismic velocity anomaly (Figs. 4.11–4.14) (Tsikalas et al. 1998c). It appears that the central traveltime anomaly has a slightly elongated shape in the SW-NE direction and its top is slightly offset by ~2 km to the WSW from the geometric crater center (Fig. 4.29b) (Tsikalas 2005).

4.8.5 Impact Direction and Angle

The Mjølnir crater lies on the Bjarmeland Platform in the central Barents Sea, which represents a stable sedimentary platform since Late Paleozoic times without evidence for considerable tectonism during its evolution (Faleide et al. 1993, 2008; Gudlaugsson et al. 1998). At the time of impact, ~142 Ma (Jurassic-Cretaceous transition), the region contained Upper Paleozoic strata, mainly carbonates and evaporites, overlain by 4–5 km thick Mesozoic siliciclastic sediments (Worsley

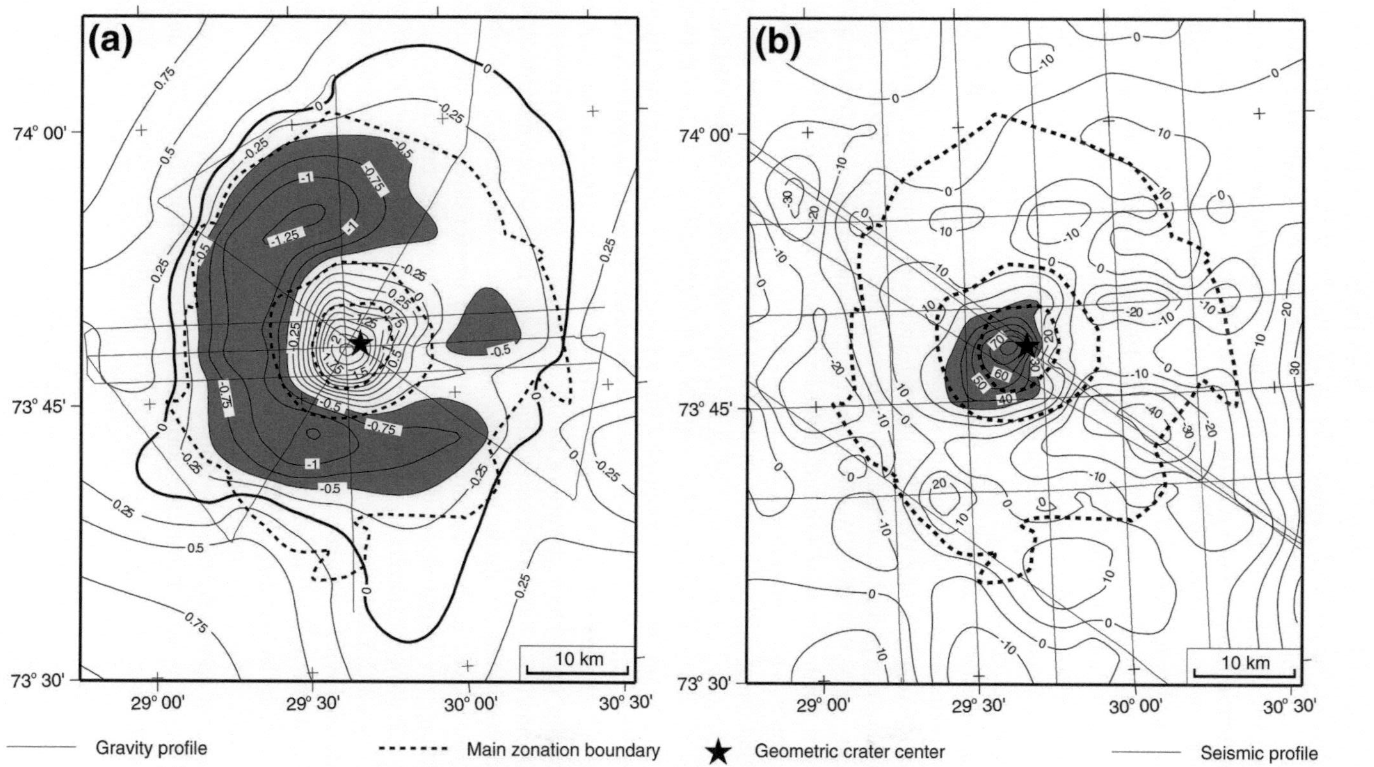

Fig. 4.29 (**a**) Residual free-air gravity anomaly map with 0.25 mGal contour interval, (**b**) Residual traveltime anomaly map at the top Permian reflector (confer Figs. 4.9 and 4.14) with 10 ms contour interval

et al. 1988; Gabrielsen et al. 1990). As indicated by the regional sequence thicknesses of Triassic and Jurassic strata (Fig. 4.27) the Bjarmeland Platform at the time of impact exhibited a minor dip of 0.5° towards the south, and the pre-impact Mesozoic successions appear sub-horizontal without presence of any angular strata or angular unconformities (Gabrielsen et al. 1990; Gudlaugsson et al. 1998). Furthermore, the post-impact sedimentary succession initially filled the subtle impact-generated crater relief on the sea bottom and subsequently deposited uniformly above the entire structure (Tsikalas et al. 1998a; Tsikalas and Faleide 2007). Therefore, we exclude the possibility that the observed asymmetric structural and geophysical patterns may be caused by any pre-impact irregularities in the regional sea-bottom topography or sequence stratigraphic relations of older strata, as well as by post-impact activity.

The detailed analysis in this study has revealed several evidences that, when combined with the established diagnostic structural and geophysical asymmetries of oblique impacts (e.g., Schultz 1992; Schultz and Anderson 1996; Sugita and Schultz 2002), can be directly related to impact direction and angle estimates for the Mjølnir impact. First, the N-S/NNE-SSW elongated crater diameter translates into impact from a similar northward or southward direction without, however, definition of the explicit azimuth. The same impact direction is further supported by the observed slight elongation of crater diameter in the ESE-WNW direction as such elongation, approximately traverse to the impact trajectory, has been reported for both the Manson (Schultz and Anderson 1996) and Chicxulub (Schultz and D'Hondt 1996) craters. A similar northward or southward impact direction is also supported by the elongated shape of the 0-mGal gravity anomaly contour. Second, the asymmetry and elongation in both the crater radius and the shallow broad-brim part of the impact-induced seismic disturbance towards a northward direction is the first evidence for an impact from the southward direction. Third, the breached and open towards N/NE peak ring in combination with the horseshoe shape of the annular gravity low which remains open in NE, further support a southwestern impact direction. The shape of the annular gravity low is connected with intense up-range brecciation and a down-range shallower excavation. Fourth, the transient cavity maximum-depth lateral offset by ~2–2.5 km towards the south-southwest from the geometric crater center, combined with a similar structural uplift lateral offset towards the south, a similar central gravity high offset towards southwest, and an elongated traveltime central anomaly offset towards WSW, all demonstrate an up-range highest elevation offset and, thus, support an impact direction from south-southwest.

Two different approaches, a quantitative and a qualitative one, were followed to define the angle, measured from the horizontal, of the Mjølnir impact. The quantitative approach is based on laboratory experiments and crater studies on Venus relating the effect of impact angle on the ratio of crater diameter to central high/peak ring diameter (Schultz and Gault 1990; Schultz 1992), as well as insights from structural and shock asymmetries (Schultz and Anderson 1996; Dahl and Schultz 2000, 2001). A possible impact angle ranging 30°–45°, is found to be representative for the Mjølnir dimensions. The qualitative approach uses the results obtained by the laboratory experiments of Gault and Wedekind (1978) and the numerical computations of O'Keefe and Ahrens (1985) for oblique impacts at various angles. For all resulting crater geometries, crater radii asymmetry factors are calculated by fitting a parabolic-shaped transient cavity and the results are extrapolated to the full range of impact angles. A possible impact angle of ~45° was estimated, ranging 30°–60°, based on the crater radius/seismic disturbance asymmetry factors for Mjølnir (Table 4.4).

Planetary crater studies and laboratory experiments showed that the oblique penetration phase of the impactor appears to be at least in part preserved within the central crater: The up-range offset of transient cavity, structural uplift and central gravity peak, and the down-range breach of the peak-ring zone are consistent with the extended region of energy transfer created during the early penetration stages by an oblique impact (Schultz and Gault 1990; Schultz 1992). A similar down-range shallower excavation and up-range offset of the central gravity high connected with oblique impacts are revealed at detailed gravity studies on lunar complex impact craters (Zuber et al. 1995). However, based on a comprehensive statistical approach for Venusian oblique impact craters, Ekholm and Melosh (2001) argued recently that the preserved transient cavity asymmetry in the final crater, which is observed in simple craters, does not necessarily hold in complex craters. Although the clarity of this issue is still debatable, the various lines of evidence presented in this study for Mjølnir all point towards a similar impact direction from the south/southwest.

4.8.6 Mjølnir Impact Obliquity Constrains Models for Near-Field Perturbations

Numerical simulations and experimental analogues have shown that obliquity is accompanied by less energy transfer from the projectile to the target (e.g., Gault and Wedekind 1978; Hayhurst et al. 1995; Schultz 1996; Burchell and Mackay 1998; Ivanov and Artemieva 2002). The Mjølnir energy release estimates of Tsikalas et al. (1998b) were made considering an elevation impact angle of 45° based, at that time, on well-known probability arguments (Shoemaker 1962; Shoemaker et al. 1990). The energy release was estimated to be in the order of 16×10^{20} J (range of 2.4–53×10^{20} J), translating into 3.8×10^{5} megatons TNT equivalent (range of $5.7 \times 10^{4} - 1.2 \times 10^{6}$) (Tsikalas et al. 1998b). An oblique impact at a ~45° (possibly 30°–45°) angle, as estimated in this study, is expected to have resulted in similar energy release.

Energy release dissipation at the proposed trajectory and angle for the Mjølnir impact may have a direct consequence on the distribution of proximal ejecta and tsunami-waves following the cessation of the impact-related processes at the impact-site. This is because the oblique impact most probably has created a down-range sector/corridor of thicker ejecta deposits and greater water column disturbance (Fig. 4.30). Such a sector/corridor may have been responsible for a geographic variation of short-term perturbations/environmental stress magnitude on the Barents Sea and adjacent regions, as it may have intensified the stress at a specific location and left the others almost unaffected.

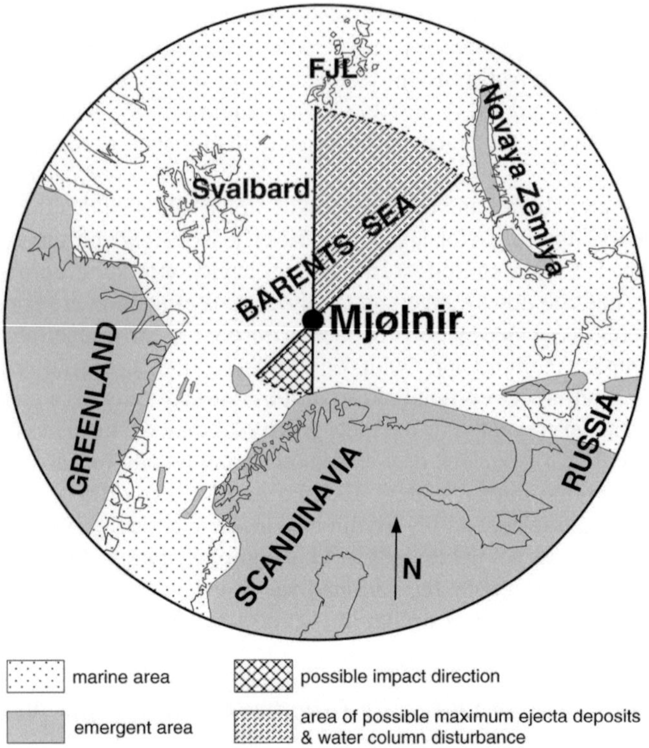

Fig. 4.30 Mjølnir impact location with possible range of impact direction azimuth and down-range area of possible maximum ejecta deposits and water column disturbance, shown at a ~142 Ma plate reconstruction based on Lawver et al. (1999) and overlaid on a simplified paleogeographic synthesis based on Brekke et al. (2001) approximately at the time of impact. FJL, Franz Josef Land

4.8.6.1 Nature and Distribution of Proximal Ejecta

Theoretically, the volume of material displaced from the crater equals the volume of excavated cavity (Croft 1985; Melosh 1989). Geophysical observations constrain the volume of a parabolic excavated cavity to 180 km^3 (Tsikalas et al. 1998b), whereas numerical simulations indicate a volume of ~230 km^3 (Shuvalov et al. 2002). The ejecta layer is expected to be thickest close to the crater rim, decreasing rapidly with distance from the crater center (Tsikalas et al. 1998b; Shuvalov et al. 2002). Accepting an oblique Mjølnir impact from the south/southwest, ejecta iso-thickness contours will probably not be circular around the crater site, but rather elongated towards the north/northeastern direction (Fig. 4.30). This issue is further discussed in detail together with Mjølnir numerical simulations (confer Chap. 8).

Borehole evidence for the Manson crater case (e.g., Anderson et al. 1996) substantiate that ejecta deposits are thinnest in the up-range direction and that only the top target layers are ejected due to shallower excavation as a result of oblique impact (Schultz and Anderson 1996). At Mjølnir, possible shallower penetration and excavation may be the reason for the small discrepancies in the estimated transient cavity depth of 4–5 km versus 6 km and the excavated/ejected volume of 180 km^3 vs. 230 km^3, based on geophysical observations and numerical simulations, respectively (Tsikalas et al. 1998b; Shuvalov et al. 2002). Note that numerical simulations have for simplicity considered a vertical incidence for the Mjølnir impact (Shuvalov et al. 2002). Furthermore, the fact that the shocked quartz grains and the iridium anomaly peak are located at the base and top, respectively, of the 80-cm-thick ejecta deposit at borehole 7430/10-U-01 (Fig. 4.1) (Dypvik et al. 1996) may have possibly resulted from a multistage ejecta emplacement, similar to those attributed to oblique impacts as revealed by laboratory experiments and planetary impact crater studies (Schultz and Gault 1990; Schultz 1992; Schultz and D'Hondt 1996). Again, this issue is further discussed in detail together with Mjølnir numerical simulations (confer Chap. 8).

Magnetic modelling has indicated that only low quantities of dispersed-character melts localized in the crater periphery may have been produced during the water-covered, sedimentary target Mjølnir impact (Tsikalas et al. 1998c). A similar absence of considerable impact glass and melts at Manson crater has been also attributed to the sedimentary target and, more importantly, to the obliquity of the impact as this results in shallower target penetration and less direct energy transfer from the projectile to the target (Izett et al. 1993; Schultz and Anderson 1996). Recent geochemical analyses of samples from the Mjølnir central crater core (borehole 7329/03-U-01, Fig. 4.1) have showed the absence of (Cr, Co, Ni) or weak (Ir) siderophile-element anomalies (Sandbakken 2002). This translates into a low abundance or total absence of projectile material in the crater itself, being consistent with oblique impact models. In such models a large fraction of the projectile material retains a net down-range motion and fragments of it may survive the impact, due to higher ejection velocity and lower shock compression, and may be deposited outside the crater proper (Schultz 1996; Pierazzo and Melosh 2000; Artemieva and Shuvalov 2001; Dypvik et al. 2004a, c, 2006; Shuvalov and Dypvik 2004).

4.8.6.2 Tsunami-Wave Distribution

The growing crater rim and ejecta curtain following the Mjølnir impact form a water surge that eventually breaks up and causes the formation of several waves that, in turn, together with reflected waves, will generate tsunamis. Tsunami wave heights resulting from a vertical incidence Mjølnir impact at various radii from the impact site have been calculated based on different approaches (Tsikalas et al. 1998b; Shuvalov et al. 2002) (see Chap. 10). Although the impact tsunami theory for vertical incidence impacts is well understood (e.g., Ward and Asphaug 2002), there is an almost total absence of computational experiments of tsunami-wave distribution resulting from oblique impacts.

Due to the relatively shallow water-target depth, we visualize an oblique Mjølnir impact (south/southwest azimuth at ~45° angle, possibly 30°–45°) to have generated a greater down-range water column disturbance, probably giving rise to faster travelling tsunami-waves at the down-range rather than the up-range region

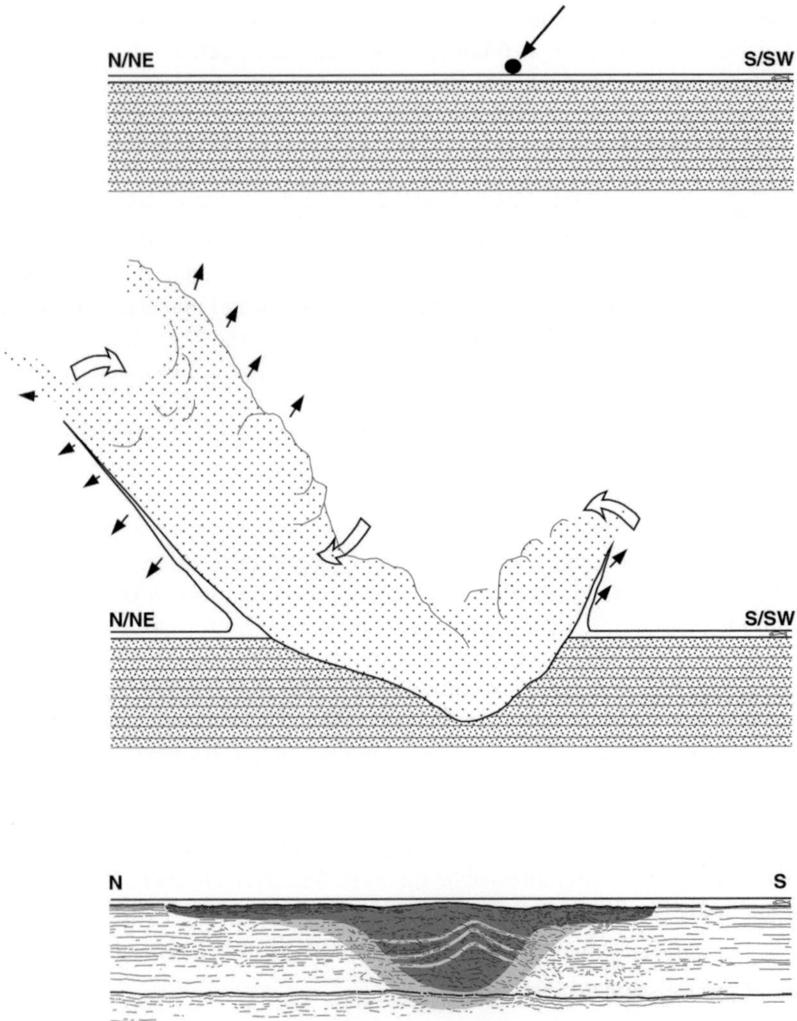

Fig. 4.31 Schematic cross-section diagrams showing the proposed formation sequence of the Mjølnir crater, with focus on the ejecta distribution and water-column disturbance, resulting from an oblique impact from south/southwest. Detailed hydrocode simulations of a vertical-incidence Mjølnir impact are provided by Shuvalov et al. (2002)

(Fig. 4.31) (Tsikalas 2005). Such scenario is better approximated with the non axis-symmetrical propagation of tsunamis resulting from submarine slides (Ward 2001). Additional and more advanced tsunami propagation scenarios are discussed in Chap. 10.

The ~80-cm-thick ejecta layer at borehole 7430/10-U-01 (Fig. 4.1) (see Chap. 6) is the thickest Mjølnir ejecta detected so far. The minor thickness ejecta detected on Svalbard ($\leq$1 cm; Dypvik et al. 2004b) and the absence (undetected so far despite the intense efforts) of tsunami-deposit signatures on NE Greenland, are additional evidence for the obliquity of the Mjølnir impact and possible geographic indications in ejecta and tsunami-waves distribution patterns. The proposed conceptual model (Figs. 4.30 and 4.31) envisages thickest ejecta distribution and faster travelling (thus most devastating) tsunami waves concentrated in the area between Svalbard and Novaya Zemlya. The analysis clearly shows the importance of impact direction and angle in the distribution pattern of ejecta and tsunamis, and further research must, therefore, focus on the proposed down-range region (Fig. 4.30).

Chapter 5
Impact Cratering and Post-impact Sedimentation

Henning Dypvik, Morten Smelror, Atle Mørk, Stephanie C. Werner, and Trond H. Torsvik

5.1 Introduction

The impact origin of the Mjølnir structure has been confirmed by detailed studies of lithologies from two shallow boreholes; one close to the crater center (7329/03-U-01) and one 30 km NE from the crater periphery (7430/10-U-01) (Fig. 1.7). The boreholes revealed brecciated sediments containing shocked quartz grains. In 7430/10-U-01 a prominent ejecta layer with strong iridium enrichment has been recovered (Dypvik et al. 1996; Dypvik and Ferrell 1998; Dypvik and Attrep 1999; Sandbakken 2002) (see Figs. 6.2 and 6.3). In this chapter the Mjølnir core (7329/03-U-01) will be presented, whereas core 7430/10-U-01 is presented in Chap. 6.

Recent studies of marine impact targets and systematic compilations of their structural and morphological features have shown significant differences among impact craters formed on land and those formed at sea (Gersonde and Deutsch 2000; Ormö and Lindström 2000; Dypvik and Jansa 2003; Dypvik et al. 2004a; Gohn et al. 2008). The primary cause for several of these differences is the high water content in the sediments themselves, as well as the overlying sea water. These characteristics explain the rheological differences, and trigger different erosional and depositional processes that normally do not occur on land. During a marine impact, a water cavity is formed and modified by the growing crater rim and the ejecta curtain pushing the water outwards. Collapse of this water cavity starts at its base and causes a water flow towards the crater (Shuvalov 2002b). When the water depth is sufficient to overflow and cut through any uplifted rim, characteristic erosional/depositional resurge gullies can be formed, acting as inlets of water and material flowing back into the crater. This flow greatly affects the crater rim and leads to extensive infilling of the crater. In addition to the slumping, slides, avalanches and density currents are formed by the collapsing water-saturated sediments of the rim, the peak rings and

H. Dypvik (✉)
Department of Geosciences, University of Oslo, NO-0316 Oslo, Norway
e-mail: henning.dypvik@geo.uio.no

H. Dypvik et al. (eds.), *The Mjølnir Impact Event and its Consequences*, Impact Studies, DOI 10.1007/978-3-540-88260-2_5,

the central high. In the Mjølnir case, such complex sedimentation processes mixed sediments of variable lithologies from different stratigraphic levels of the pre-impact Upper Paleozoic and Mesozoic succession of the impact area. A detailed description of these rocks is given below.

5.2 The Mjølnir Crater Core (7329/03-U-01)

A shallow borehole (7329/03-U-01) was drilled into the Mjølnir crater at a water depth of 350 m in late August 1998 by the drillship M/S Bucentaur (Figs. 1.2, 1.6, and 1.7). A soil-drilling pipe was used to penetrate the 50-m-thick Quaternary sediment and till package (Fig. 5.1), and no samples were taken from this interval. An ordinary diamond drilling system placed inside the soil drilling pipe was used to core the bedrock. After every three meters of drilling, the core was retrieved using a wire-line system, then carefully extracted from the core barrel and a preliminary description made. Sonic velocity measurements and spectral gamma radiation analyses on the core were run where possible.

The coring operation went smoothly. The first 7 m of the marly limestones of the Klippfisk Formation were sampled, followed by 17 m of the grey, laminated shales of the Hekkingen Formation (Table 5.1 and Fig. 5.2). The rocks retrieved between 74.05 m below seabed and the core base at 171 m differed from anything else obtained in the Barents Sea. The core, however, had very low internal strength and pieces could not be lifted out of the core barrel without falling apart. The sound velocity could, therefore, not be measured regularly and only very low values were obtained, except in the few well-cemented intervals. The lack of core-strength prevented measurements of natural gamma activity aboard, and a technical breakdown that terminated the drilling operation made logging of the hole by ordinary petrophysical wire-line tools impossible.

After the core had been transported to the laboratory, and allowed to dry slowly, about 62 m had sufficient strength to be slabbed, and these parts were mounted in alumina trays as display cuts (Figs. 5.3–5.12). A spectral gammalog was recorded 4 years after drilling, using laboratory analyses analyses directly on the core (Fig. 5.13). Macrofossils, mainly from the Hekkingen Formation, were carefully extracted, together with samples for palynological and micropaleontological analyses (Fig. 5.14).

5.2.1 The Ragnarok Formation

The disturbed rocks cored between 171 and 74.05 m in borehole 7329/03-U-01 forms a mapable unit recognized within the Mjølnir crater, the so-called Ragnarok Formation formally described by Dypvik et al. (2004b) (Fig. 1.4). The occurrences of this unit are diagrammatically displayed in Figs. 5.2, 5.5, 5.6, and 5.8. They

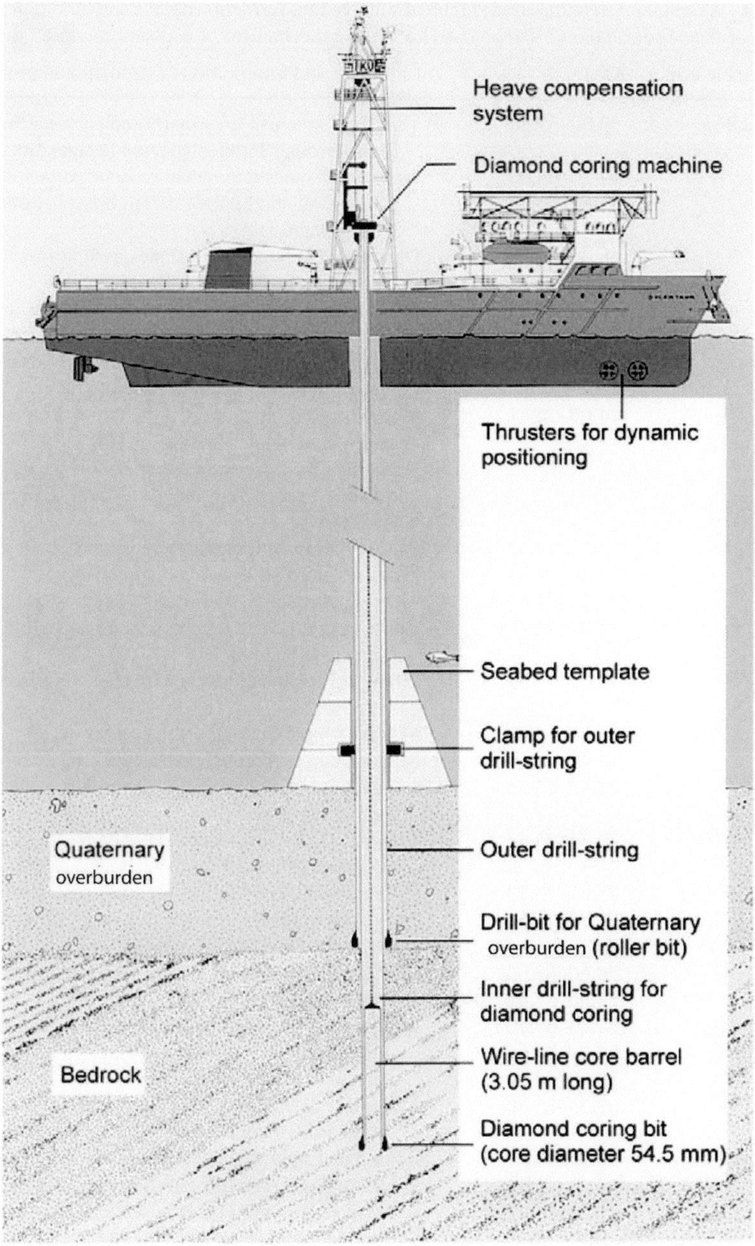

Fig. 5.1 The drilling device at Bucentaur RS, illustrating both overburden penetration and sediment coring

Table 5.1 A compilation of the stratigraphical subdivision, formation depths and lithological composition of the Mjølnir core (7329/03-U-01). Possible mechanism of deposition written in italics

Stratigraphic unit	Depth m	Lithology and interpration of depositional process
Klippfisk Fm	50.00–57.20	Very light greenish grey marls and carbonates, fossiliferous. Hemipelagic and pelagic rain
Hekkingen Fm	57.20–68.00	Light grey shales laminated, in some parts no fossils. From 63 m and up core rich in bioturbation. Hemipelagic rain
	68.00–73.78	Dark grey/black org.-rich shales, rich in fossils. Parallel lamination. Hemipelagic rain
	73.78–73.89	Interval with conglomerates and sandstones in mudstones. Storm-triggered turbidity currents
	73.89–74.05	Dark grey/black org.-rich shales, rich in fossils. Parallel lamination. Hemipelagic rain
Ragnarok Fm. subunit IIc	74.05–74.31	Matrix- and grain-supported, pebbly mudstone, faintly laminated. From level 74.05 to 75.73 mainly en-masse sediment transport possibly related to debris flow and/or turbidity current transport
	74.31–74.37	Parallel laminated, medium sandstones
	74.37–74.71	Massive, grain-supported conglomerate
	74.71–74.81	Medium sandstone, planar cross bedding, well sorted
	74.81–75.27	Clast- to matrix s.cgl., general inverse grading with faint cross bedding
	75.27–75.33	Fining-upwards sandstone with clay clasts, matrix supported
	75.33–75.42	Matrix-supported conglomerate, clay clasts
	75.42–75.48	Matrix-supported conglomerates with thin clay beds
	75.48–75.54	Mudstone with outsized clay clasts, matrix-supported
	75.54–75.61	Thin clay bed overlain by homogenous, clast-supported cgl
	75.61–75.73	Homogenous, clast-supported conglomerate
Ragnarok Fm. subunit IIb	75.73–79.50	Dark grey to brown, olive green, normally graded mudstone. Suspension fall-out
	79.50–83.00	Contains levels that are in situ brecciated (jig-saw
	83.00–87.43	puzzle fabric)
Ragnarok Fm.	87.43–87.70	Matrix supported conglomerate beds. Debris flow
subunit IIa	87.70–88.09	(mud flow)
	88.09–88.30	
	88.35–88.30	Siderite bed, septarian lower part and laminated darker brown upper part
Ragnarok Fm. unit I	88.35–171	Folded and fractured bedded sandstones, siltstones and shales with a few alternating carbonate beds. 120 cm monomict breccia at top of interval. Slumping and sliding of large sediment slabs

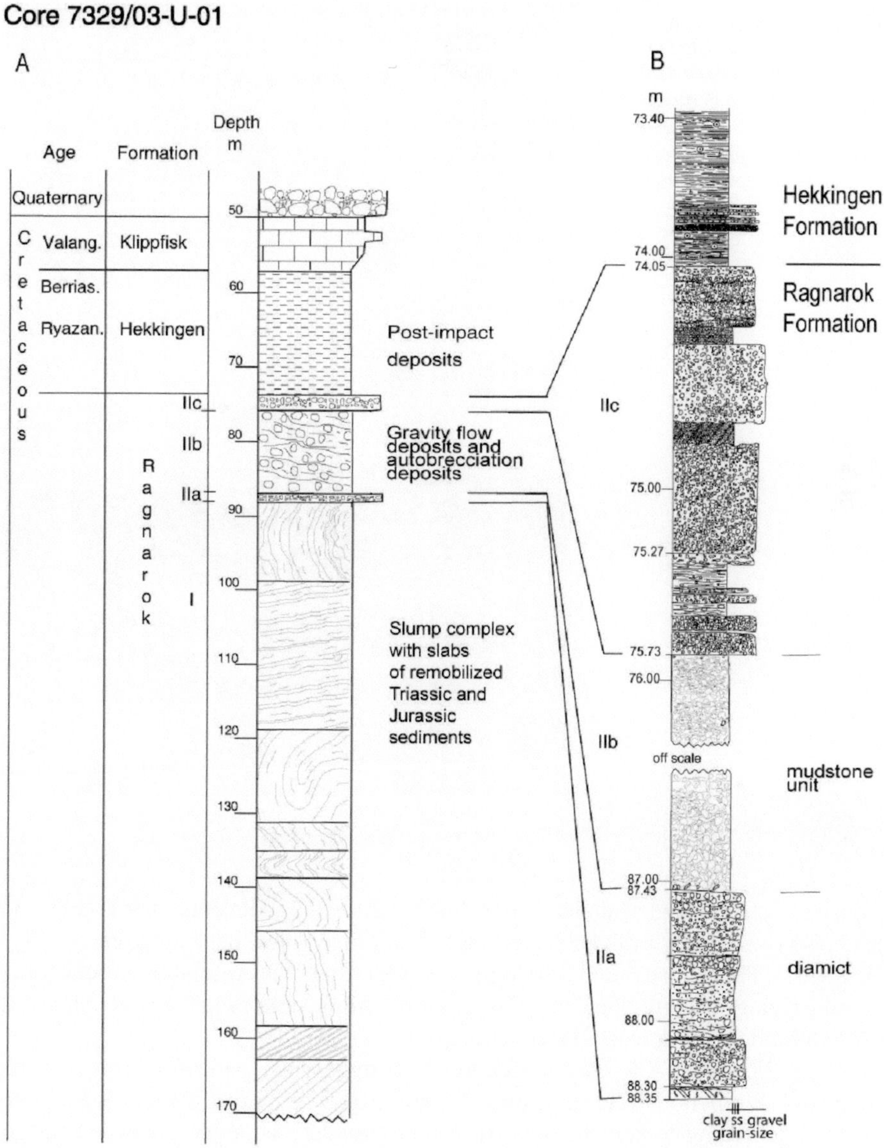

Fig. 5.2 The general core log of the Mjølnir core (7329/03-U-01) and detailed logs of the upper sedimentary formations. In the grain-size scale at the base of the sedimentary column, s s = silt and sand. Further details are presented in Figs. 5.5, 5.6, and 5.8, and in Chap. 8

Fig. 5.3 Overview core photos of cut slices, the Mjølnir core (7329/03-U-01), depth level of top part is given at the cores. See the logs of Fig. 5.2 and Table 5.1 for stratigraphical positioning

consist of chaotic and avalanche- and slump-dominated sediments (unit I) (Figs. 5.3 and 5.4) overlain by avalanche and mass- and gravity-flow deposits (unit II) (Figs. 5.3–5.10). The units contain lithologies and rock fragments resembling rocks found in underlying stratigraphical units in the surrounding areas of the Barents Shelf and on Svalbard (Dypvik et al. 2004b, 2004c).

Core 7329/03-U-01 only penetrates the uppermost part (96.95 m) of the Ragnarok Formation (Table 5.1, Figs. 5.2 and 5.3). The base of the unit is presently defined according to seismic data. Its lower boundary is defined between the so-called autochthonous and allochthonous to parauthochtonous breccias, reaching 1.3 km depth below sea bed (Tsikalas et al. 1998a). This level represents a recognizable horizon on seismic data. The Ragnarok Formation consequently includes part of the structurally uplifted and slump-back deposits (parautochtonous breccias) (Figs. 5.3 and 5.4) and the fall-out/back and reworked/retransported resurge deposits (allochtonous breccias) (Figs. 5.3, 5.6, 5.9, and 5.10). The insitu and highly fragmented brecciated rocks formed during the excavation and modification stages in the transient crater, the so-called authochtonous breccias, are found below and were not reached by the core. The Ragnarok Formation is succeeded by the Hekkingen

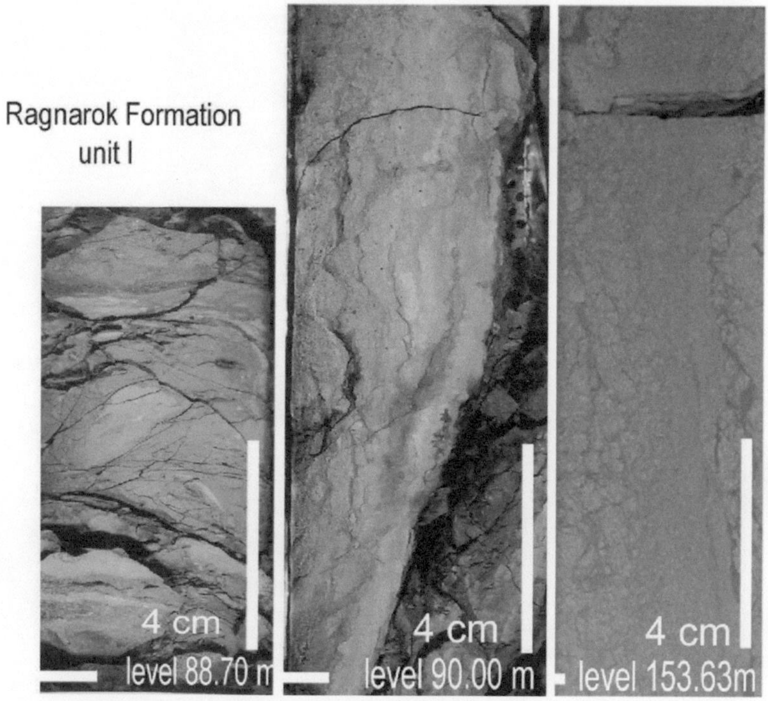

Fig. 5.4 Core photos of unit I, fractured and folded sedimentary successions. Details on their formation are presented in Chap. 8

and Klippfisk formations below a 50-m-thick Quaternary cover (Figs. 5.3, 5.11, and 5.12) (Mørk et al. 1999; Dypvik et al. 2004b, c).

Dating of the underlying and overlying sediments show the Ragnarok Formation was formed close to the Volgian-Ryazanian boundary time (Smelror et al. 2001a, b; Bremer et al. 2004; Smelror and Dypvik 2006). The formation, however, carries re-sedimented fragments, dominantly of late Early to Late Triassic ages, and to a lesser extent, of Jurassic age (see below).

Outside the Mjølnir crater rim, equivalents to the Ragnarok Formation are recognized on seismic data (Tsikalas et al. 2002a). As expected, the Hekkingen Formation is found above and below the Ragnarok Formation. The boundaries of this wedge in near crater areas have so far not been cored or sampled. The Sindre Bed, the ejecta unit and lateral equivalent of the Ragnarok Formation, represents the impact-related material outside the crater.

5.2.2 Ragnarok Formation, Unit I

Unit I (171–88.35 m) (Table 5.1, Figs. 5.2, 5.3, and 5.4) consists of strongly folded and fractured clay-, silt- and sandstones. These are described in detail and discussed

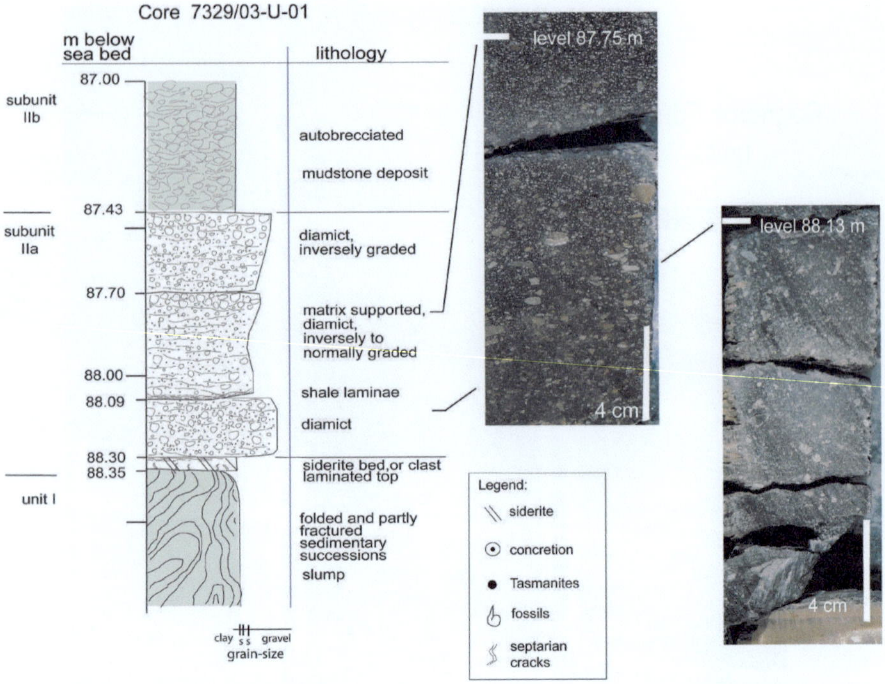

Fig. 5.5 Detailed core log of Unit IIa of core 7329/03-01. The matrix supported diamicts are illustrated as are the lower boundary to the siderite bed (88.13 m). In the grain-size scale at the base of the sedimentary column, s s = silt and sand

in Chap. 8. The palynomorph assemblages in this unit comprise a mixture of taxa derived from deposits of Spathian (late Early Triassic) to Late Jurassic age (see below, Fig. 5.14).

The degree of folding and faulting varies throughout the core and in some parts both vertical and horizontal bedding can be observed, often in combination with soft sedimentary deformation and water escape structures (see Chap. 8). Unit I is interpreted to represent reworked, folded and fractured pre-impact sediments, that were deposited as scree, avalanches, and slumps along the central peak of the crater (Dypvik et al. 2004a, b).

The uppermost 120 cm of unit I (Fig. 5.5) contains a fragmented, brecciated bed (89.55–88.35 m). The bed has a sharp upper boundary and an angular unconformity with an overlying siderite layer (Fig. 5.5) and the succeeding diamict of unit IIa. This 120 cm thick bed consists of partly folded clay clasts, grain-supported with a clayey matrix. The palynomorphs recorded from this unit are of the same ages as in the deposits below (Figs. 5.14 and 5.15). Lithologically, the clasts are similar to those just below this top part, dominated by grey claystones with a low content of silt and sand.

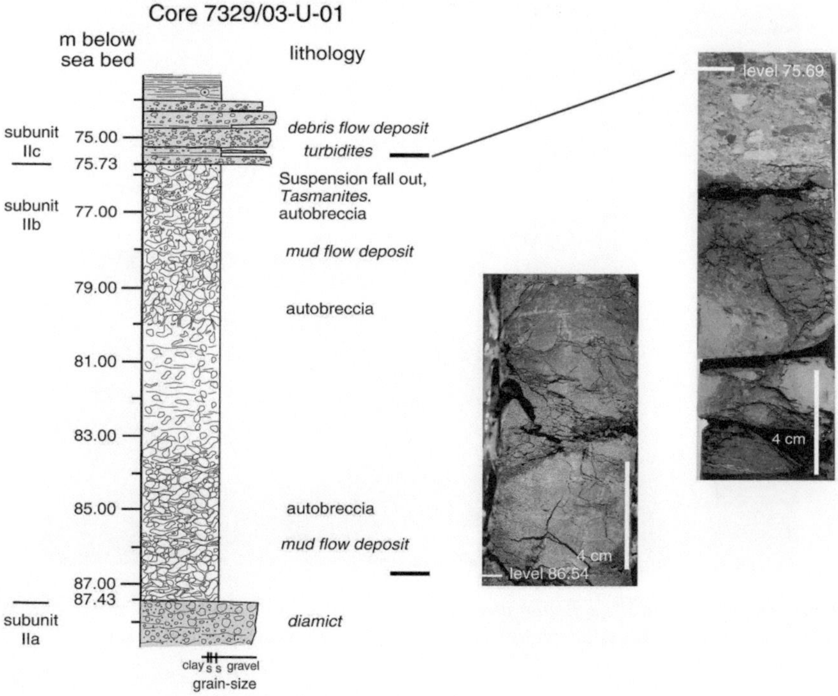

Fig. 5.6 Detailed core log of autobrecciated beds in Unit IIb of core 7329/03-01). In the autobrecciated part (between levels 75.73 and 87.43 m) the clasts and matrix have the same composition. In the grain-size scale at the base of the sedimentary column, s s = silt and sand

5.2.3 Ragnarok Formation, Unit II

Unit II (88.35–74.05 m) generally consists of poorly sorted conglomerates and can be divided into three subunits IIa, IIb and IIc (Figs. 5.2–5.10; Table 5.1).

Subunit IIa (88.35–87.43 m) consists of two well-defined parts: A lower, light brown, 5 cm thick, dense siderite bed or concretion (88.35–88.30 m), and an upper 87 cm thick (88.30–87.43 m) diamict with a dark grey, sandy clay matrix (Table 5.1 and Fig. 5.5). Septarian cracks are typically found in the lower 3.5 cm of the siderite bed. The upper 1.5 cm of the siderite shows a faint parallel lamination and is sharply separated from the overlying conglomerates but, shows similar orientation of bedding. The cored siderite may represent a siderite concretion, but the bedding being similar to the overlying sediments, indicates a possible in-situ origin associated with the diamicte above.

The major part (87 cm) of unit IIa is a homogeneous and poorly sorted diamict composed of a matrix-supported, dark grey, pebbly mudstone with subrounded to subangular clasts (Fig. 5.5). The intraclasts consist of sandy silt and claystone fragments up to about 2 cm in size, but with 2 mm as an average. The lithological

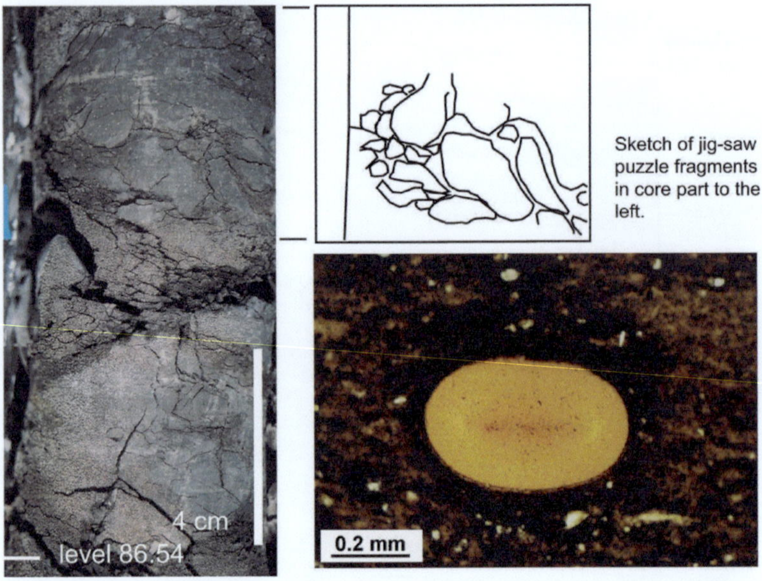

Fig. 5.7 Details of Unit IIb; *left core piece* with jigsaw puzzle clay clast, *upper right* a sketch of the fit of these clasts in the upper part of the core. The *lower right* shows a tasmanites fossils from Unit IIb

compositions of the clasts resemble the reworked Triassic and Jurassic sediments also found in unit I. Only faint clast-orientation has been observed in the pebbly mudstone (diamict). In Fig. 5.5 they seem to have their long axis parallel to horizontal, possibly indicating some shear movement. Glass/melt or glass/melt fragments have not been found so far. The sediments are poorly cemented and disintegrate easily in water, as does most of unit II.

The pebbly mudstones of unit IIa may be further subdivided into three minor parts that are separated by two cm-thick layers containing irregular shear laminae. The lowermost layer (88.30–88.10 m) is fairly homogeneous. The overlying section (88.10–87.70 m) shows faint layering and contains more lightly coloured, subrounded clasts especially in its upper part (87.85–87.70 m). This change in colour and roundness may reflect an additional input from a different source, but presently we do not have any good candidate. The uppermost section (87.68–87.43 m) shows some clast enrichments along its uppermost layers, but no changes in clast size (Fig. 5.5).

In the upper part of subunit IIa, the alga *Leiosphaeridia* is present in great abundance, as is also the case in the overlying Hekkingen Formation (Bremer et al. 2004). An algal bloom of *Leiosphaeridia* is also recorded in other correlative sediments and is attributed to increased nutrition of the seawater caused by the impact (Smelror et al. 2002). This indicates that the finest grained gravity deposited material of this subunit was deposited some time after the impact.

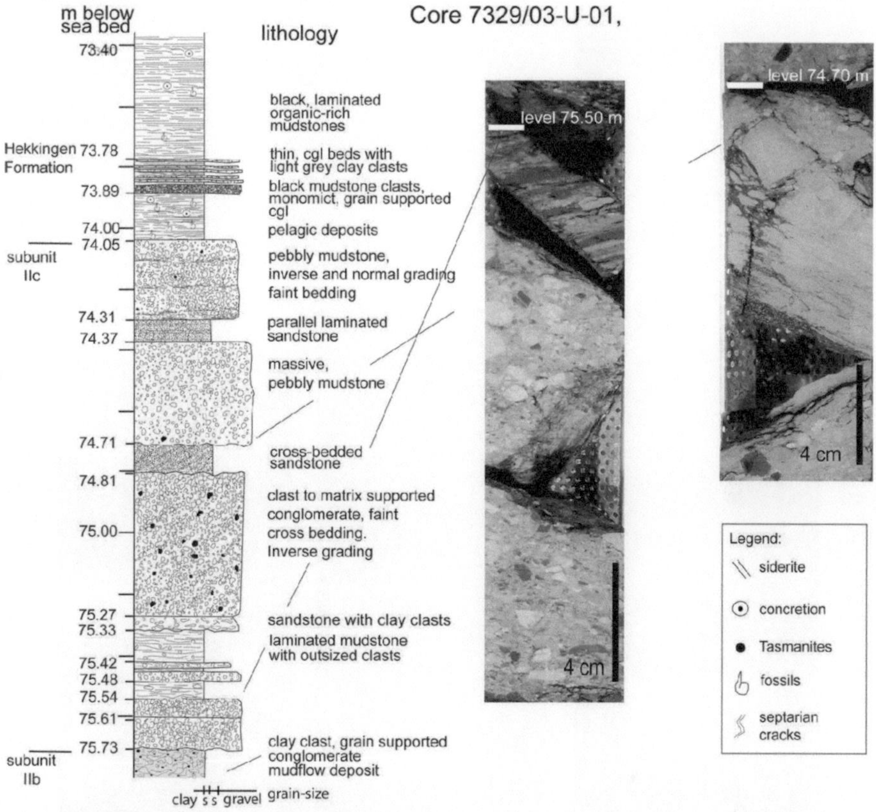

Fig. 5.8 Detailed core log of Unit IIc of core 7329/03-01. The unit consists mainly of conglomerates, sandstones and laminated mudstones. Detailed core photos in Fig. 5.9. In the grain-size scale at the base of the sedimentary column, s s = silt and sand

Subunit IIa most likely was deposited from debris or mud flows along the central peak, while the siderite bed may represent a diagenetic alteration product (Dypvik et al. 2004a, b). Unit IIa appears to represent the first collapse phase of the central peak, possibly tsunami related.

Subunit IIb (87.43–75.73 m) is an 11.70 m thick, homogenous unit consisting of dark grey to brown and olive green, highly fragmented claystones (Figs. 5.5, 5.6, and 5.7). Some few cm-sized siderite concretions are also present. The clast- and matrix-supported unit IIb displays clay clasts in a matrix of mud. The clay clasts are composed of similar material as the matrix and are difficult to distinguish macroscopically, but are easily distinguished in thin section with a more consolidated appearance than the matrix. The angular to subrounded clay clasts range in size from a few millimeters to 5 cm, indicating that the clay must have been somewhat compacted/consolidated before they were reworked.

The disintegrated clay clasts and matrix together consist of more than 90% coarse clay and fine silt, with average grain sizes between 3 and 13 μm (Dypvik et al.

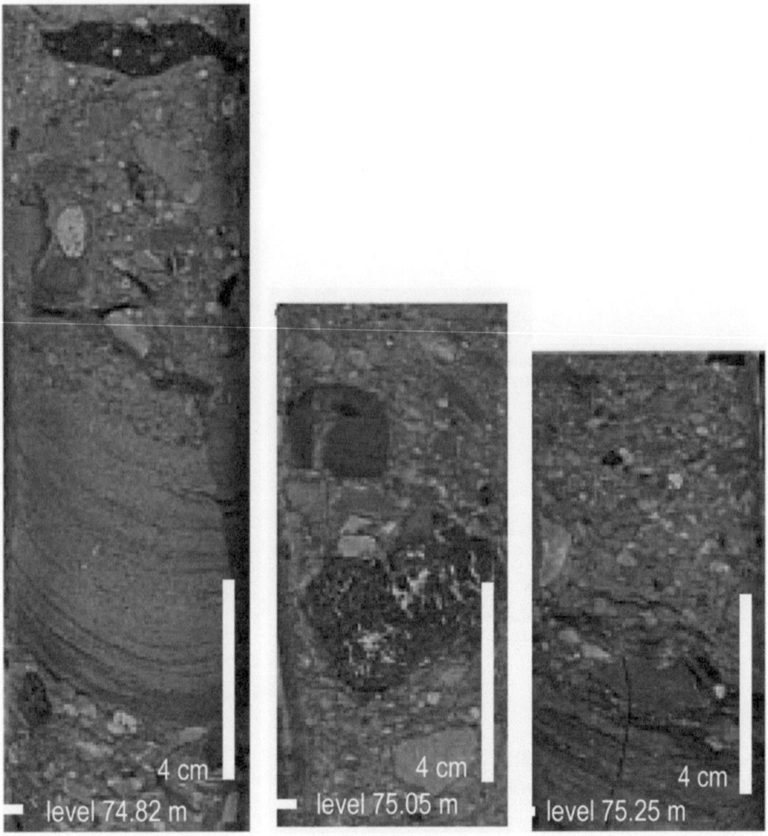

Fig. 5.9 Core photos Unit IIc; displaying conglomeratic and laminated sandstone sections. The location in core is shown in Fig. 5.8

2004b, c). In the uppermost parts of the unit indications of bimodal grain size distribution are seen, with minor enrichments at 40–50 μm grain size. The grain-size distribution in unit IIb can generally be characterized as unimodal, with a slightly fining upward trend. The subunit does not show any obvious bedding, except for a faintly laminated and apparently almost clast free interval between 83 and 81 m (Fig. 5.6). At few levels (incl. 79.4 m) chemical alteration structures are seen as somewhat lighter bands that cut the core. Brownish alteration bands and spots are found sporadically, and are most likely secondary diagenetic feature.

Recycled palynomorphs in the subunit are of mixed Early Triassic (Spathian) to Jurassic age (Fig. 5.14). In the uppermost 2.5 m of subunit IIb, the prasinophyte alga *Tasmanites* is found together with 1–3 cm large pyrite concretions. The recorded *Tasmanites* are around 0.5 mm black spheres. In thin sections, they have a well-defined orange (light) to green appearance under crossed polars (Fig. 5.7). *Tasmanites* algae of similar size are abundant in Middle Triassic sediments on

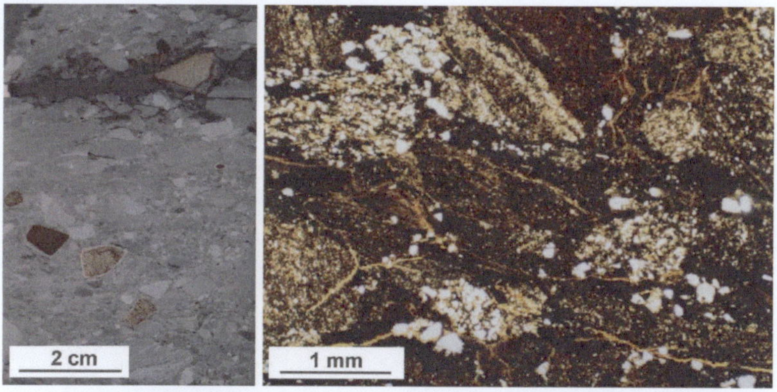

Fig. 5.10 Core photo (*left*) and thin section photo (*right*) of Unit IIc. The light oxidation rims on a couple of grains are well developed in the *left photo*

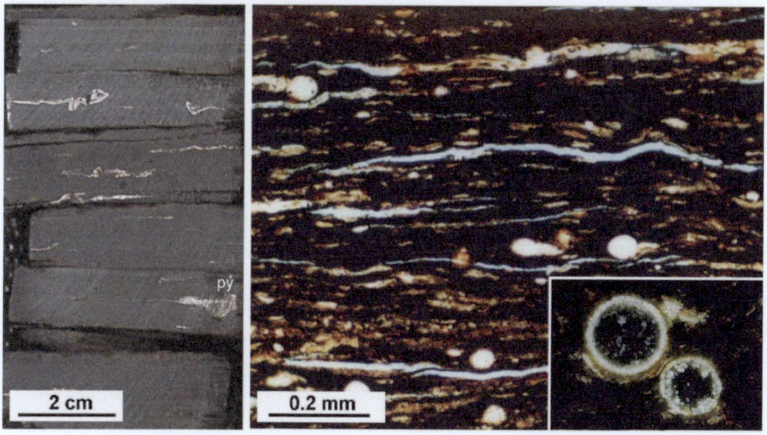

Fig. 5.11 Core photo (*left*) and thin section photos (*right*) of the Hekkingen Formation. In the *left photo* fine laminated black shales with light Buchias and pyrite concretion (py) is shown. In the *right photo* the thin section of this finely laminated shale show a large content of spheridal algae (*inset*)

Svalbard and in Northern Barents Sea cores (A. Mørk and J.O. Vigran, personal communication).

Subunit IIb most likely represents mud flow deposits formed by autobrecciation, possibly in relation with the tsunami reworking and the major modification phase of the crater (Dypvik et al. 2004b). The upper boundary towards subunit IIc is a sharp, well-defined erosional surface, which may have been caused by the succeeding density flows of subunit IIc (Figs. 5.6 and 5.8).

Subunit IIc (75.73–74.05 m) has a composition characterized by well-defined diamict, conglomerates and minor sandstone beds (Figs. 5.8, 5.9, and 5.10,

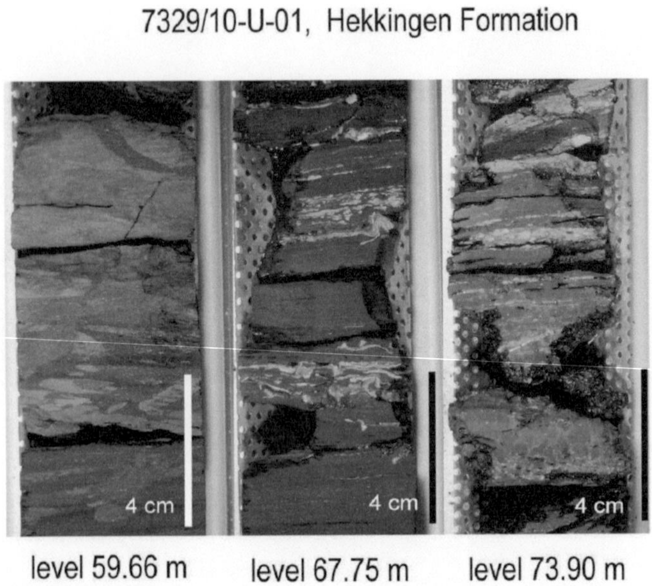

Fig. 5.12 Core photos of Hekkingen Formation; the *left* show bioturbated grey shales, the *middle* one show finely laminated black shales with macrofossils, while the *right photo* show the black shales with this, distal turbidite laminae (sedimentological logs in Figs. 5.2 and 5.8)

Table 5.1). The three thickest conglomerate beds consist of mainly dispersed subangular to subrounded clasts (Fig. 5.8). These clasts are in thin section seen to be composed predominantly of greyish dark brown clay- and siltstone with few sandstone clasts (Fig. 5.10). The texture is most commonly grain-supported with a poorly sorted grey, silty to clayey sandy matrix and a few dispersed *Tasmanites* algae. Sorting is poor to moderate and clasts are distributed throughout the bed, whereas a grain-supported appearance increases near the top. The clasts are commonly around 3 mm in size and mainly randomly oriented. The largest clast found was a 4 cm long sandstone pebble. Rare brown colored sandstone clasts show light greyish-brown coloured alteration rims, whereas no comparable rim has been observed in the much more common grey clasts (Fig. 5.10).

Thinner sandstone beds occur in between the conglomeratic beds. These are generally well sorted and decrease in grain size upwards, forming an overall fining-upwards succession. The sandstones show parallel lamination and rare cross-bedding, typical for tractive sediment transport. These conglomerates and sandstone beds are interpreted as turbidite and debris flow beds based on grain size and sorting, as well as on clast- and matrix-support (Figs. 5.8 and 5.9). Such density flows may have been formed along the central peak and the crater rim, representing the late modification or early postimpact phase of the crater (Dypvik et al. 2004c).

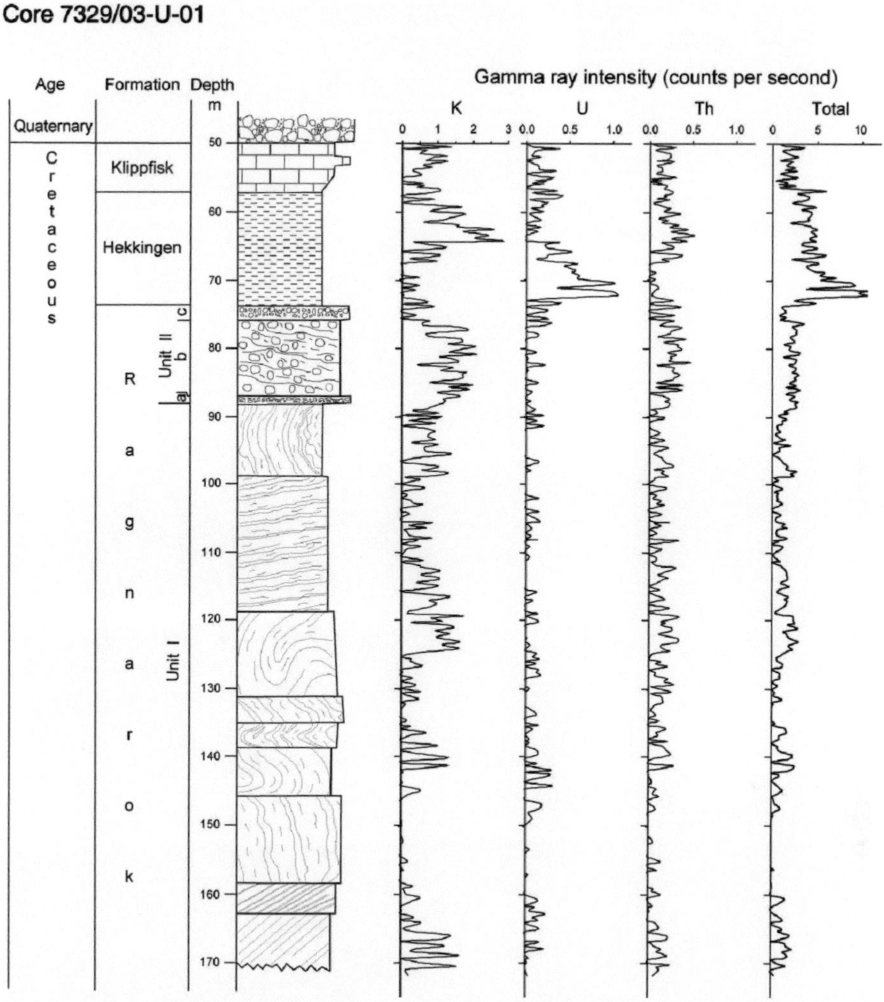

Fig. 5.13 Total gamma log of core 7329/03-01, along with the K, U and Th contributions

The samples have been analysed by standard total organic carbon (TOC) and Rock Eval analysis. Rock Eval is a standard pyrolysis method for source rock characterization and evaluation (Espitalié et al. 1977). In the pyrolysis the measured hydrogen index (HI) (mg hydrocarbons/g organic carbon) and oxygen index (OI) (mg CO_2/g organic carbon) characterizes the kerogen: high HI values (> 100) are typical for good petroleum source rocks. The organic maturity is measured by the T_{max} value in the Rock Eval analysis: the temperature when hydrocarbon production is at its highest in the pyrolysis (Espitalié et al. 1977). The total organic carbon (TOC) content and Rock-Eval hydrogen index values (HI) of the rocks from subunits

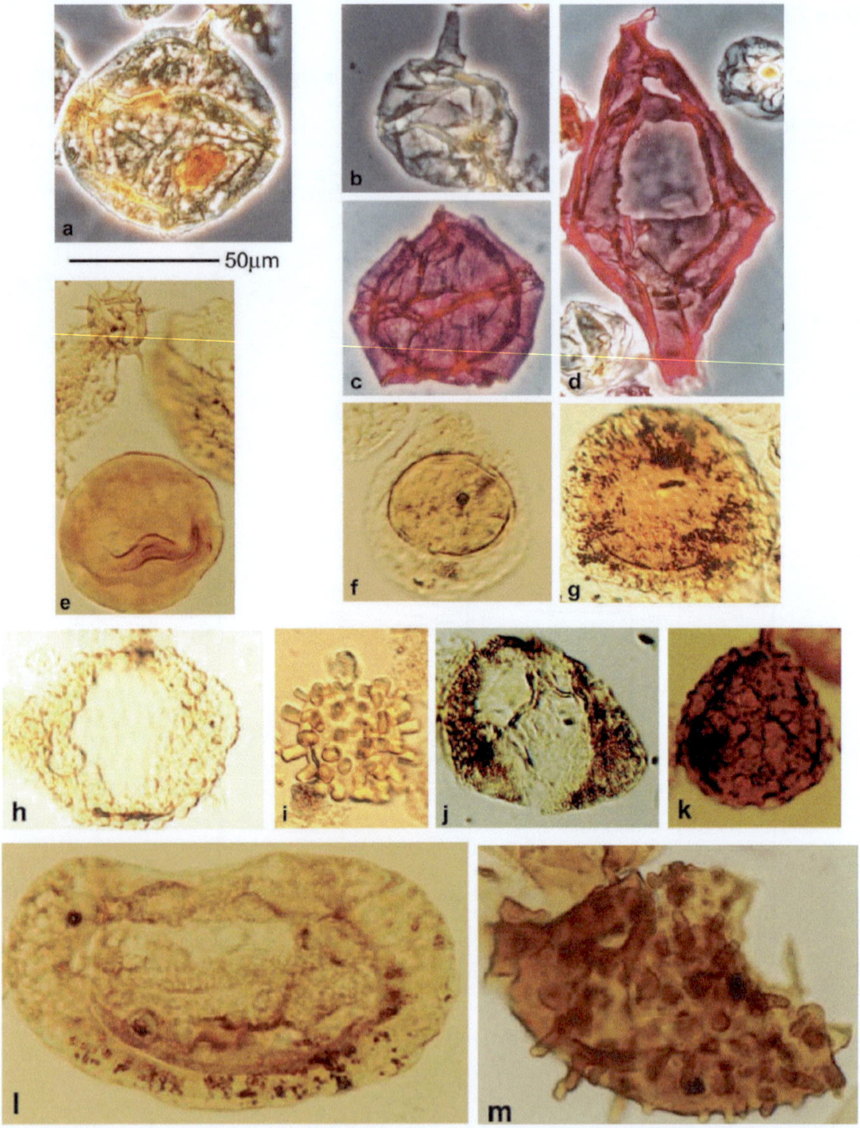

Fig. 5.14 Palynomorphs found in Ragnarok Formation of core 7329/03-01. The 50 μm scale indicates the enlargement of the palynomorphs. Details and references for the different palynomorphs can be found in Fig. 13 in Dypvik et al. (2004b): **(a)** *Cribroperidinium globatum*, **(b)** *Pareodinia ceratophora*, **(c)** *Sirmiodinium grossii*, **(d)** *Tubotuberella*, **(e)** *Tasmanites* sp. and *Microhystridium* sp. (*the upper left corner*), **(f)** *Aulisporites astigmosus*, **(g)** *Doubingerispora filamentosa*, **(h)** *Protodiploxypinus macroverrucosus*, **(i)** *Echinitosporites iliacoides*, **(j)** *Triadispora obscura*, **(k)** *Rewanispora foveolata*, **(l)** *Illinites chitonoides*, and **(m)** *Jerseyiaspora punctispinosa*

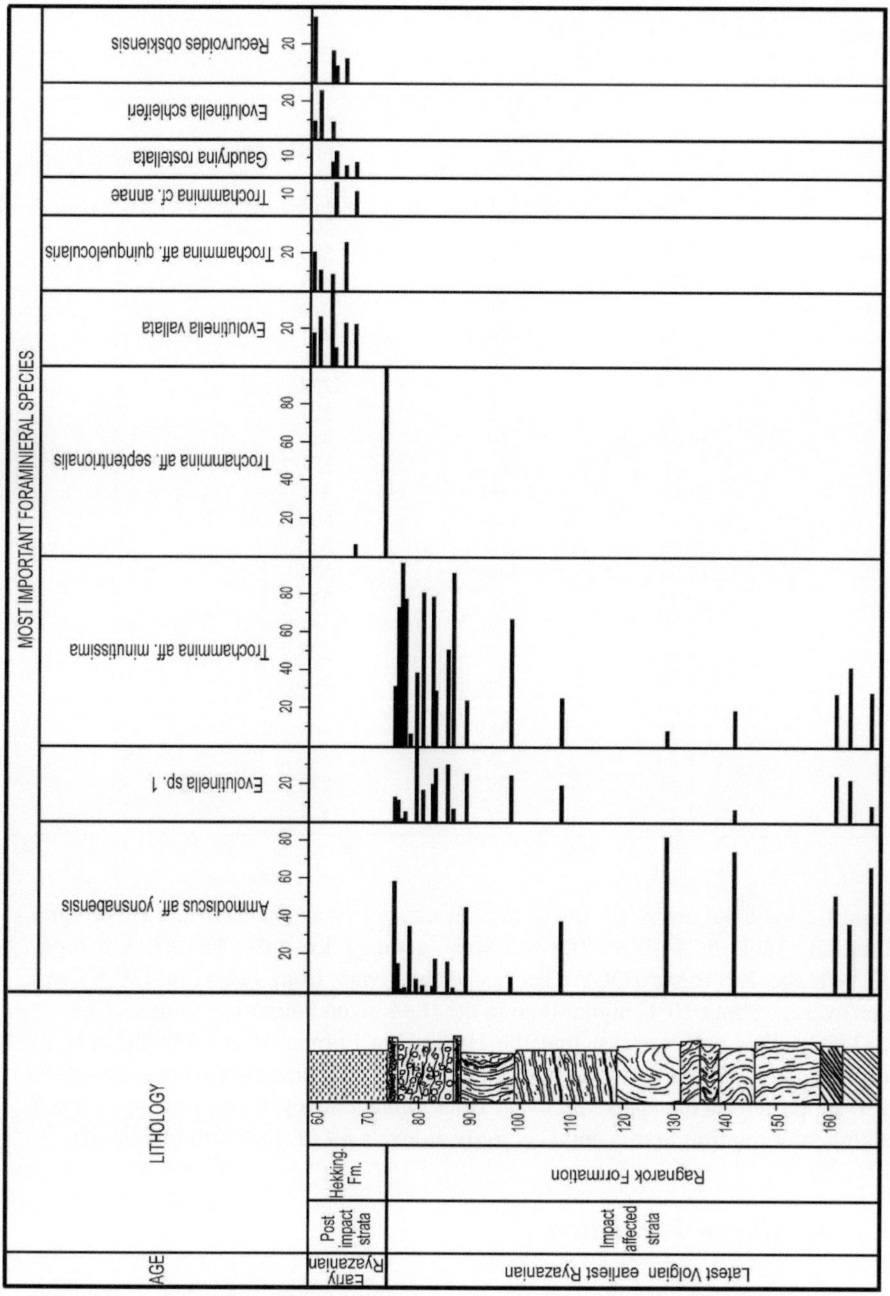

Fig. 5.15 The most important foraminiferas found in core 7329/03-01 (Bremer et al. 2004)

Table 5.2 Geochemical data from the Barents Sea. TOC analysis (weight %) and Rock Eval analysis of core 7329/03-U-01. Sample number is m in depth below sea floor. According to Ohm et al. (2008), the average lower Hekkingen contain 9.7 weight % TOC and HI of 303, while the upper Hekkingen contain 4% TOC and HI of 288

Core 7329/03-U-01

Sample (m depth)	TOC %	T_{max} °C	HI mgHC/g C	OI $mgCO_2/gC$
58.5	2.03	421	125	4
59.5	7.13	420	446	4
60.6	8.17	416	505	12
61.5	9.98	425	539	7
62.5	3.33	421	276	5
63.5	4.06	420	291	4
64.5	6.89	418	455	6
65.5	4.00	415	298	4
66.5	18.42	412	504	5
67.5	23.24	410	449	4
68.5	21.18	425	591	2
69.5	21.07	419	351	4
70.5	22.12	416	500	3
71.5	25.48	417	522	4
72.5	16.84	420	504	3
73.5	20.21	424	547	4
74.5	0.63	428	51	22
75.5	4.46	427	328	6
76.5	1.67	434	327	10
77.5	1.50	422	63	5
78.5	1.32	434	104	7
79.5	1.19	438	146	87
80.5	1.11	437	120	152
81.5	1.18	437	121	133

IIb and IIc vary, but are much lower than in the overlying shales of the Hekkingen Formation (Table 5.2) (TOC: 0.6–4.5 wt%, average 1.6 wt%; HI: 51–328 mg/g TOC, average 130 mg/g TOC). The T_{max} values vary from 422°C to 439°C, and are on average about 10°C higher than in the Hekkingen Formation samples (432°C vs. 422°C). This may suggest that the Hekkingen Formation contributed only to minor amounts of reworked sediments to the Ragnarok Formation. This assumption is also supported by the sparse recovery of dinoflagellate cysts from the pre-impact Hekkingen Formation in this unit (i.e., only at levels 89.11, 88.20 and 97.93 m).

5.2.4 Hekkingen Formation

In the Mjølnir crater core the Hekkingen Formation comprises black to medium grey, organic rich and laminated shales (74.05–57.10 m; Figs. 5.3, 5.8, 5.11, and 5.12). The bivalve *Buchia* is abundant throughout the formation, and some specimens have even preserved their nacreous aragonite inner coatings. The shales contain abundant pyrite concretions (0.2–2 cm) and occasionally pinkish brown siderite concretions. The lithology is similar to that of the upper part of the same formation in the neighbouring core 7430/10-U-01 (Fig. 2.12) and elsewhere in the Barents Sea (Worsley et al. 1988; Leith et al. 1992) (Figs. 5.3, 5.8, 5.11, and 5.12).

The interval from 73.89 to 73.78 m contains six thin conglomeratic layers, only from 5 cm to less than 1 cm thick (Figs. 5.8 and 5.12). They are clast-supported and typically display a sharp, erosional base. The conglomerates contain dark grey clasts reworked from Hekkingen Formation shales, and clasts of a lighter grey colour, which have a composition similar to those in the conglomerates of unit IIc of the Ragnarok Formation.

The interval between 73.78 m and the top of the Hekkingen Formation (57.10 m) is dominated by parallel laminated shales. These are dark grey up to about level 68 m and lighter grey in the remaining upper part of the interval (68–57.10 m) (Fig. 5.12). Thin sections display remnants of Prasinophytae algae *(Leiosphaeridia,* 20–50 μm in diameter) and enrichments in *Buchia* shell-fragments. Between 65 and 63 m no macrofossils were found, but they are again present in the upper part of the Hekkingen Formation from 63 to 57.10 m. Trace fossils (dominantly *Chondrites* and *Planolites)* and carbonate concretions are common at different levels in the upper part of the interval. Based on the macrofossils the post-impact Hekkingen Formation is dated as latest Volgian to earliest Ryazanian (Smelror et al. 2001a, b).

The organic matter is thermally immature, as seen in Rock-Eval, T_{max} values of typically 420°C–432°C (Tables 5.2 and 5.3) and production index (S1/(S1 + S2)) values typically less than 0.05. The uranium distribution curve derived from spectral gamma ray measurements (see below) shows a maximum at 67–74 m depth and correlates with the organic richness (Tables 5.2 and 5.3, Fig. 5.13). The total organic carbon (TOC) content in the lower, laminated part of the Hekkingen Formation varies between 17 and 34%, which is similar to, and partly exceeds, the organic richness reported for this formation elsewhere in the northeastern Barents Sea (e.g., Leith et al. 1992; Bugge et al. 2002). High TOC values and hydrogen index values of 351–715 mg/g TOC indicate that the kerogen belongs to mostly type II and locally type I, deposited in an anoxic environment. Spectral gamma values for uranium in the upper part of the section (~68–57.10 m) are rather low.

Organic richness and hydrogen index of these rocks are more variable and generally lower than in the underlying interval (1.1–21.6 wt% TOC, HI = 49–574 mg/g TOC) (Tables 5.2 and 5.3). This is consistent with a less oxygen-deficient depositional environment as suggested by the variable, generally lighter colours and the occasional bioturbation of these shelf deposits.

5.2.5 Klippfisk Formation

The 7.1 m thick interval representing the Klippfisk Formation consists of heavily bioturbated, light greenish-grey, argillaceous carbonates (marls) (Figs. 5.2 and 5.3). The lithology is similar to that of the Klippfisk Formation in type well 7430/10-U-01, described by Smelror et al. (1998) as representing a condensed carbonate

Table 5.3 Geochemical data from the Barents Sea. TOC analysis (weight %) and Rock Eval analysis of core 7430/10-U-01. Sample number is m in depth below sea floor. According to Ohm et al. (2008), the average lower Hekkingen contain 9.7 weight % TOC and HI of 303, while the upper Hekkingen contain 4% TOC and HI of 288

Core 7430/10-U-01				
Sample (m depth)	TOC %	T_{max} °C	HI mgHC/g C	OI $mgCO_2$/g C
45.1	12.90	420	517	14
45.3	7.52	424	537	13
45.4	3.90	424	288	12
45.6	2.15	423	174	14
45.7	2.99	425	218	10
45.9	6.48	421	392	12
46.2	2.88	422	305	28
46.3	5.45	419	330	11
46.6	17.90	416	462	13
46.8	16.40	416	456	12
47.0	19.80	416	300	10
47.4	9.13	421	491	11
47.5	4.07	421	389	13
47.6	5.30	415	235	16
47.8	17.70	426	572	8
48.2	15.40	415	314	10
48.4	13.60	419	510	8
48.5	27.80	423	451	8
48.6	13.30	425	496	7
48.9	19.60	423	423	8
49.1	17.10	425	497	8
49.2	7.72	421	306	11
49.4	15.50	425	467	7
49.5	14.30	424	446	8
49.6	15.00	422	350	8
49.9	16.70	425	475	8
50.0	23.40	422	319	7
51.0	13.80	419	374	10
52.0	12.10	419	338	12
52.5	18.30	420	330	11
52.9	16.00	417	321	12
53.5	8.91	413	317	14
53.7	10.00	414	323	14
54.0	8.89	417	377	15
54.5	6.03	412	274	22
55.0	5.50	418	294	22
55.3	6.39	418	194	21
55.5	6.21	417	284	21

platform. The top of the formation is not preserved at this locality in the Mjølnir crater as the overlying Quarternary till rests directly on the marls of the Klippfisk Formation.

5.2.6 Spectral Gamma Results

Natural spectral gamma radiation (K, U, Th, total gamma) was measured on the core in the laboratory (Fig. 5.13). Both the total gamma and the spectral gamma readings for the Klippfisk and Hekkingen formations in core 7329/03-U-01 are similar to those found in the neighbouring core 7430/10-U-01. The total gamma measurements in the cored Ragnarok Formation are significantly lower than those of the Hekkingen Formation, and the spectral intensities for uranium are close to the detection limit. Uranium is known to be associated with anoxic depositional conditions and organic material. Gamma readings in homogeneous lithologies therefore commonly correlate with variations in the contents of organic matter (e.g., Supernaw et al. 1978; Schmoker 1981; Fertl and Rieke 1980; Dypvik and Eriksen 1983; Dypvik 1993). The low uranium gamma values in the Ragnarok Formation suggest that the Hekkingen Formation, which was still unconsolidated when the impact occurred, contributed little sediment to the Ragnarok Formation, in agreement with the palynological observations (see below). In the target area most of the Hekkingen Formation was blown away or went up in fire (Dypvik et al. 2008b).

The pronounced increase in K-activity at 65 m in the Hekkingen Formation may reflect a decrease in smectitic fractions and an increase in illitic components (Fig. 5.13). This is also indicated by Dypvik et al. (2003), who demonstrated an increased in the illite content in the one sample (60.00 m) they analyzed from the Hekkingen Formation interval shallower than 65 m.

5.2.7 Paleontology of the Ragnarok Formation

The marine macro- and microbiotas of core 7329/03-U-01 have been described by Smelror et al. (2001b, 2002), Bremer et al. (2004), Dypvik et al. (2004b) and Smelror and Dypvik (2005, 2006).

Below follows an overview of the palyno-flora found in the Ragnarok Formation. The main purpose of these analyses has been to document the ages of the rocks from which the impact breccia was derived. Altogether 117 palynomorph taxa have been recognized in core 7329/03-U-01. Palynomorphs derived from terrestrial sources dominate and indicate that the formation comprises rocks of different ages. Figure 5.14 present some of 39 stratigraphically important taxa.

Only age-significant taxa are commented in the following:

- *Illinites chitonoides* and *Jerseyiaspora punctispinosa,* which occur regularly, range in age from the Late Spathian to the Carnian. *J. punctispinosa* is restricted upwards to Middle Anisian deposits. Both *Densoisporites nejburgii* and *Rewanispora foveolata* have ranges restricted upwards to the Lower Anisian deposits (Fig. 5.14).
- *Aratrisporites macrocavatus* and *Conbaculatisporites hopensis* are characteristic for the Upper Anisian to Ladinian of the Sassendalen Group on the Svalis

Dome. On Svalbard these taxa continue into the Carnian of the Kapp Toscana Group (J.O.Vigran, unpublished data). Bjærke and Manum (1977) illustrate them from Hopen and Kong Karls Land (Kapp Toscana Group; De Geerdalen and Wilhelmøya formations).

- *Echinitosporites iliacoides* is restricted to the Ladinian on Svalbard and the Barents Sea Shelf (Vigran et al. 1998). *Ovalipollis pseudoalatus*, *Schizaeoisporites worsleyi* and *Sellaspora rugoverrucata* have been recorded from the Lower Ladinian on Svalbard and the Barents Sea Shelf, Svalis Dome (Vigran et al. 1998).
- An Upper Ladinian-Lower Carnian association is recognized on the basis of *Aulisporites astigmosus*, *Camerosporites secatus*, *Chasmatosporites major*, *Doubingerispora filamentosa* and *Paracirculina tenebrosa*. It should, however, be noted that most of the evidence described by Bjærke (1977) and Bjærke and Manum (1977) from the Upper Triassic deposits of Hopen and Kong Karls Land has not been recorded in this core.
- *Cerebropollenites macroverrucosus* has the oldest appearance in lower Lower Jurassic rocks. Poorly preserved specimens recorded as Lower Jurassic dinocysts (Fig. 5.14, Nos 16–23) confirm that Lower Jurassic deposits are represented throughout the Ragnarok Formation.

The palynomorph content suggests that the scree and slump complex of sediments forming the Ragnarok Formation in core 7329/03-U-01, interval 170.94–89.11 m (most of unit I) have their major source in the Middle Triassic deposits of the Sassendalen Group. Some of the palynomorphs probably represent the Lower Carnian, i.e., the oldest deposits of the Wilhelmøya Subgroup, but there is no palynological evidence allowing recognition of a Carnian-Norian association of the Kapp Toscana Group. Lower Jurassic palynomorphs are generally only present in minor proportions.

Samples at 88.20–77.80 m, representing the uppermost part of the Ragnarok Formation (uppermost part of unit I, subunits IIa and most of subunit IIb) contain a stratigraphic mixture of pollen and spores from the Middle to Upper Triassic and Lower Jurassic. The dominance of *Botryococcus* and marine plankton of presumed Jurassic age distinguishes these samples from those in the lower part of the core, containing dominantly terrestrial material.

One sample at 86.20 m contains dinoflagellate cysts of late Middle to early Late Jurassic affinity (i.e., *Dichadogonyaulax* sp. and *Chytroeisphaeridia hyalina*). Three samples from 89.11, 88.20, and 87.93 m include Late Jurassic dinoflagellate cysts in addition to the Late Triassic to Early Jurassic terrestrial palynomorphs. Characteristic species found in these samples are *Atopodinium haromense, Paragonyaulacysta borealis, Cribroperidinium* spp. (including *C. globatum*), *Senoniasphaera jurassica,* and *Tubotuberella apatela.* They are derived from the pre-impact deposits of the Hekkingen Formation.

The foraminiferal zonations are displayed in Fig. 5.15 (Bremer et al. 2004), which are presented in more detail in Sect. 5.4.5.

5.2.8 Paleontology of the Hekkingen Formation

Overlying the Ragnarok Formation there are 17 m dark shales of the Hekkingen Formation. The Hekkingen Formation contains a low-diversity macrofauna with species of the bivalve genus *Buchia*, and low diversity assemblages of palynomorphs and microfossils (Smelror et al. 2001b, 2002; Bremer et al. 2004; Smelror and Dypvik 2005, 2006).

The most prolific feature of the post-impact sediments in the Mjølnir crater core is the very distinct bloom of the prasinophycean alga *Leiosphaeridia* in the oldest post-impact sediments of the Hekkingen Formation. This impact-induced algal bloom is described in more details in Sect. 5.4.5.

At the base of the post-impact succession *Buchia unschensis* appears at 73.97 m (Smelror et al. 2001a, b). It is the first identifiable species to occur above the disturbed strata of the Ragnarok Formation and therefore is most important for identifying the date of return to normal sedimentary conditions after the Mjølnir impact. Specimens of *Buchia unschensis* are found at several levels upwards in the core, with the uppermost occurrence at 67.14 m. According to Zakharov et al. (1981), the *Buchia unschensis* Zone can be correlated to the upper part of the *Craspedites okensis* and the lower part of the *Chetaites sibericus* ammonite zones in Siberia. In North Greenland, *Buchia uschensis* is found in strata of the *Crappedites okensis* to *Hectoroceras kochi* ammonite zones and in East Greenland from the *Virgatosphinctes tenuicostatus* to *Hectoroceras kochi* zones (Surlyk and Zakharov 1983). *Buchia unschensis* has also been recovered from the *Craspedites nodiger* ammonite Zone (Yershova 1983).

Further upwards in the Mjølnir crater core *Buchia okensis* is found between 66.80 and 60.13 m. This species is known to range from the upper part of the *Chetaites sibericus* ammonite Zone to the *Bojarkia mesezhnikowi* Zone in the central part of the Russian Platform (Zakharov 1981). *Buchia okensis* is also known from the Ryazanian of Spitsbergen (Yershova 1983).

At 60.37 m in the Mjølnir crater core 7329/03-U-01 an ammonite identified as *Borealites* sp. has been described (Smelror et al. 2001a). This species is closely comparable to *Borealites* sp. aff. *fedorovi* as illustrated by Håkansson et al. (1981) from the Early Ryazanian *Hectorocera kochi* ammonite Zone of Peary Land on Northern Greenland.

The present biostratigraphic evidence from the Mjølnir crater core supports the data from borehole 7430/10-U-01 that the oldest post-impact sediments correspond in age to the uppermost Volgian – lowermost Ryazanian *Buchia unschensis* Zone. The age-range of this zone corresponds to the *Subcraspedites primitivus* to *Runctonia runctoni* ammonite zones of the standard Boreal succession (Rawson et al. 1999).

The marine microfloras found in the post-impact Hekkingen Formation of the Mjølnir crater core are described in details by Smelror and Dypvik (2005). The marine microfloras of the oldest post-impact deposits (74.05–64.9 m) are totally dominated by the *Leiosphaeridia* bloom, while dinoflagellate cysts are most common in the overlying deposits of the Hekkingen Formation (64.9–58.5m). The diversity of marine microplankton is relatively moderate throughout the formation, varying from 4 to 18 species per sample. Few age-diagnostic species are recovered, but the presence of *Gochteodinia villosa* at 62.0 m allows a correlation to the *Gochteodinia villosa* (Gvi) Interval Biozone of Riding and Thomas (1992). The Gvi-zone spans the Portlandian to Ryazanian in the British Jurassic.

5.2.9 *Magnetic Properties and Densities of the Mjølnir Crater Core (7329/03-U-01)*

Here we present measurements of magnetic properties and densities for two sample sets. Susceptibilities were measured using a Bartington MS2c sensor in a core-scanning set-up. Sampling was limited by the physical condition of the core, as the instrument set-up requires intact core-sections of at least 10 cm in length. For this sample-set only total susceptibilities were measured (Fig. 5.16). The second dataset consists of selected samples investigated in a laboratory environment (Torsvik and Olesen 1988), and both volume-specific susceptibilities and densities were determined (Fig. 5.16). Susceptibility values ranged between zero (i.e., below the instrument noise level) and 200×10^{-5} SI, common values for marine sedimentary rocks. The sample selection is biased to competent samples. Overall, maximum susceptibility values are found in samples with rusty colours. Typical magnetic sources in such a sedimentary environment aside the bedrock are deposited detrital magnetic minerals and siderite-cemented beds or nodules aside.

The measured densities range between 1,500 and 2,900 kg/m^3 and mostly around 2,400 kg/m^3 (Fig. 5.16). This selection is also biased to more competent core sections because of the density measuring procedure, i.e., the "Archimedes" principle that does not allow measurements of unconsolidated sediments. The densities determined here are dry densities, because any persistent exposure to water would have dissolved most samples. The two extreme end-members are related to materials such as coal (sample 12) or siderite nodules (sample 6). Generally, the densities increase with depth likely due to compaction. During the drilling, the first 60 m were penetrated without recovering the core. These first layers are unconsolidated Quaternary sediments, which have densities less than 1,700 and as low as 1,300 kg/m^3. Following the density measurements of the core samples, densities for the uppermost layers of the subsurface model of Mjølnir were derived and are plotted in Fig. 5.16.

5.3 The Mjølnir Impact Event in a Sequence Stratigraphical Framework

The Mesozoic formations can be correlated in detail across the Arctic, and within the Jurassic and Cretaceous several stratigraphical sequence boundaries can be followed (Mørk and Smelror 2001), as already discussed in Chap. 2. Embry (1989) presented

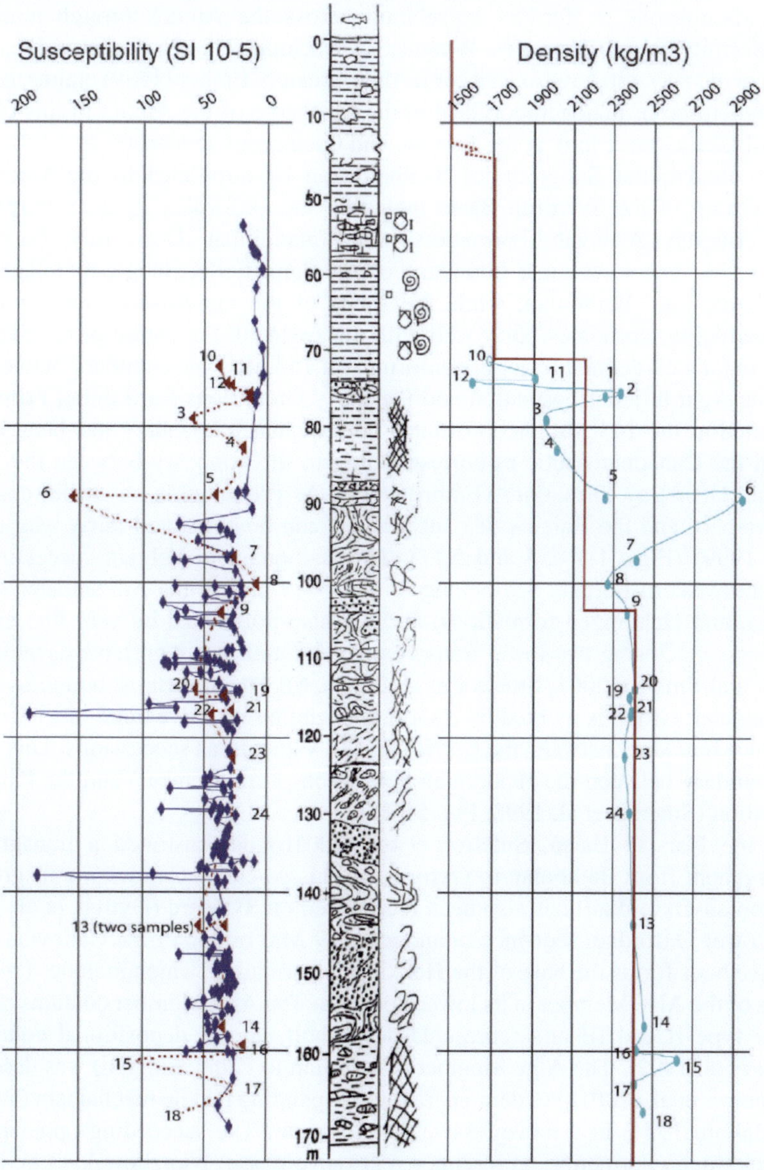

Fig. 5.16 Stratigraphy of core 7329/03-U-01 (Mørk, personal communication 2006) and measurements of susceptibilities (*left*) and densities (*right*) of competent samples. The *red curve* indicates the density variations with depth as used for the *upper* most layers of the density model. Degrees of fracturing and folding indicated in- and along the central sketched geological column

the Carboniferous to Tertiary correlation across the Arctic, through numerous unconformity bound sequences. Whether this correlative development is due to tectonics or eustasy will have to wait for further research. Embry (1989) claims tectonic influence to be of major importance in the Mesozoic of the Arctic. In this chapter we will take a closer look at the Jurassic and Cretaceous sequences.

The Janusfjellet Subgroup of Svalbard can be correlated to the Mackenzie King Group of the Sverdrup Basin including the Oxfordian Ringnes Formation, Kimmeridgian Awingak Formation and Valanginian Deer Bay Formation (Fig. 5.17). The lowermost boundary of the Janusfjellet Subgroup falls within the Middle-Late Bathonian, while the sands of the Oppdalsåta Member (about Kimmeridgian) correlates fairly well with the Awingak Formation of the Sverdrup basin and a well defined lower unconformity at 153 Ma. The boundary between the Jurassic Agardhfjellet Formation and the Early Cretaceous Rurikfjellet Formation is located at the 143 Ma unconformity. This sequence boundary has been identified in the Canadian Arctic as corresponding to the boundary between the Awingak and Deer Bay formations (Embry 1989). As presented in an earlier chapters, the Svalbard and the Barents Sea successions can be correlated fairly well (Mørk et al. 1999) (Figs. 1.3, 2.5 and 5.17). The Bathonian to Volgian successions of the southwestern Barents Sea belong to the lower part of the Adventdalen Group (Fuglen and Hekkingen formations) and are also dominated by very fine grained sediments disclosing two lower transgressive and an upper regressive development (Mørk and Smelror 2001; Smelror et al. 2001a, b). The uppermost boundary of the Adventdalen Group is marked by a short and subtle regressive trend just before the sequence boundary between the Berriasian and Valanginian successions. This marks the boundary between the Hekkingen Formation (Krill Member) and the Klippfisk Formation (Smelror et al. 1998; Fig. 5.17).

In the Harstad Basin, Smelror et al. (2001b) demonstrated a transgressive development from Bathonian to Oxfordian. This so-called Callovian transgression (Bathonian-Oxfordian) has also been recognized on Svalbard (Dypvik et al. 1992). The Lower Oxfordian sequence boundary (159 Ma) on top of the Callovian transgressive beds forms the base of the Hekkingen Formation, which includes the black shales of the Alge Member in its lowermost part. The Alge Member contains organic matter type II and III and represents deep shelf, anoxic depositional conditions (Leith et al. 1992). The Alge Member (Oxfordian to Kimmeridgian) was described by Smelror et al. (2001b) to demonstrate the expanding puddle mechanism (Wignall and Hallam 1991) in a transgressive development. The succeeding upper part of the Hekkingen Formation represents a regressive phase. Wierzbowski et al. (2002) studied the Upper Oxfordian to Kimmeridgian in Nordland VII area to the south of Barents Sea; focusing in particular on the ammonite and dinoflagellate faunas. The studied units from Nordland VII show the same transgressive developments and can be stratigraphically tied to the Barents Sea, Svalbard and East Greenland (Fig. 1.3).

The uppermost part of the Hekkingen Formation is made up of the Krill Member, which generally represents anoxic to dysoxic deep shelf conditions with a mixture of terrestrial and marine organic matter (Leith et al. 1992). The greenish, partly nodular marls of the Klippfisk Formation represent deposits from local highs with reduced

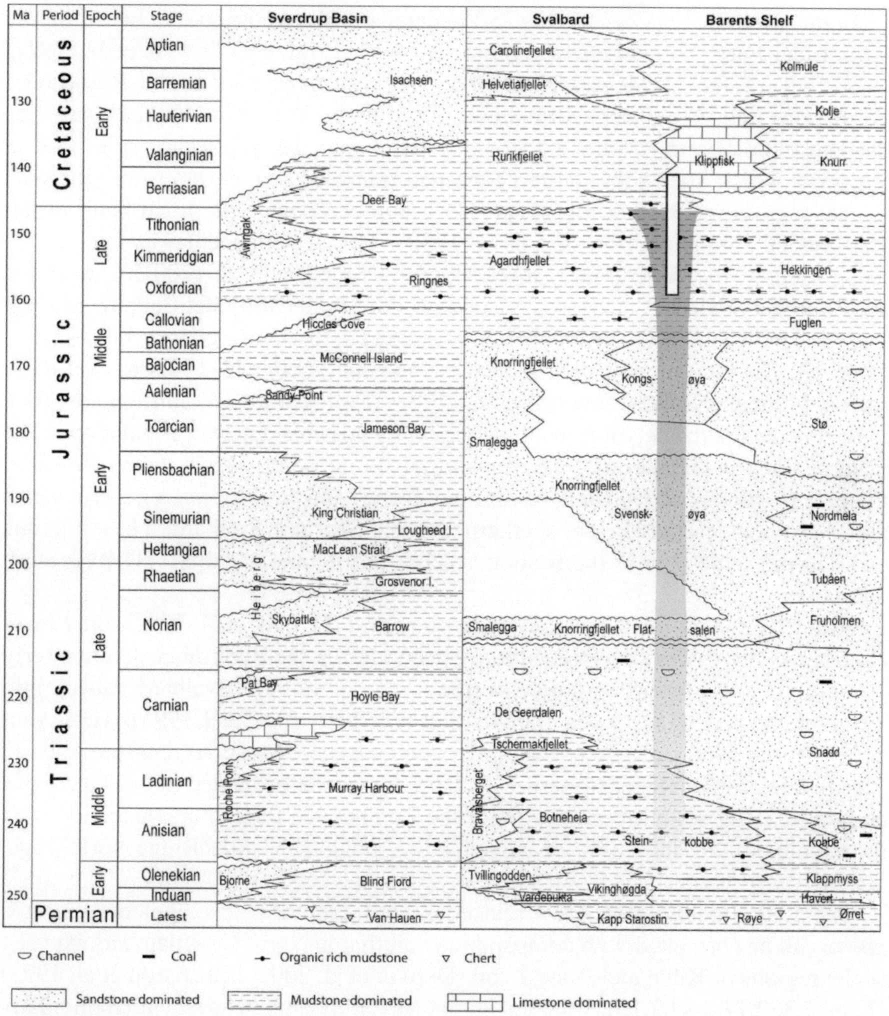

Fig. 5.17 The correlation scheme illustrates the stratigraphical correlation in the Arctic between the Sverdrup Basin (Canada) and the Barents shelf region. The Mjølnir crater erosinal cut is illustrated, while the *white, solid block* (Mjølnir core) indicates that only the upper part was drilled. The *shadowed area* displays that the ejected/reworked crater material found in the breccias of the crater core are dominately by Triassic and early to mid Jurassic age clasts. Fossils from ejecta material indicate disturbance down to the Lower Triassic, while the Permian-Triassic bounday was not penetrated. The Figure is base on Embry and Beaucamp (2008), Mørk et al. (1999) and Mørk and Smelror (2001)

clastic sedimentation and increased biological production in the Barents Sea region (Smelror et al. 1998, 2001b). This Lower Cretaceous unit is confined by a sequence boundary succeeded by the dark claystones of the Kolje Formation. Further, into the Cretaceous, the basin development was characterized by improved ventilation in the water masses and fading out of the anoxic/dysoxic domain.

In the Svalbard succession (Fig. 5.17) details in the transgressive Late Jurassic (Oxfordian-Early Volgian) development was demonstrated by Dypvik et al. (1992) in a study of the Myklegardfjellet Bed (the lowermost part of the Rurikfjellet Formation) and its adjacent formations. They found the major Volgian part, i.e., the latest Jurassic to display a regressive trend into the Myklegardfjellet Bed at Svalbard, which after a thin and fine-grained transgression, develops into the well known regressive Lower Cretaceous succession composing the main body of the Rurikfjellet Formation (Worsley et al. 1988). This development was later partly recognized in the Barents Sea successions by Smelror et al. (2001b), as stated above. In the Svalbard succession, just below the Myklegardfjellet Bed, the Mjølnir impact is indicated by ejected material recognized in the Janusfjellet succession (see Chap. 6). The Late Jurassic regressive development (about 143–142 Ma) is of more regional/global character. Smelror et al. (2002) displayed the Mjølnir impact effect to be reflected in the palynoflora of the region by possible renewed circulation in the water masses and temporarily increased supply of nutrients. This is demonstrated in the prolific bloom of *Leiospheridia* (green algae) and *Botryoccoccus* (fresh water algae) over wide areas in the southern Barents Sea and Svalbard. These blooms occur at the same time as the Ir anomaly, i.e., at the time of impact (Dypvik et al. 2006) (see e.g., Fig. 6.3).

After the Early Cretaceous transgressive period (Dypvik et al. 1992), the Lower Cretaceous regressive development took over. During the Barremian, southeasterly prograding deltas from an uplifted northern region North of Svalbard, moved into the northern part of the paleo Barents Sea (Gjelberg and Steel 1995). This event partly overlaps and strengthens the more regional Early Cretaceous regressive phase. The deltaic progradation across the Svalbard region can be related to local tectonics and uplift along the northern margin of the Barents Sea Region in connection with the opening of the Nansen Basin and spreading along the Gakkel Ridge North (Figs. 2.1 and 2.2).

The Late Jurassic to Early Cretaceous stratigraphical developments outlined above can be correlated with comparable evolution in North Greenland, as exposed in the regions of Kilen and Peary Land (Dypvik et al. 2002; Håkansson et al. 1993) (Figs. 1.3, 2.14, and 2.15). Both the so-called Callovian transgression (Bathonian–Oxfordian) with a maximum extent (flooding?) in Late Kimmeridgian and the following regressive phase are seen (Dypvik et al. 2002; Håkansson et al. 1993; Heinberg and Håkansson 1994). Within the Callovian North Greenland transgressive successions several sand units are present and commonly display coarsening upwards developments (Dypvik et al. 2002).

According to Nagy et al. (1988) and Dypvik et al. (1992) the Ryazanian flooding succeeding the Myklegardfjellet Bed on Svalbard (Fig. 1.3) can be interpreted to represent a maximum flooding situation. This would be comparable to the middle part of the Krill Member (Hekkingen Formation, Barents Sea), Middle Part of Deer Bay Formation (Sverdrup basin) and lower part of the Dromledome Formation (N Greenland, Figs. 1.3, 2.14 and 5.17).

5.4 The Evidence for Impact Crater Formation

The identification and study of potential impact structures formed by asteroid or comet collisions is complicated, and includes integration of several different types of information in order to achieve the most reliable interpretations. Consequently the identification and presentation of the impact evidences are of great importance and will normally be met with major interest and a large portion of skepticism by the impact community. In this compilation we therefore present some of the main impact evidences we have been able to gather for the Mjølnir structure. For further studies of the more general aspects of these arguments the reader is referred to French (1998) and Montanari and Koeberl (2000).

5.4.1 The Crater: Its Structure and Shape

The first recognition of impact structures is often based on convincing geomorphologic observations of the structure and its more or less circular configuration. Surface structures of circular shape can be formed in several different ways and asteroid and comet impacts are only one of the many explanation alternatives for the circular shapes. Structures with a diameter of 40 km could be formed by, e.g., extensive karstification, halokinesis, volcanic activity, and structural geological activity such as folding. Consequently the identification of a circular structure is not enough, additional evidence is needed.

In the Mjølnir case, the first descriptions and the reconstructions of Gudlaugsson (1993) formed the crucial starting point for further investigations. The structure has been portrayed as a complex crater, 40 km in diameter, with a central high that is 8 km wide along the base, surrounded by a 12 km wide annular basin (Gudlaugsson 1993; Tsikalas et al. 1998a) (e.g., Figs. 3.4, 3.10, and 3.14). A closer description of these characteristics and other geophysical information (e.g., gravimetric and magnetic) is presented in detail in Chap. 4.

The annular basin of the Mjølnir crater has been estimated to have a diameter of 16 km. The dimensions of annular basin and the transient crater, are crucial parameters in understanding crater mechanisms, and are vividly discussed these days (see Turtle et al. 2005). This is in particular the case for complex craters with a brim and an inverted sombrero shape (Melosh 1989), just like the Mjølnir structure (e.g., Figs. 1.10, 3.8, and 3.19). The Mjølnir structure has a peak-ring geometry as should be expected of impact structures larger than 26 km in diameter (Melosh 1989; French 1998). The well preserved peak-ring configuration is additional impact evidence. The excellent preservation of the Mjølnir Crater is the result of an early burial in the marine environment (e.g., Tsikalas and Faleide 2007). The relations between peak height and diameter of the Mjølnir depression are well within the scaling relations of Melosh (1989) (e.g., Tsikalas et al. 1998a–c) (see also Chap. 4).

5.4.2 Fracturing and Conglomerates

In the seismic lines crossing the Mjølnir structure extensive fracturing and brecciation can be observed (Chaps. 3 and 4). This intense fracturing is also highly evident in the Mjølnir core (Figs. 5.3 and 5.16). Several structural phases can be recognized and soft sediment deformation and water escape phenomenon have been demonstrated. This is discussed in detail in Chap. 8. The fracture patterns, observed in the core and of large scale in the seismic lines, display a general circular orientation around the center associated with outwards decreasing intensities. Such observations are very important in impact crater identification (Tsikalas et al. 1998a) and fracture patterns like this have been identified in other impact structures (Chesapeake Bay-Virginia, USA and Neugrund, Estonia). Impact structures may often be deeply eroded and the original fracture patterns can be difficult to detect. In other cases the structural geometry may have been completely eroded, and only remnants of a circular development of brecciaed and partly melted rocks may be present (French 1998).

In the seismic lines and in the few meters of sediment penetrated by the Mjølnir core, several breccias are present (Dypvik et al. 2004b). So-called monomictic authochtonous breccias were formed by crushing the target area and only moving the clasts around locally for short distances. This resulted in a jigsaw puzzle texture, which typically is found in the deeper and central parts of impact craters (French 1998). In a standard impact setting polymict, allocthonous breccias will follow on top of these conglomerates. These polymict breccias are made up of ejected material and sedimentary conglomerates/breccias with clues of traction and sediment transportation (French 1998). In the cores from the Mjølnir Crater only this kind of polymictic breccia were recovered by the drilling, comprising the lowermost part of the Ragnarok Formation (Figs. 5.2 and 5.3). The monomictic authochonous breccias have not been encountered in the Mjølnir core, but most likely make up the deeper central, un-cored parts of the structure. They are most likely represented in the chaotic reflections of the seismic lines in those deeper sections (Tsikalas et al. 1998a–c) (e.g., Fig. 4.4).

Covering the impact breccias, commonly in several craters, glass- and melt-rich formations and so-called suevites follow. They may in turn be succeeded by various sedimentary conglomerates. In the case of Mjølnir no melts or suevites have been found so far, but various types of sedimentary conglomerates are present in the parts of the Ragnarok Formation, while the melt problem of Mjølnir is discussed in the next paragraph.

Shatter cones are a typical structural feature so far only found in impact structures (Montanari and Koeberl 2000). However, no such structures have been found in the Mjølnir core and would be hard to see at such scale.

5.4.3 Mineralogical Evidence of Impact Cratering

Several mineralogical impact indicators have been discovered through the years of impact research (Montanari and Koeberl 2000). The major evidence and most commonly used mineralogical arguments are fragments of meteorites, shocked minerals (quartz, feldspar, and zircon), high pressure polymorphs (coesite, stishovite), Ni-rich spinels, diamonds, meltphases of various types, including diaplectic glasses and tektites.

In the studies of the Mjølnir impact, analysis of cores from the impact crater, analyses of the ejecta beds and field observations have been carried out. Shocked quartz have been found in both the crater and in the ejecta beds deposited 30 km outside the crater (Sindre Bed), as seen in core 7430 (level 47.6 m see Figs. 6.2 and 6.3) (Dypvik et al. 1996, 2006). The few shocked quartz grains (about 1 by ‰ by number) found in the 7430 core are dominated by PF structures, but also some few PDF have been found (Dypvik et al. 1996; Langenhorst and Dypvik 1996). The shocked quartz from the central part of the structure (Mjølnir core) has been studied and described in detail, with a particular focus on planar microstructures (planar fractures -PF and planar deformation features-PDF) by Sandbakken (2002) and Sandbakken et al. (2005) (Figs. 5.18, 5.19, and 5.20). These shocked grains are dominated by planar fractures and only a few of good planar deformation features have been found (Figs. 5.18 and 5.19). This may largely be attributed to the great difference in compressibilities of the pores and sand grains, resulting in large shear stresses that have been accommodated mainly by fracturing. The different shock features indicate possible formation pressures between 5 and 20 GPa (Fig. 5.20). The samples from the lower part of the core typically show the lowest pressures, whereas the uppermost part of the succession displays the highest pressures, a development comparable to that of the Ries crater (Sandbakken et al. 2005).

In some cases, Ni-spinels (Byerly and Lowe 1994; Robin and Molina 2006), their possible derivatives and even parts of the meteorite may be found (Kyte 2002). In the Mjølnir case we have looked for such impact evidence, but no good examples have been discovered, so far. In a study of samples from Janusfjellet section at Svalbard, Robin et al. (2001) found microscopic grains of nickel-rich iron oxides. These may be remnants of the altered bolide. However, Ni enrichments are commonly found in several of the Upper Jurassic formations, and likely represent enrichments of sulphide phases, mainly demonstrating the consistent anoxic depositional conditions of that time (Dypvik et al. 2006; Dypvik and Harris 2001).

The discussion of impact glass formation, in particular in marine impacts, is going strong within the impact community. It turns out that most marine impact craters are poor in impact melt rocks, and it has been suggested that the production of impact glass normally could be smaller in marine targets (Ormö and Lindström 2000; Dypvik and Jansa 2003). The presence of water results in more violent impacts and possibly more wide-spread distribution of ejecta (Shuvalov et al. 2002; Shuvalov and Dypvik 2004; Tsikalas 2005). So far no impact glass or melt rocks have been found during the detailed petrographic and geochemical studies of the Mjølnir drill-core. In this connection it should, however, be mentioned that the amounts of smectite is rather high in the impact related sediments of Mjølnir (Dypvik and Ferrell 1998, Dypvik et al. 2003). Smectite is regarded to be a most likely alteration product of the impact melt, but it can be formed by several different mechanisms. In this respect the appearance of smectite enrichments can be taken as a weak indication of an original presence of impact glass. No mineralogical or

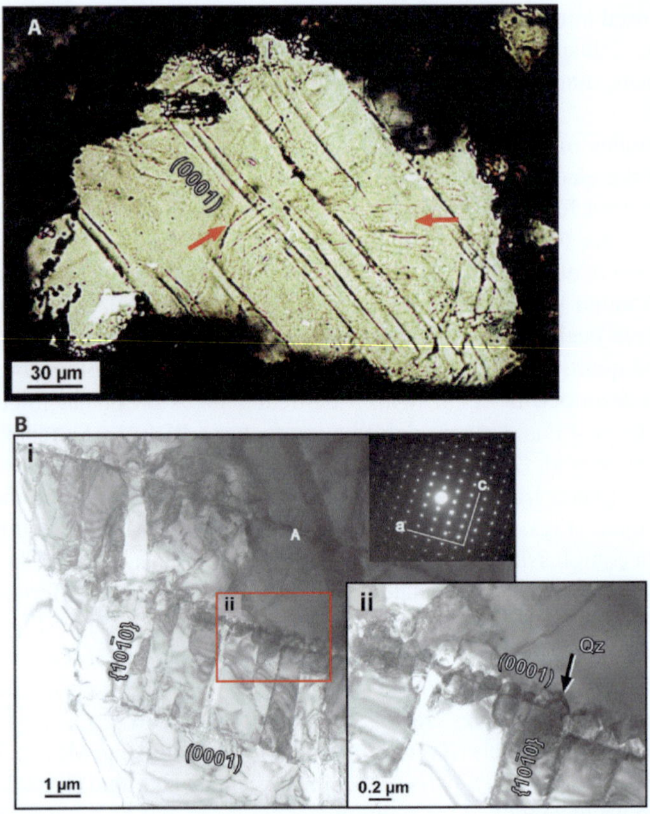

Fig. 5.18 **(a)** (*Upper photo*) Photomicrograph of a shocked quartz grain from unit IIc (74.05 m) with well-developed, irregularly spaced (4–20 µm) PFs (~1 µm broad) parallel to (0001). Two sets of thin (<1 µm) and closely spaced (2–4 µm) PDFs terminate towards the PFs (*red arrows*). On the *left side* of the grain there are indications of chemical etching (plane-polarized light). Shocked quartz grains from 7329. **(b)** (*lower three photos*) Bright field micrograph (TEM) of quartz from unit I (PDF/PF) (100.70 m). (i) Two sets of planar microstructures are shown. The partly open (0001) PFs terminate the thin, sub-planar and sub-parallel {10$\bar{1}$0} PFs. A corresponding diffraction pattern (*inset*) was used to index the planar micro-structures. The *rectangle* (ii) marks the enlarged area in (ii), showing microquartz (Qz) that has crystallized in an open part of the PF (From Sandbakken et al. 2005)

geochemical characteristics of the smectite have, however, been found, which could pin-point an impact related origin.

5.4.4 Geochemistry

Platinum Group Element (Ru, Rh, Pd, Os, Ir, Pt) (PGE) enrichments are well known and commonly used as indications for an impact event (Montanari and Koeberl 2000; Koeberl 2007). In particular Ir and partly Os have been in focus.

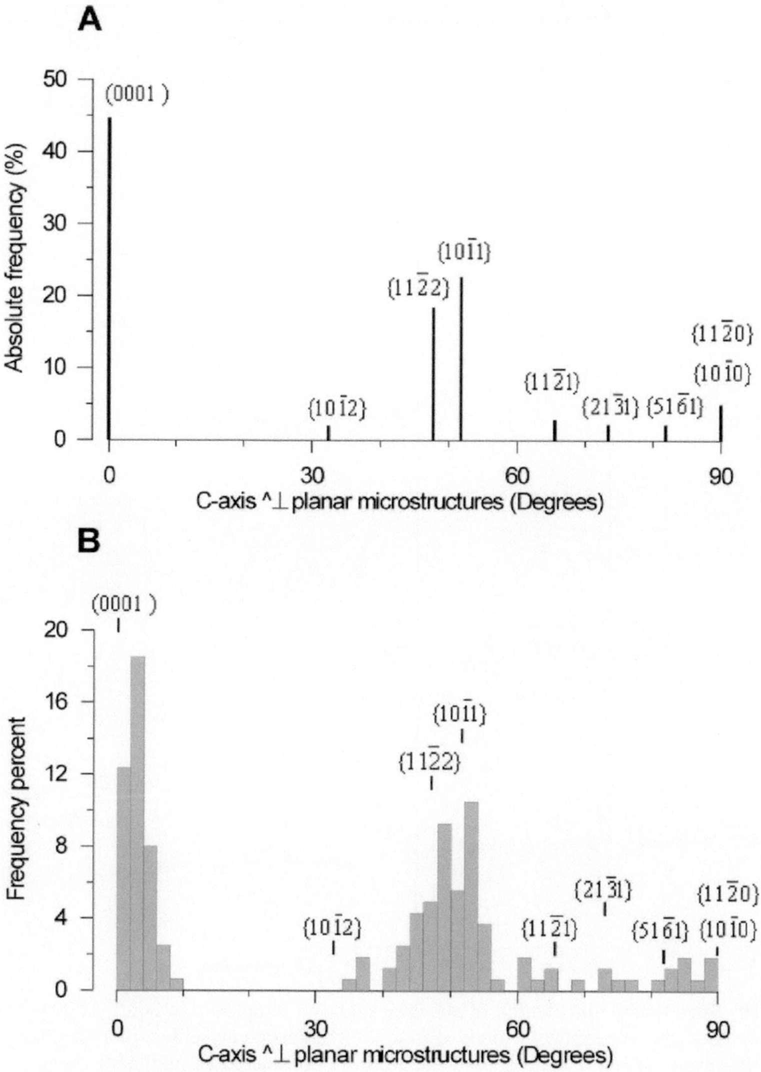

Fig. 5.19 **(a)** Histogram showing the crystallographic orientations of PFs in 100 quartz grains (141 sets, 18% unindexed planes) from core 7329/03-U-01. The planes have been uniquely indexed using a template containing known planar microstructure orientations (From von Engelhardt and Bertsch 1969). **(b)** Histogram showing the angles (binned at 2°) between the c-axis and the poles to PFs. This figure presents similar data as in Fig. 5.19a, but also includes the unindexed planes. Note the analyzed samples represent allochtonous material that represents a mixture of quartz grains recording different degrees of shock (From Sandbakken et al. 2005)

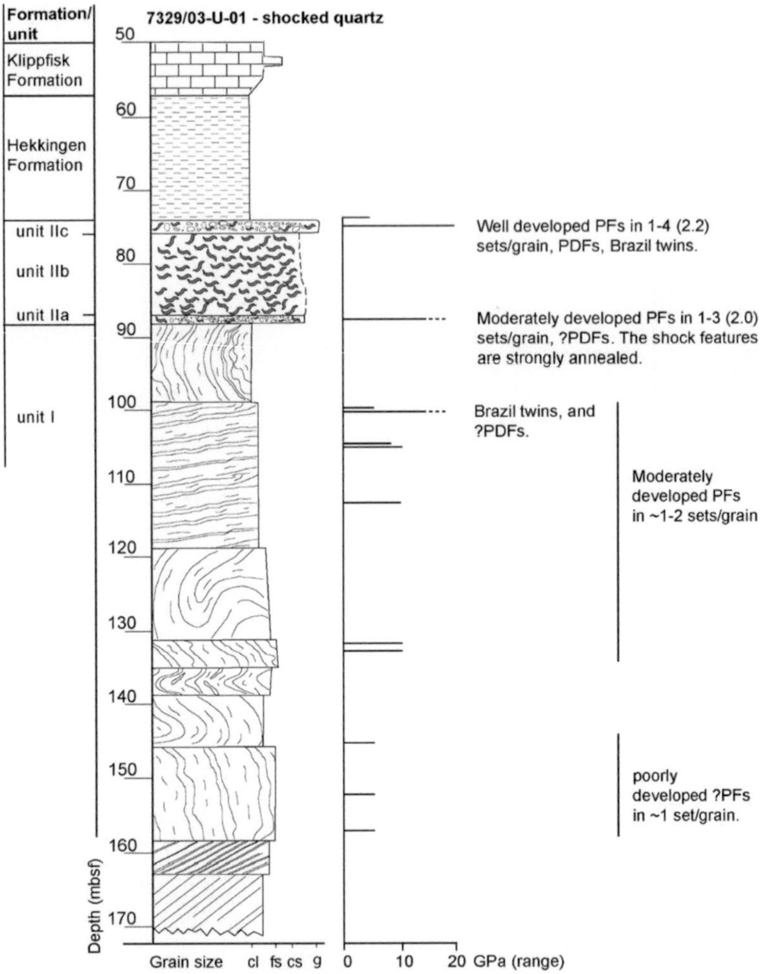

Fig. 5.20 Stratigraphic distribution of shocked quartz in samples from core 7329/03-U-01. The range of shock pressures shown is derived from calibrations on crystalline rocks by shock experiments (Grieve et al. 1996) and should therefore be considered as minimum shock pressures. Mbsf = meter below seafloor. cl = clay, fs = fine sand and silt, cs = coarse sand, g = grain size coarser than sand (From Sandbakken et al. 2005)

In the Earth's crust and surface the Ir concentrations are normally low (20–60 ppt), while Ir enrichments are found in mantle derived rocks. However, not even the largest volcanic eruptions are able to supply high enough Ir-concentrations to the surface in order to match extraterrestrial Ir-supply caused by asteroid and meteorite impacts. In chondrites and iron meteorites several hundred ppb of Ir have been detected (Koeberl 2007). The Ir contents in the samples from the Mjølnir crater core are more than 200 ppt. In the ejecta from drillhole 7430/10-U-01, located 30 km NE of the crater, a sharp peak of more that 1 ppb has been found (see Figs. 6.2 and 6.3). In this respect it should be mentioned that one sample from Svalbard have been recognized with 2 ppb Ir (Dypvik et al. 2006) at a horizon correlative with the Mjølnir event.

Os- and Cr-isotope analyses can be used in order to confirm the impact origin and a determination of the possible bolide type (Koeberl 2007; Montanari and Koeberl 2000). Recent preliminary Os-isotope analyses of the impact horizon at Janusfjellet, may point towards an iron oxide (possible nickel) rich bolide (S. Graham, personal communication, 2006).

The Cr and Ni distributions have been analyzed in most of the available Svalbard sections and in cores from the Barents Sea region (Dypvik and Attrep 1999; Dypvik et al. 1996, 2006). In the Ir-enriched beds of the 7430 drillhole Dypvik and Attrep (1999) found Ni and Cr values that were interpreted to indicate an iron rich meteoritic composition. This is in agreement with the preliminary Os-isotope results mentioned above and along with the Ni-rich iron oxides discovery from the impact beds at Janusfjellet, Svalbard (Robin et al. 2001). It should be mentioned that mutual Ni and Cr enrichments maybe be partly related to the anoxic/dysoxic deposition conditions at the time of the Mjølnir impact.

5.4.5 Paleontological Evidence of Impact Cratering

The Late Jurassic and earliest Cretaceous on the Barents Shelf were dominated by fine-grained clay sedimentation, with mostly anoxic to hypoxic depositional conditions. The stratified water masses contained typically relatively rich, but low diversity, nektonic faunas and marine microfloras above the pycnocline. In contrast, the benthic faunas contained only a few bivalve species (dominantly *Buchia*) and low diversity communities of foraminifera.

There is no evidence of any major biotic extinction or changes in diversity related to the impact event, but the overall composition of the microfossil assemblages show a significant turnover within the impact influenced strata (Bremer et al. 2004; Smelror and Dypvik 2006). Of particular interest is the previous described sudden bloom of algaes *Leiosphe ridia* and *Botryoccoccus* just after impact and in the same beds with the Ir-enrichments (Smelror et al. 2002; Smelror and Dypvik 2005) (see Fig. 6.3). In the oldest post-impact sediments of the Mjølnir core (i.e., at level 74 m) the acme of *Leiosphaeridia* reach 513,000 specimens/gram sediment (post-compacted) and remains at abundances around 450,000 specimens/gram sediment up to about the 71 m level. Continued up to 68.5 m the abundance varies between 320,000 and 360,000 specimens/gram sediment, and from 68 to 66 m the abundance drops to between 107,000 and 152,000 specimens/gram sediment. From 65.5 m and up to the uppermost studied sample at 58.2 m the abundance drops further and is reduced to between 50,000 specimens/gram sediment (at 64.5 m) to around 500 specimens/gram sediment (at 59.0 m). This prolific acme is interpreted to reflect a dramatic change in environmental conditions in relation with the impact and just thereafter. Smelror et al. (2002) suggested that the algal bloom was induced by the enormous amounts of nutrients released into the water column by the impact and the following tsunami. The brief, dramatic change in the depositional conditions and new environmental setting created a unique opportunity for opportunistic and disaster species such as the *Leiospheridia*.

In the post-impact "algal bloom" interval only a monospecific assemblage, with few foraminifera (i.e., *Trochammina* aff. *Septentrionalis*) is found (Bremer et al. 2004) (Fig. 5.15). Above 67 m the diversity and abundance of foraminifera increase somewhat, but in spite of this faunal expansion the diversity is still low. The assemblages are dominated by agglutinated taxa but there are also a few calcareous forms present in the samples at 66.0 and 60.0 m. The faunas are dominated by *Evolutionella vallata*, *Gaudryina rostellata*, *Recurvoides obskiensis* and *Trochammina* spp. (Bremer et al. 2004) (Fig. 5.15). These assemblages closely resemble those found in the Agardhfjellet Formation on Svalbard. These observations of the benthic foraminifera faunas suggest that "normal" (pre-impact) oceanographic and depositional conditions of the Hekkingen Formation environments were restored at the time of deposition of the sediments above 67 m in the Mjølnir core 7329/03-U-01.

The duration of the aftermath, covering the time of the prasinophycean bloom and ocean eutrophication is hard to determined precisely, but most likely took place during a relatively short time. This means we have to assume a very high sedimentation rate to account for the about 7–8 m of post-impact dark shales containing the *Leiosphaeridia* bloom and the monospecific foraminifera assemblage. This depositional rate is extreme for such fine-grained sediments, but not unlikely given the fact that about 233 km^2 of sea-bottom sediments and underlying bedrock were thrown up and spread in the water columns and air in a few seconds (Shuvalov et al. 2002). From the subsequent fall-back and resurge much of the sediments were brought back in the crater, including some organic-rich sediments from the target area and more distal areas agitated by the tsunamis.

Only a sparse macrofauna has been recovered from the Mjølnir crater core and adjacent boreholes. This is because the cores are only 5 cm in diameter, and because the anoxic to hypoxic depositional environments of the Hekkingen Formation contained relatively sparse benthic communities. Based on the available material no dramatic changes have been registered in our collections. This is, however, to be expected for an impact of this size (Smelror et al. 2001a; Bremer et al. 2004).

Chapter 6
Ejecta Geology

Henning Dypvik, Morten Smelror, Atle Mørk, and Filippos Tsikalas

6.1 The Identification of Ejecta Beds

6.1.1 Introduction

Ejecta recognition is an important factor in impact research and has been a key element in the Mjølnir impact studies. The characterization of ejecta covers several different geological and geophysical topics as summarized in French (1998) and Montanari and Koeberl (2000). In particular in studies of marine impact events (submarine craters) or in cases where the impact site has not been recognized, ejecta recognition is crucial. This is well exemplified in the K/T investigations and the Chicxulub impact structure (Alvarez et al. 1995; Smit 1999), in the investigations of the North American tektite (melt particles formed by impact, partly melt of target material, immediately before crater formation) strewn fields and the Chesapeake Bay crater (Koeberl 1989; Poag et al. 2004) as well as in the investigations of the Precambrian spherule beds of Australia and South Africa, where still no impact sites have been recognized (Simonson et al. 1998).

The so-called ejecta beds may consist of material deposited from both the ejecta curtain (a curtain of particles ejected from the forming crater) and the fireball (a spherical mass of fire with melt and clastic material incorporated, formed at impact). Its composition will, therefore, vary according to impact configuration/dimension/location and geological composition of the target area. While the ejecta curtain will contain crushed rocks, melt material, spherules, and shocked minerals, the fireball related ejecta will be more enriched in the finer fractions and gas. This fraction normally also carries the PGE enrichments (e.g., Ir-enrichments). The ejecta composition will vary accordingly and make it possible to identify the ejecta source. This has been done in the case of the North American tektite strewn field and for the moldavites of Europe, i.e., the Chesapeake Bay crater and Ries crater, respectively.

H. Dypvik (✉)
Department of Geosciences, University of Oslo, NO-0316 Oslo, Norway
e-mail: henning.dypvik@geo.uio.no

H. Dypvik et al. (eds.), *The Mjølnir Impact Event and its Consequences*,
Impact Studies, DOI 10.1007/978-3-540-88260-2_6, © Springer-Verlag Berlin Heidelberg 2010

The ejecta distribution may vary in composition from locally distributed ejecta being characterized by rock fragments and breccias of target rocks, to globally spread ejecta and fireball material being recognized by thin layers enriched in platinum group elements (PGEs) (Ir included), siderophile elements and possibly fine-grained shocked quartz (Alvarez et al. 1995; Smit 1999; Montanari and Koeberl 2000; Claeys et al. 2002). In both the Chicxulub and the Mjølnir cases soot has been found related to the impacts. In this chapter we have chosen to include the soot particles in the ejecta category.

6.1.2 The Ragnarok Formation and Sindre Bed

The Ragnarok Formation consists of various conglomerates, breccias and chaotic organized sediments deposited in and just around the Mjølnir crater structure (Fig. 1.4 and Fig. 1.12). Mainly airborne or reworked ejecta are included in the deposits outside the Mjølnir crater rim and away from the deposits of the Ragnarok Formation (see Chap. 5). These beds, named the Sindre Bed, consist of a large selection of lithologies and with considerable lateral variation in composition, reflecting mixing combinations of both local background sedimentation and the exotic Mjølnir ejecta. In the case of major impact, deposits can be expected to be found thousands of kilometers from the impact site – spread all around the globe (Melosh 1989; French 1998; Montanari and Koeberl 2000; Claeys et al. 2002; Shuvalov and Dypvik 2004).

Ejecta from the Mjølnir impact has been identified as separate beds within the Hekkingen Formation in the western Barents Sea and possibly within time-equivalent units in more distal areas such as the Norvik Peninsula in West Siberia (Dypvik et al. 2006; Zakharov et al. 1993). At a proximal position only 30 km away from the Mjølnir structure (core 7430/10-U-01) the Sindre Bed, forms tens of cm to a few meter thick sedimentary unit consisting of dark grey shales and conglomerates (Figs. 1.3, 1.4, 1.12, 1.13 and 6.1).

The conglomerates of the Sindre Bed were first described from the shallow core 7430/10-U-01 by Dypvik et al. (1996) (Figs. 1.4, 6.1, and 6.2). In that paper the bed was not formally named, but its content of shocked quartz and iridium enrichments on top of claystones highly enriched in smectite, was presented and taken as strong evidence that the time equivalent and nearby located Mjølnir structure had an impact origin (Figs. 6.1, 6.2, and 6.3). Later, Dypvik and Ferrell (1998) discussed in detail the considerable smectite enrichments from possible altered impact glasses in the same unit; a specific unit name forced its way (Figs. 1.4, 6.3, and 6.4). This forms the base for the definition of the Sindre Bed, as first described in the borehole 7430/10-U-01 (Dypvik and Ferrell 1998; Dypvik et al. 2002, 2004b, c 2006).

Borehole 7430/10-U-01 was drilled 30 km NNE of the rim of the Mjølnir crater (Figs. 1.2, 1.7, and 6.3). From core-base (67.6 m) to the base of the Sindre Bed (52.0 m) the core consists of finely laminated, dark grey, organic rich shales, which are typical for the Hekkingen Formation in the Barents Sea (Worsley et al. 1988). From 52 to 46.5 m, the core contains sediments defined to belong to the Sindre Bed (Fig. 6.3). The interval from 52 to 47.60 m consists of dark grey, smectitic

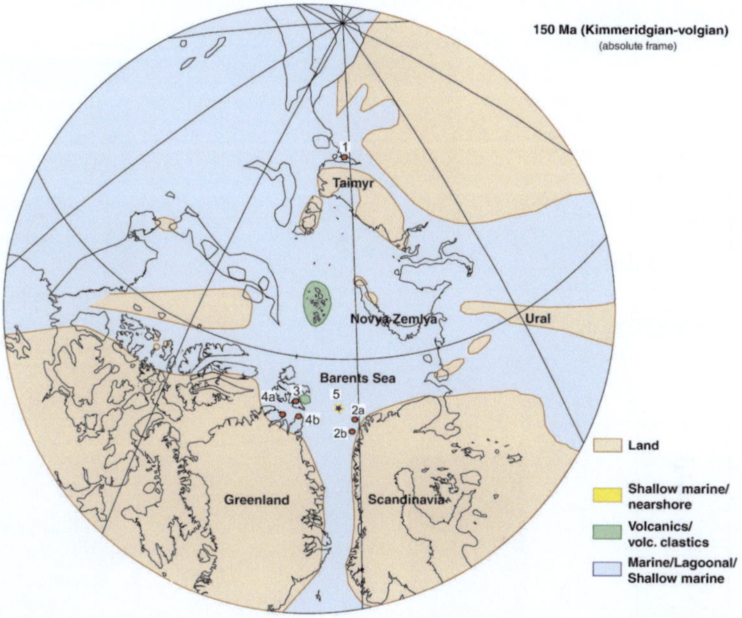

Fig. 6.1 Kimmeridgian-Volgian paleogeographical map, showing the locations (1–5) where ejecta and possible ejecta have been recovered 1 = The Nordvik section, 2a = 7018/05-U-01 core, 2b = 6814/04-U-02 core, 3 = Janusfjellet section, 4 = North Greenland sites; 4a = East Peary Land, 4b = Kilen, 5 = The Mjølnir crater site

claystones and shales, with dispersed mudflake clasts (Figs. 6.2 and 6.3). This unit, which is rich in smectite, is embedded within the organic rich, finely laminated, dark grey claystones of the Hekkingen Formation (Dypvik et al. 1996; Dypvik and Ferrell 1998). The smectite enrichment was most likely derived from altered impact glass in this lower part of the Sindre Bed (Fig. 6.4). At the 47.60–47.40 m level, three coarse, sand- to gravel-sized, upwards fining conglomeratic units are found, separated by mm to cm thick claystone/shale laminae (Fig. 6.2). This 20 cm thick conglomeratic package is succeeded by 120 cm of sandy and silty shales (47.40–45.20 m), which at 46.85 m has a well-defined iridium peak (Figs. 6.2 and 6.3). The very last geochemical ejecta-signals are found at level 46.5 m (Ir, Ni and smectite enrichments), consequently defined as the top of the Sindre Bed (Fig. 6.3). Enrichments of *Leiospheridia* and soot particles are also found in the uppermost parts of the Sindre Bed (see Figs. 6.13 and 6.14). The succeeding cemented siltstones of Hekkingen Formation (above 46.5 m) are heavily bioturbated and contain fossil fragments and insitu bivalves of the genus *Buchia* (Smelror et al. 2001a). The smectite found in the Sindre Bed, especially enriched in its lower part of core 7430/10-U-01, has been interpreted to represent altered impact glass (Dypvik and Ferrell 1998). Glass alters quickly to smectitic clays and several million years old glasses are rarely found unaltered, being most commonly completely changed (Declercq et al. 2009). Therefore, the lower boundary of the Sindre Bed is defined by the first well-defined glass/smectite appearance.

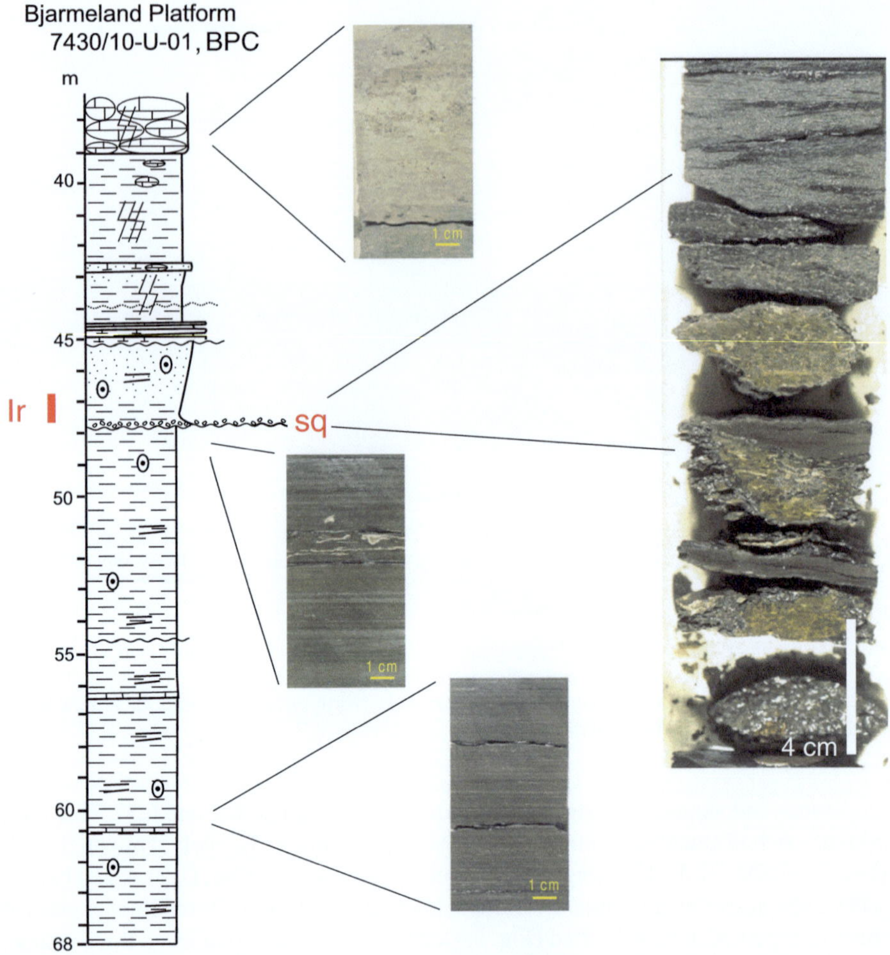

Fig. 6.2 Core photo of the Sindre bed in core 7430/10-U-01 (BPC = Bjarmeland Platform Core). The shocked quartz grains (sq) are present in the conglomeratic bed at 47.6 m, while the Ir enrichments have been detected from 47.6 and 0.8 m above. Depth values in m below sea floor

Sindre Bed equivalents have been recognized hundreds of kilometers away from its type section of core 7430/10-U- 01, in a section on Svalbard and in other cores from the Barents Sea (7018/05-U-01, 6814/04-U-02) (Dypvik et al. 2006) (Figs. 6.1, 6.2, 6.5 and 6.6). Borehole 7018/05-U-01 core was drilled 500 km and corehole 6814/04-U-02 around 800 km south-west of the impact site (Figs. 1.2 and 6.1). Dypvik et al. (2006) presented stratigraphical, palynological, and geochemical information, which made it possible to indirectly trace the level of the Sindre Bed in core 7018/05-U-01. In the 7018/05-U-01 core the mutual peak in the *Leiospheridia* distribution and the environmental changes as reflected in the trace element distributions have been interpreted to mirror the effects of the Mjølnir impact (Fig. 6.7). Recirculation and reworking of nutrient- rich bottom waters and sediments may have

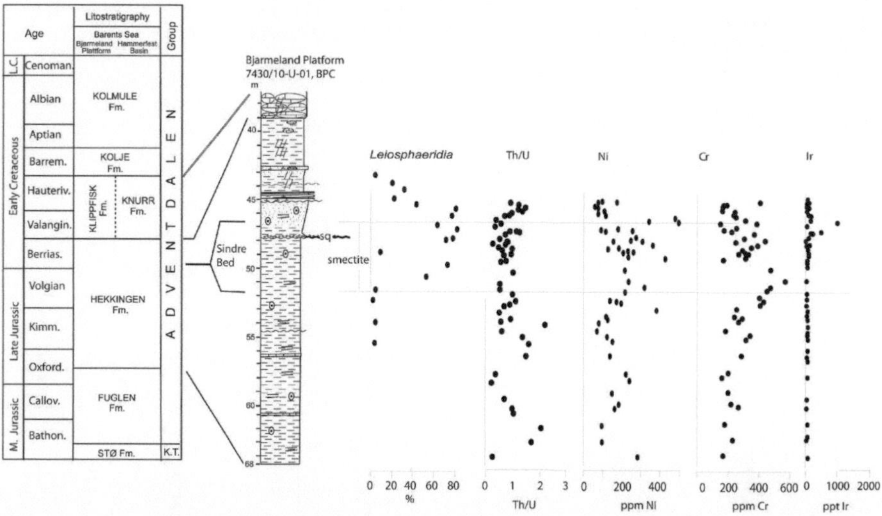

Fig. 6.3 The sedimentological log for core 7430/10-U-01 is correlated to the stratigraphical scheme of the Barents Sea and showing the position of the smectite anomaly recovered and shocked quartz grains (sq). The distributions of *Leiosphaeridia*, the Th/U ratios and the Ni, Cr and Ir concentrations are plotted

resulted in the large algal bloom, which, in turn, terminated in new anoxic period with low Th/U values and increased accumulations of Ni and Cr.

Borehole 6814/04-U-02 (Fig. 6.8) was drilled close to the paleocoastline of the paleo-Barents Sea and 300 km farther southwards from the Mjølnir crater than 7018/05-U-01 (Fig. 6.1). In the 6814/04-U-02 core case, no clear-cut geochemical or palynological evidences were evident, but the core contains a sandy unit with a mud-flake conglomerate and zones of parallel lamination (Fig. 6.8). This relative coarse-grained unit time-wise matches the Mjølnir impact well, and may represent related coastal erosional wave and possible tsunami effects along the paleocoastline of the Barents Sea at that time. Another effect of the Mjølnir event may be the possible collapse structures at the irregular Jurassic-Cretaceous boundary off Mid Norway as described by Rokoengen et al. (2005). They suggested these structures have been caused by the Mjølnir tsunami.

In the Janusfjellet section of Svalbard possible Sindre Bed equivalents (Figs. 6.6, 6.9 and 6.10) have been identified by palyological and geochemical evidence; enrichments of *Leiospheridia* and mutual shifts in the Th/U ratios and the Ni distributions (Dypvik et al. 2006) (Figs. 6.9 and 6.10). In addition, Robin et al. (2001) analyzed samples from well 7430/10-U-01 and Svalbard with respect to nickel-enriched iron oxides. However, no Ni-spinels were found in the Ni-rich iron oxides in the Svalbard section at level 76 m (Fig. 6.10), which is the stratigraphical level equivalent to the units enriched in shocked quartz in core 7430/10-U-01 (Fig. 6.5). One single sample from the Janusfjellet section (Svalbard), level 79 (33.70 m) did also show high Ir values (2 ppb) a result that was replicated in separate analyses (Dypvik et al. 2006).

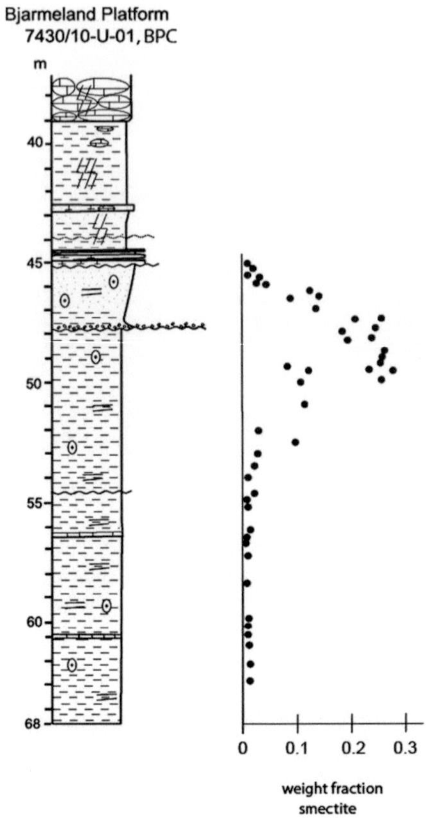

Fig. 6.4 The depth in m are given below seafloor in core 7430/10-U-01. Smectite (SI85RD) values are from Dypvik and Ferrell (1998)

The search in North Greenland for Mjølnir-related tsunamite locations at Kilen and East Peary Land have not yet resulted in any positive findings. The Ladegårdsåen and Kugleleiet formations, from East Peary Land and Kilen, repectively, have been investigated in great detail, both in the field and by thin sections analysis. Some possible reworked coarse grained beds can be picked out, but the uncertainties in dating due to poor fossil contents make a correlation to Mjølnir difficult.

The Sindre Bed ejecta can be correlated over very long distances. Probably the extreme Ir-enrichments published from time-equivalent beds in Nordvik, Siberia, about 2,500 km to the northeast (Zakharov et al. 1993) represents one such example (Figs. 1.3 and 6.12). The Nordvik section of north western Siberia is characterized by black organic rich shales with dispersed phosphate concretions. In one of these concretions Zakharov et al. (1993) found up to about 7 ppb Ir, a remarkable enrichment (Fig. 6.12). The stratigraphical timing is also exact when the Mjølnir event took place; close to the Volgian-Ryazanian boundary or in the Berriasian *Berriasella jacobi* Zone when correlated to the Tethyan stratigraphy (Smelror et al. 2001a). We

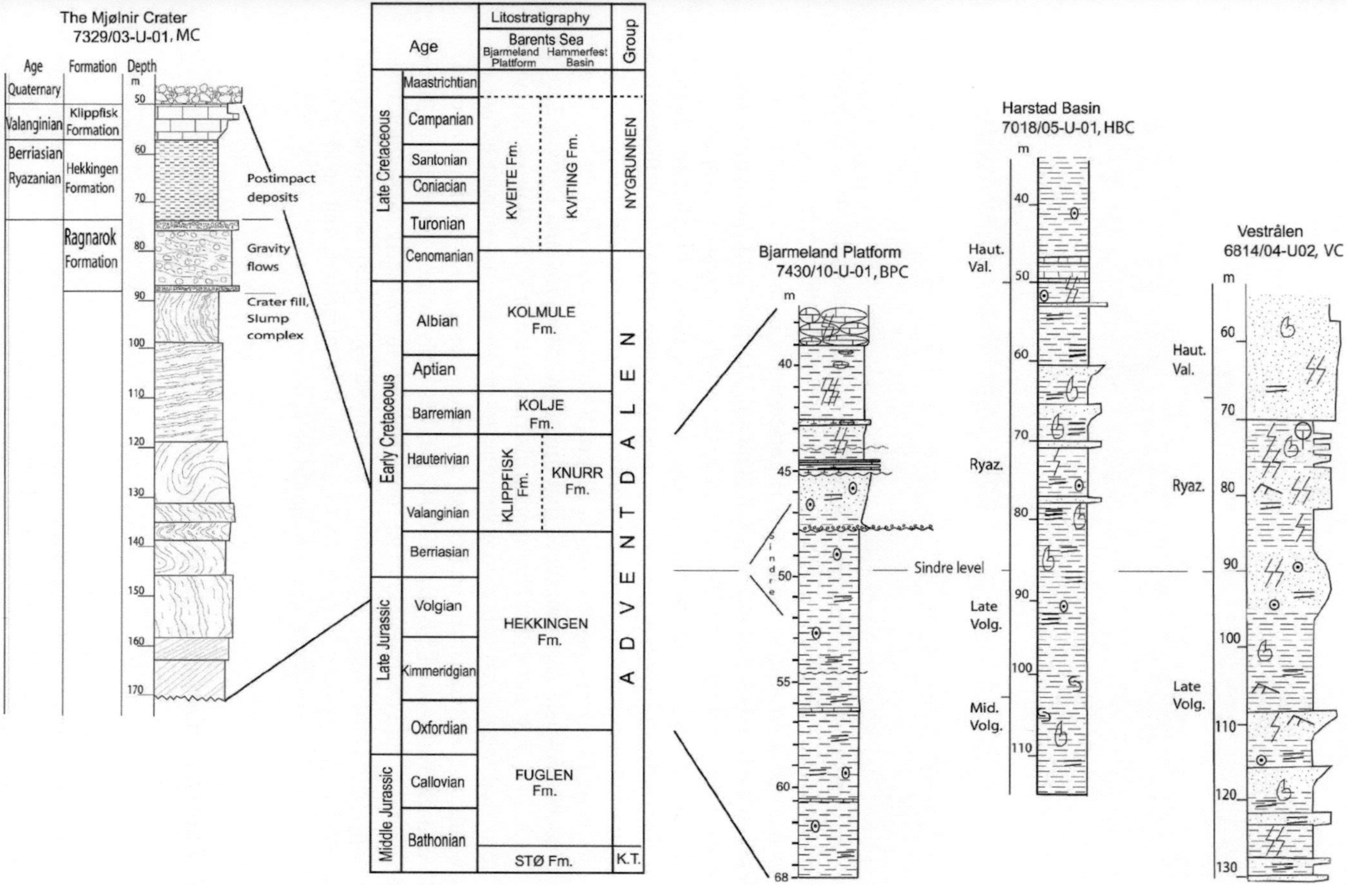

Fig. 6.5 Correlation of the Sindre Bed in cores 7430/10-U-01 from the Bjarmeland Platform and core 7018/05-U-01 (HBC = Harstad basin core) in the Harstad Basin, and the estimated position of the Volgian-Ryazanian boundary in core 6814/04-U-02 (VC = Vestrålen core) off Vesterålen. MC = Mjølnir core)

Fig. 6.6 The location of the possible Sindre Bed at Janusfjellet, Svalbard

have recently analyzed a new sample batch from this Siberian section, to get more detailed information about this distant ejecta formation. In these later analyses, we have, so far, not been able to detect any Ir-enrichments in the phosphates and the highly organic- rich shales of the Nordvik peninsula (Dypvik and Zakharov 2010).

Applying the estimates of Tsikalas (2005), which suggest an impact from southwest at an angle of about 45°, along with the ejecta distribution simulations of Shuvalov and Dypvik (2004), the most probable area to track the thicker developments of the Sindre Bed will be northeast of the Mjølnir crater (Fig. 1.12).

In this way all arrows point towards Siberia (Nordvik Peninsula), but with an uneven ejecta thickness distribution in that direction (Shuvalov and Dypvik 2004) (Figs. 1.12, 7.1, 7.8 and 7.9). The spatial distribution of the Mjølnir ejecta and the Sindre Bed were even more complex due to probable post–impact action of heavy winds, waves, tsunamis, and ocean currents. Based on the Shuvalov et al. (2002) and Shuvalov and Dypvik (2004) models only minor amounts of fine-grained ejecta should be expected to be found towards in southerly (SW) directions (see Chap. 7). Macroscopic, stratigraphical, mineralogical/geochemical evidence of the Sindre Bed would not be expected in that direction, but evidence of Mjølnir-triggered waves and tsunamis should be found.

6.1.3 The Discoveries of Large Amounts of Soot in Mjølnir Related Sediments

The most exotic ejecta found so far, being derived from the Mjølnir impact are possibly the large amounts of soot particles observed in the analyzed cores from the Barents Sea (7430/10-U-01 and 7329/03-U-01) and the Svalbard section (Wolbach et al. 2001; Dypvik et al. 2008b) (Figs. 6.13 and 6.14). The soot is characterized as aciniform type (bunches of grapes), and soot grains inspected so far have an appearance comparable to soot formed by combustion of hydrocarbons (Fig. 6.13)

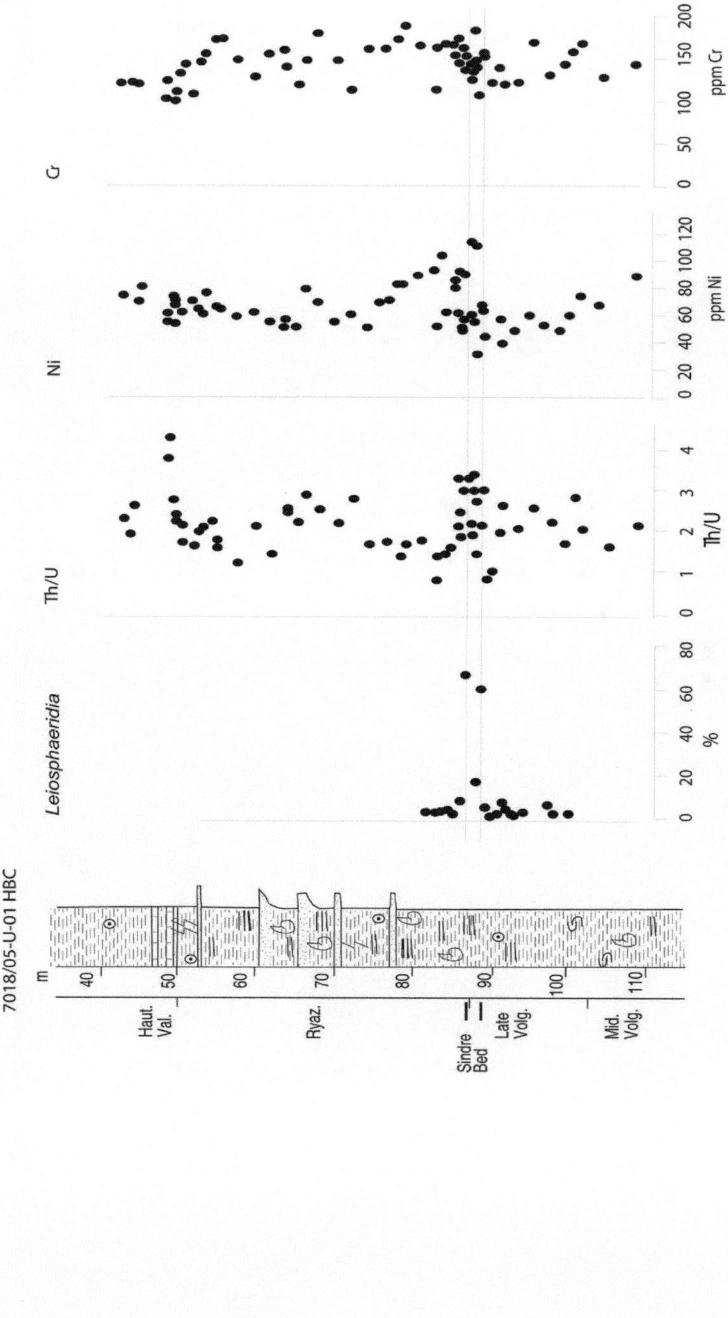

Fig. 6.7 Inferred position of the Sindre Bed in core 7018/05-U-01, determined by the distributions of the algae *Leiosphaeridia*, together with measured values of Ni, Cr and Th/U

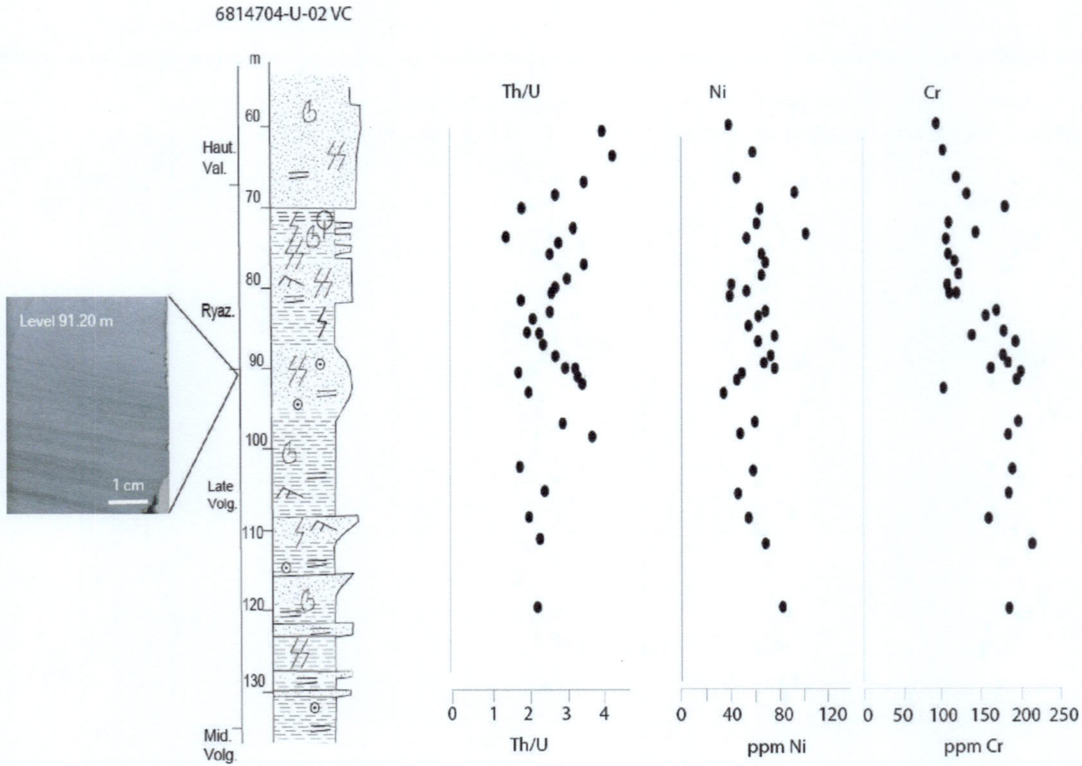

Fig. 6.8 The sand bed at around level 90 possibly represents the time of the Sindre Bed, a possible reworking phase in this, also at that time, coastal locality of core 6814/04-U-02 off Vesterålen

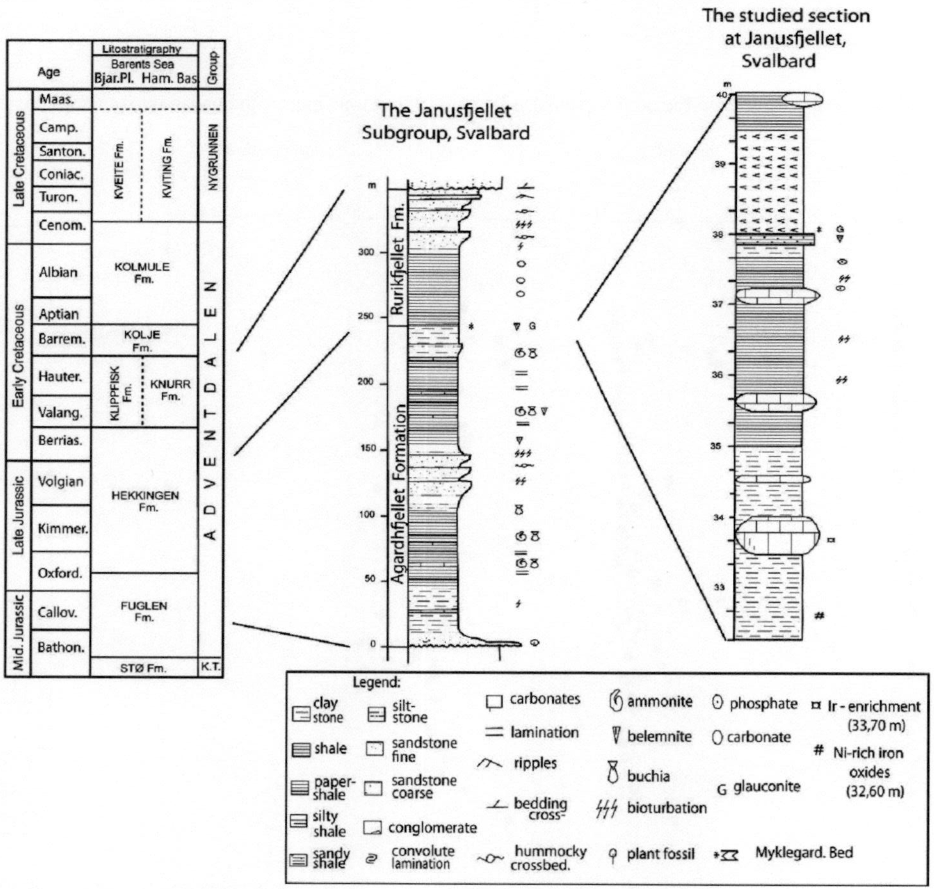

Fig. 6.9 Lithostratigraphic log of the Janusfjellet Subgroup, with details around the possible equivalent to the Sindre Bed at Janusfjellet, Svalbard

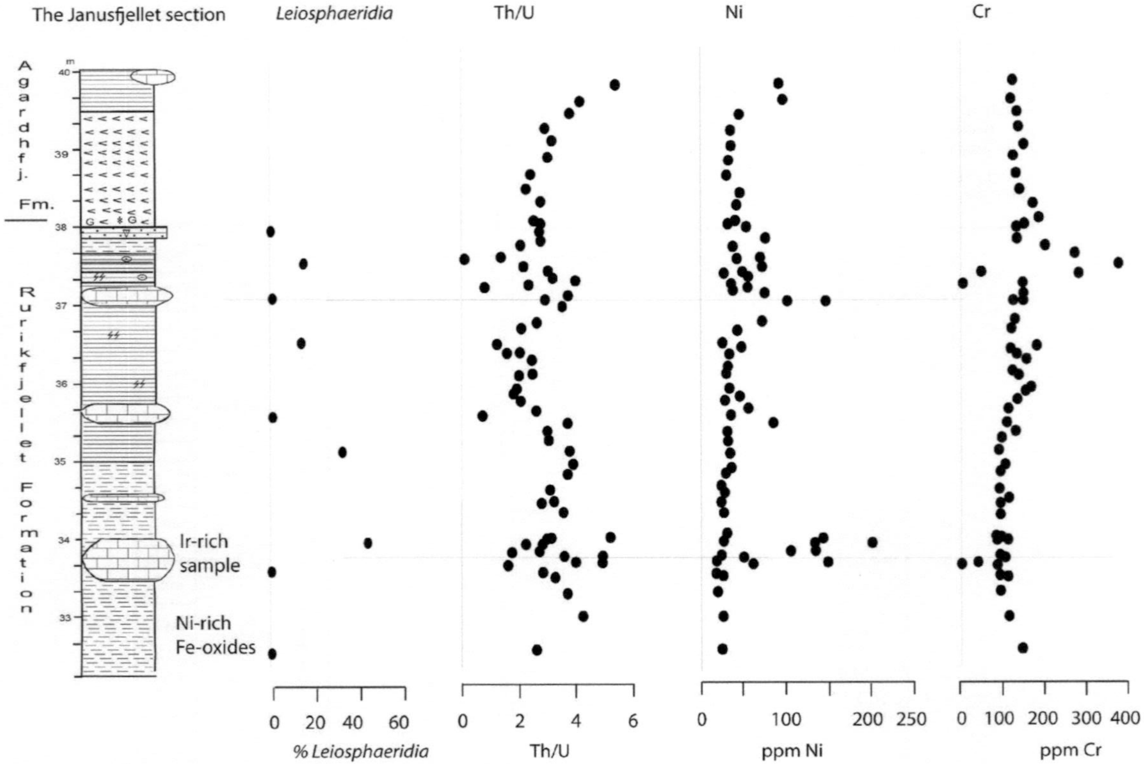

Fig. 6.10 Distribution of paleontological and geochemical indicators suggesting a possible distal occurrence of the Sindre Bed at Janusfjellet on Spitsbergen, Svalbard

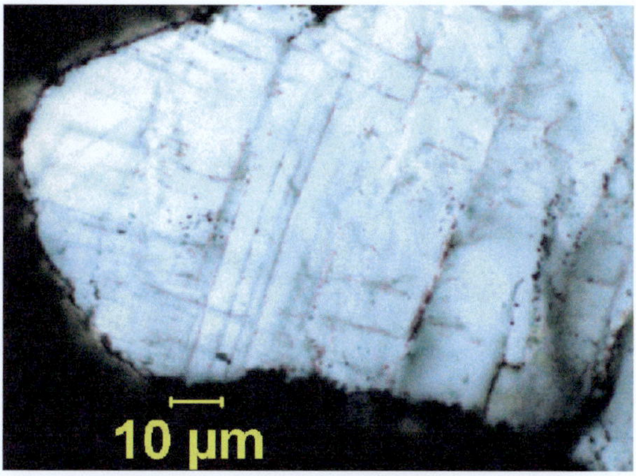

Fig. 6.11 Part of a shocked quartz grain, displaying well developed planar fractures (PF). Sample from the ejecta bed level 47.60 m in core 7430/10-U-01

Fig. 6.12 The Jurassic/Cretaceous boundary section at Nordvik (Siberia), with phosphate concretionary beds with possible ejecta deposits exposed between the flags. The cliff hight (photos (**a**) and (**b**)) is about 12 m. **c** is a closeup of one phosphate concretion from the section (Photos: Victor Zakharov)

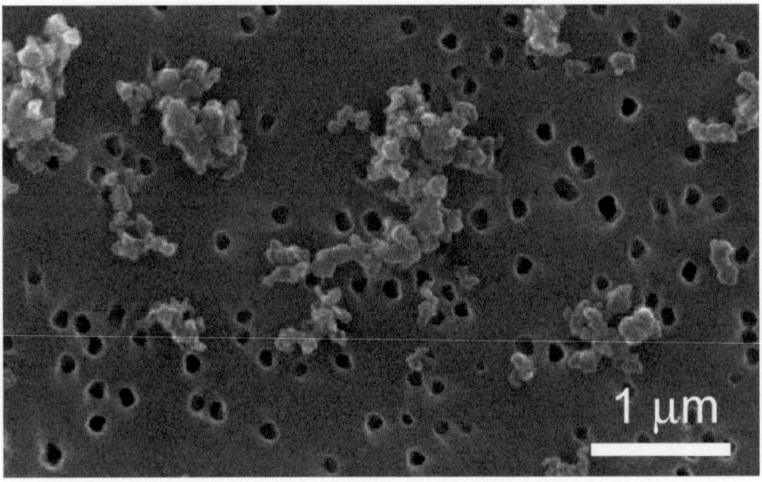

Fig. 6.13 The soot particles have been separated out Agardhfjellet Formation shales from Janusfjellet (Svalbard) and are dispersed on a filter paper with dispersed pores (*black*)

(Fernandes et al. 2003). The soot found in the Sindre Bed displays no microscopic features from higher plants, such as remnants of cell structures.

Dypvik et al. (2008b) considered the soot formation to have occurred just after impact due to shock heating of target material and later probably due to ejecta heating in the very hot impact plume. Most likely the soot originally had an initial distribution similar to that of other ejecta with similar aerophysical properties. The soot generation was first tied to an early phase of pyrolysis and succeeded by a combustion stage, but both the size and distribution of the soot particles found are consistent with a combustion origin (I. Gilmour, personal communication 2005). Winds and turbulence may to some extend make the distribution of this fluffy material less clear than theoretically expected.

At the time of impact the paleogeographic position of the impact site was hundreds of km from land and the closest forested area, making huge wildfires on surrounding land not very probable (Fig. 6.1). Spontaneous ignition of vegetation would also need higher temperatures and create soot with different appearance than the one observed in the studied samples. An extraterrestrial carbon source seems less likely, since the geochemical analysis indicate an iron-nickel rich impactor, which normally is very poor in carbon (Dypvik and Attrep 1999; Robin et al. 2001; Stuart Graham, personal communication 2006). The soot particles, therefore, most likely came from ignition (pyrolysis and combustion heating) of the organic rich, partly volatile, clays of the sea floor (Figs. 5.11, 5.12, and 6.15). Presently these claystones and shales make up the more than 100-m-thick Hekkingen Formation, which at the time and place of impact were able to feed the blazing fire with huge amounts of burnable oranic (kerogen) material.

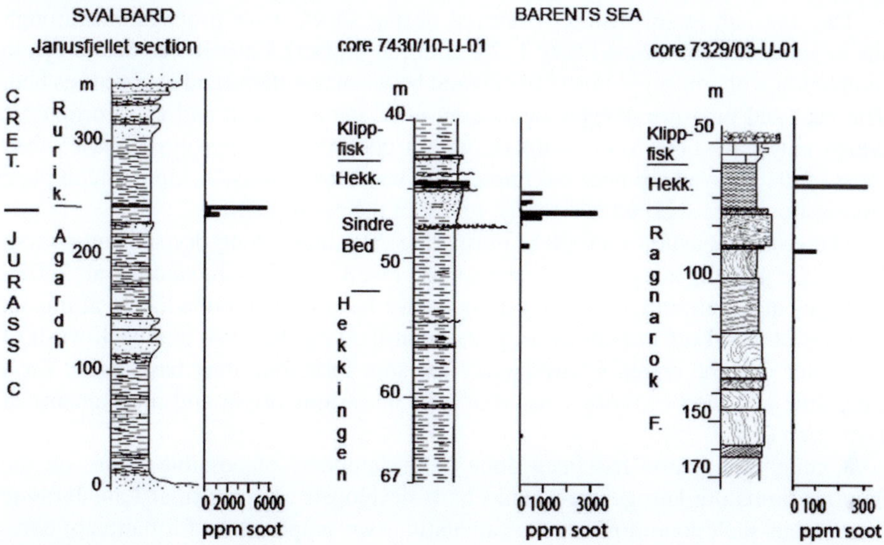

Fig. 6.14 The distribution of soot particles in the studied section at Janusfjellet and in cores 7430/10-U-01 and 7329/03-U-01

Fig. 6.15 An artistic expression of the Mjølnir fire. Painting (water-color) by Jon Reierstad

The first and major heating occurred during shock wave propagation through the target sediments (most likely 1–2 s after the impact). Later it was due to ejecta interaction with impact plume (fireball) and between the plume and exposed sea bed. The heat and pressure developments controlled the alteration and soot formation, which may have been very complex, with a possible early pyrolysis phase when most of the atmosphere near the impact site was blown away. A rapid, incomplete combustion stage succeeded after the return of the atmosphere.

The last combustion took place during the 20 minutes-long dry sea bed period, before the pre-impact sea level was restored (see Chap. 7) (Shuvalov et al. 2002). The fires in the air and along the seafloor must have started immediately at impact and were on and off in periods, in parts controlled by the wash and back-wash of sea water into the crater. Consequently the soot formation may have lasted for a long time and the beds may consist of soot of various origin and mechanisms of formation (Fig. 6.15).

A rough estimation has been done of the amounts of possible future oil the Mjølnir immature kerogen fire could have developed under normal Late Jurassic Norwegian-shelf conditions. In the calculations we supposed the formation of common Norwegian shelf oil with averagenC10-oil composition and 30 API (American Petroleum Institute) gravity. It is supposed, as an average, that about 4% of the kerogen is converted to oil which after migration is found as oil in-place (D.A. Karlsen, personal communication 2005). During the 20 min-long Mjølnir fire, kerogen pyrolysed and combusted were equivalent to source rocks capable of forming about 30 million standard m^3 oil in place (Dypvik et al. 2008b). This amount represents a peak year production from a giant Norwegian oilfield (e.g., Statfjord – one of the world's largest offshore fields). Today the tiny, fluffy grains of soot (Fig. 6.13) are the only traces left of this conflagration.

6.2 The Stratigraphical Distribution of the Ejecta Beds

The stratigraphical position of the ejecta beds provides key information on the age of the Mjølnir impact. Based on seismic correlation from the Mjølnir structure to borehole 7430/10-U-01 and biostratigraphical data from this corehole, Dypvik et al. (1996) suggested a general Volgian-Ryazanian age for the impact. Since then more accurate age-determinations of the impact have been provided from more detailed shallow seismic reflection profile correlations (see Sect. 3.2.2., Fig. 6.16) (Tsikalas et al. 2002a) and biostratigraphic analyses of the ejecta beds (i.e., the Sindre Bed) and underlying pre-impact and overlying post-impacts deposits in borehole 7430/10-U-01(Smelror et al. 2001a; Bremer et al. 2004; Smelror and Dypvik 2005).

In addition new biostratigraphic evidences have been obtained from the Mjølnir crater core 7329/03-U-01. As described in Sect. 6.1.2 possible traces of the Mjølnir impact have been found farther off the crater in the offshore Troms III area, on central Spitsbergen and on the Nordvik Peninsula in Western Siberia (Smelror et al.

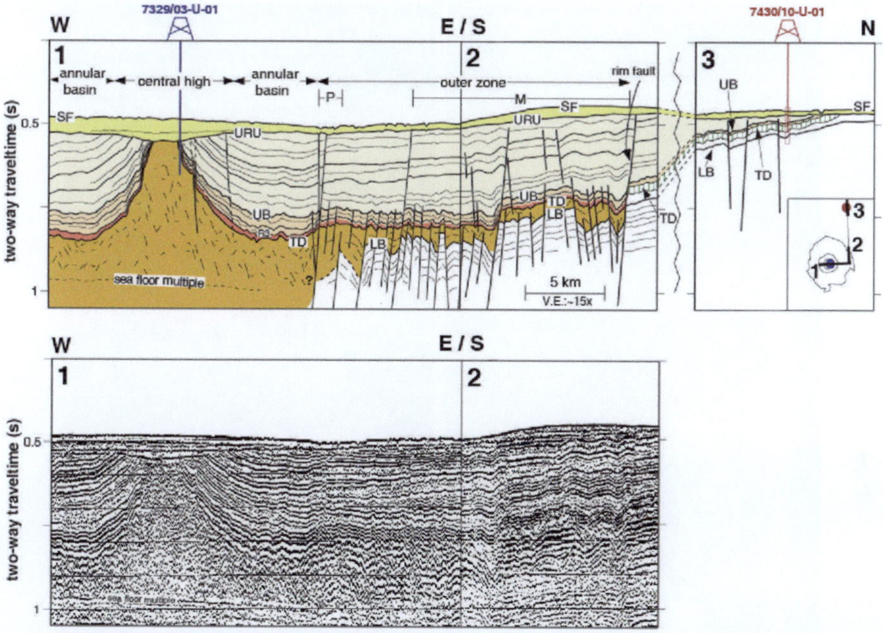

Fig. 6.16 Interpreted (*top*) and uninterpreted (*bottom*) examples of shallow multichannel seismic reflection profiles crossing the Mjølnir crater and providing the seismic reflection correlation between the two shallow boreholes. Reflectors UB (*upper boundary*) and LB (*lower boundary*) constrain the time of impact. SF, sea floor; URU, late Cenozoic upper regional unconformity; UB, Late Ryazanian-Early Barremian; TD (top disturbance, impact horizon), the first continuous reflector above the seismic disturbance; R3, top of gravity flows imposed by the Mjølnir impact; LB, base Upper Jurassic. *Vertical bars* (raster) denote the uncertainty in the seismic tie of the impact horizon. P, peak ring; M, marginal fault zone

2001b; Dypvik et al. 2006). In the following section the biostratigraphic data used for age-determination of the Mjølnir impact are presented in detail.

6.2.1 Borehole 7430/10-U-01

In borehole 7430/10-U-01(Figs. 3.12, 6.2, 6.3, and 6.17), both pre-impact, post-impact as well as ejecta-bearing strata assigned to the Sindre Bed are preserved (Dypvik et al. 1996, 2004b, c; Smelror et al. 2001a; Tsikalas et al. 2002a). In this borehole the base of the Sindre Bed is recognized at level 52.0 m. Below this depth the borehole comprises dark shale assigned to the Hekkingen Formation. The record of the bivalve *Buchia mosquensis* at 57.48 m indicate an age not younger that Middle Volgian at this stratigraphic level. Age-diagnostic Volgian foraminifera are recovered from above 56.7 m, while Middle or younger Volgian ammonites related to *Craspeditidae* are found at 53.7 m (Smelror et al. 2001a).

Fig. 6.17 The figure shows two sections of fine, laminated, *dark grey* shales from the Hekkingen Formation in core 7018/05-U-01 from the Barents Sea

Bivalves of the genus *Buchia* are also found between 50–46.5 m in the Sindre Bed (Århus 1991; Smelror et al. 2001a, b). The presence of *Buchia unschensis* between 49.95 and 46.45 m is of importance, because this species also is found in the oldest post-impact deposits of the Mjølnir crater core (Sect. 6.2.2). In borehole 7430/10-U-01 impacts indicators such as shocked quartz and an Ir-anomaly, have been found in the interval 47.65–46.85 m in the Sindre Bed (Fig. 6.3) (Dypvik et al. 1996). The uppermost geochemical ejecta signal is found at 46.5 m, and, consequently, this defines the top of the Sindre Bed. The recovery of *Buchia* cf. *volgensis* at 46.9 and 46.45 m is of importance since this species is previously reported as an indubitable Ryazanian species (Århus et al. 1990; Århus 1991). Other bivalves referable to *Buchia unschensis* and *Buchia terebratuloides* are also present in the interval from 51.88 to 46.45 m. A foraminifera assemblage with *Recurvoides obskiensis* is found at 48.45 m. This species indicates a general Late Volgian to Ryazanian age (Nagy and Basov 1998).

In the Hekkingen Formation above the Sindre Bed at 45.5 m an assemblage with the foraminifera *Recurvoides obskiensis*, *Gaudryina gerkei* and *Gaudryina rostellata* is found. The assemblage indicated a Ryazanian age, but not older than the uppermost *Chetaites sibericus* Zone. An ammonite, i.e., *Borealites* sp., which is attributed to the Early Ryazanian *Hecteroceras kochi* Zone, is found at 44.1 m. At 42.65 m a typical Late Ryazanian marine microflora, with the dinoflagellate cysts *Systematophora palmula* and *Gochteodinia villosa* as age-diagnostic species, is found (Smelror et al. 1998; Århus 1991; Smelror et al. 2001a, b).

In conclusion, the biostratigraphic evidence from the Sindre Bed, and under- and overlying strata, in boreholes 7430/10-U-01 suggest a stratigraphical age for the Mjølnir impact approximating the Volgian-Ryazanian boundary.

6.2.2 Borehole 7018/05-U-01

Borehole 7018/05-U-01 was drilled on the margin of the Harstad Basin in the Troms III area. A brief lithological description of the Upper Oxfordian to Hauterivian succession cored in this borehole is published in Smelror et al. (2002) (Fig. 6.7). The Middle Volgian to Ryazanian deposits of the Hekkingen Formation (Krill Member) consists of dark to very dark grey claystone, which are mostly finely laminated, with abundant carbonate beds (Fig. 6.17). Bioturbation is generally absent, except for some horizons close to the Volgian-Ryazanian boundary at level 88 m. Ammonites and bivalves are found at some levels, and a few coalified fragments are also present.

The identification of ejecta deposits in this borehole is not straight forward. The uppermost Volgian to lowermost Ryazanian part of the core is made up of parallel laminated shales, only with faint traces of possible bioturbation and with no clear-cut sedimentological structures. However, a distinct peak of the algae *Leiosphaeridia* is found from 89.3 to 87.1 m, with a pronounced peak at 88 m (Smelror and Dypvik 2005; Dypvik et al. 2006). At this stratigraphic level there are also increased values in increased Th/U ratio, and the Ni and Cr distributions display dramatic excursions from the decreasing trends below and above (Fig. 6.7). The Ni and Cr concentrations are here, however; significantly lower than those found in the ejecta strata (Sindre Bed) in borehole 7430/10-U-01 and in the oldest post-impact sediments in the Mjølnir crater core.

The age of the ejecta-bearing sediments in 7018/05-U-01 is constrained by the recovery of the Middle Volgian ammonite *Laugeites* cf. *groenlandicus* at 102.11 m and the Early Ryazanian ammonite *Surites* sp. found at 85.86 m (Smelror et al. 2001a, b). Additional age determinations of the ejecta-unit can be inferred by the occurrence of the dinoflagellate *Heslertonia? pellucida* at 90.1 m which is known to have its last appearance in the Early Ryazanian. The occurrence of *Cribroperidinium globatum* through the ejecta-unit is additional evidence for an age not younger than the Early Ryazanian *Hectorocera kochi* Zone.

6.2.3 Janusfjellet, Central Spitsbergen

Upper Jurassic and Lower Cretaceous deposits of the Agardhfjellet Formation are exposed several places on Central and Eastern Spitsbergen. The outcrops at Janusfjellet are dominated by shales and siltstones, with minor carbonate beds and concretions (Figs. 6.6 and 6.9). In search for ejecta from the Mjølnir impact an interval with increased Th/U-ratios and Ni-concentrations has been found in the upper part of the Agardhfjellet Formation, at an interval from around 33–34 m on the lithostratigraphic log presented by Dypvik et al. (2006) (Fig. 6.10). This interval also displays a comparable peak in the abundance of *Leiosphaeridia* as is found in and near the Mjølnir crater and in the Troms III area. As described in Sect. 6.1.2 one single sample from this interval contained elevated Ir-abundance, as was confirmed by repeated analyses. In combination this provided reasonable evidence for the presence of ejecta-bearing deposits attributable to the Sindre Bed on the Barents Shelf in the upper part of the Agardhfjellet Formation on Svalbard.

6.2.4 Nordvik Peninsula, North-Western Siberia

In addition to the Iridium anomalies described from the Sindre Bed in borehole 7430/10-U-01 and from the contemporaneous strata on Svalbard, possible iridium peaks, which can be related to the Mjølnir impact have been described in the Jurassic-Cretaceous boundary beds on the Nordvik Peninsula in North-Western Siberia (Zakharov et al. 1993). Several later analyses have, however, not been able to confirm these enrichments (Dypvik and Zakharov, in press; Koeberl, personal communication).

6.2.5 The Mjølnir Ejecta as a Regional Stratigraphic Marker

The regional relationship and stratigraphic correlation of the Mjølnir ejecta (i.e., the Sindre Bed and its correlative units), make the ejecta an important aid for dating and correlation of the Volgian-Ryazanian boundary beds in the Arctic. This is of particular value, because both the Arctic Jurassic-Cretaceous correlations and the plate tectonic relations are rather complex.

The Ir-anomaly at the base of the *Chetaites sibiricus* ammonite zone falls between M17r and M18r on the magnetostratigraphy, which corresponds to an unknown level within the Tethyan Jacobi Zone (Zakharov et al. 2007). A rough correlation to the recent chronostratigraphic timescale of Gradstein et al. (2004) gives a rough age of 142 +/– 2.6 Ma for the Mjølnir impact (Smelror et al. 2001a).

Chapter 7
The Impact Dynamics

Valery Shuvalov, Henning Dypvik, and Filippos Tsikalas

7.1 Introduction

Together with geological/geophysical studies and laboratory-scale experiments numerical simulations of impacts contribute a great deal to our knowledge of the cratering process. Whereas field studies give information about target conditions and final crater configuration, numerical modeling allow us to follow the evolving process in time and to reconstruct several important features not surviving after the cessation of the impact event and later time.

Early numerical simulations of marine target impacts were conducted for very large (~10 km) projectiles, which are interesting from the viewpoint of global catastrophes including terrestrial life extinctions (O'Keefe and Ahrens 1982; Roddy et al. 1987). More detailed recent numerical simulations have made it possible to determine the critical sea depth at which a crater is formed and/or shock modified material can be found to identify and confirm the impact structure (Artemieva and Shuvalov 2002). Recent classification and confirmation of about 25 submarine craters (Ormö and Lindström 2000; Dypvik and Jansa 2003) have provided a renewed impulse to integrate the geological/geophysical studies of these structures with detailed numerical modeling analyses. Numerical simulations were constructed to model several craters, including Lockne (Ormö et al. 2002), Eltanin (Shuvalov 2003b), Mjølnir (Shuvalov et al. 2002), Chesapeake Bay (Crawford and Barnouin-Jha 2004), Silverpit (Collins et al. 2003), and to model also the generation (Weiss et al. 2003; Shuvalov 2003a) and propagation (Ward and Asphaug 2003; Glimsdal et al. 2007) of impact induced tsunami waves.

Impacts into sea and ocean are typical examples of impacts into a laminated (layered) target with variable rheological characteristics. At least three target layers can be distinguished: the low density water, the low strength sediments and a high-strength crystalline basement. The first qualitative description of impacts into the laminated target was given by experiments of Quaide and Oberbeck (1968), where

V. Shuvalov (✉)
Institute of the Dynamics of Geospheres, Russian Academy of Sciences, Moscow, Russia
e-mail: shuvalov@idg.chph.ra.ru

H. Dypvik et al. (eds.), *The Mjølnir Impact Event and its Consequences*, Impact Studies, DOI 10.1007/978-3-540-88260-2_7,

a layer of relatively high-strength material was covered by strengthless sand. Impact structures with different morphologies formed, depending on the impactor diameter to sand-depth-ratio.

Both numerical and experimental studies show that the process and results of marine target impact are mainly determined by a ratio of water depth *h* to projectile size (diameter) *d*. No underwater crater is formed if $d/h < 0.1$ to 0.2 (Gault and Sonett 1982; Artemieva and Shuvalov 2002). Eltanin (Gersonde et al. 1997; Kyte et al. 2000) is the only presently known deep ocean impact event, in which no impact structure has been preserved. If $d/h > 1$, there is only modest direct influence of the water column on the cratering process (Shuvalov et al. 2002). However, even in those cases the structure and morphology of the resulting submarine crater may differ considerably from the counterpart on land. This is primarily due to the different rheological properties of the target material, the immediate post-impact effects of water on the final crater morphology and ejecta deposition through resurge flow and tsunami waves (Ormö and Lindström 2000; Dypvik and Jansa 2003). Mjølnir is a typical impact structure of this kind.

In the intermediate case, when $0.2 < d/h < 1$, a well defined submarine crater is formed, but its morphology and the cratering process itself are strongly controlled by the water depth. The Lockne structure is an example of this type of impacts (Lindström et al. 2005). In this chapter we describe new numerical modeling results of the Mjølnir impact, building on the foundation set by the analyses of Shuvalov et al. (2002) and Shuvalov and Dypvik (2004).

7.2 Numerical Model

The SOVA multi-material multi-dimensional (2D and 3D versions) hydrocode (Shuvalov 1999) is used to model the main stages of the impact process, including projectile penetration through water and solid target, excavation and ejecta expansion, crater modification, and tsunami generation. SOVA is an Eulerian response code with some Lagrangian features. It allows modelling of strong hydrodynamics flows with accurate description of the boundaries between different materials (e.g. solid rock, atmospheric air, vapor, and water). The code is similar in conception to the widely used CTH hydrocode (McGlaun et al. 1990). The tracer particle technique is used to follow a motion of fixed Lagrangian particles and their pressure/temperature/velocity history.

For the description of thermodynamic properties of the projectile a tabular equation of state for granite (Pierazzo et al. 1997) is utilized obtained with the ANEOS code (Thompson and Lauson 1972). Furthermore, the Tillotson analytical formulas (Tillotson 1962; Melosh 1989) are used to describe thermodynamics of water and solid target, and table readings (Kuznetsov 1965) were used to calculate air pressure.

The approach to model material strength developed by Melosh and Ivanov (1999) and O'Keefe and Ahrens (1999) is also used. The approach is based on the "rigid-plastic" model (Dienes and Walsh 1970). For fractured rocks (i.e., loose materials with finite cohesion), the yield strength was defined as (Lundborg 1968; Zamyshlyaev and Evterev 1990):

$$Y = \min(Y_0 + kP,\ Y_{max}) \tag{7.1}$$

where Y_0 is the cohesion, k is the coefficient of dry friction, P is the pressure, and Y_{max} is the limiting yield strength of the material at high pressure. Acoustic fluidization (Melosh 1989) was also taken into account using the method described by Ivanov and Turtle (2001).

7.3 Cratering Process

The results of the initial numerical simulations of the Mjølnir event as a vertical asteroidal impact were presented by Shuvalov et al. (2002). These simulations showed that the observed impact structure could have been produced after a vertical impact of a 800-m-radius asteroid (mass $M = 5.630^{12}$ kg, volume $V = 2.1$ km^3, energy $E = 270$ Gtons of TNT equivalent) at velocity U=20 km/s into a 400-m-deep sea. The best fit to observational data was obtained for a constant cohesion, $Y_0 = 10^6$ Pa, and a composite depth-friction dependence where z is the target depth:

$$\begin{aligned} &k = 0.025 - 0.253\ z \text{ if } z > -3 \text{ km} \\ &k = 0.2 - (\mathrm{z} + 3)\, 30.1 \ \text{ if} - 3 \text{ km} > z > -6.5 \text{ km} \\ &k = 0.8 \ \text{ if } z < -6.5 \text{ km} \end{aligned} \tag{7.2}$$

The results also showed that the water layer itself had minimal influence on the cratering process, i.e., projectile penetration, excavation, and modification. In detail, the numerical simulations show that 10 s after the impact the transient cavity reached its maximum depth of about 6 km, attaining a hemisphere shape. The simulation shows that about 30 s after the impact the radius of the transient crater increased due to excavation and reached 10–11 km. Approximately at this time (30 s) collapse of the crater commences.

The asymmetry of the Mjølnir structure (Tsikalas et al. 1998a–c; Tsikalas 2005) substantiates an oblique impact angle. Therefore, further numerical simulations were performed for Mjølnir as an oblique, 45°–30° from horizontal, impact (Shuvalov and Dypvik 2004). The projectile radius must be increased to 1 km (mass $M = 1.11 \times 10^{13}$ kg, volume $V = 4.1$ km^3, energy $E = 530$ Gtons of TNT equivalent) in order to obtain a crater approximately of the same size. The initial cratering stages, for a 45° oblique Mjølnir impact are shown in Fig. 7.1. These results are comparable to those presented by Shuvalov and Dypvik (2004), but were obtained in later, more advanced simulations with higher spatial resolution (301 × 151 × 181 cells instead of 181 × 91 × 121 cells in x, y, and z directions, respectively). The projectile penetrates through the water layer without any deceleration and deformation, strikes the solid target and generates strong shock waves in both water and soil. The passage of the shock wave results in extensive in situ fracturing, partial vaporization,

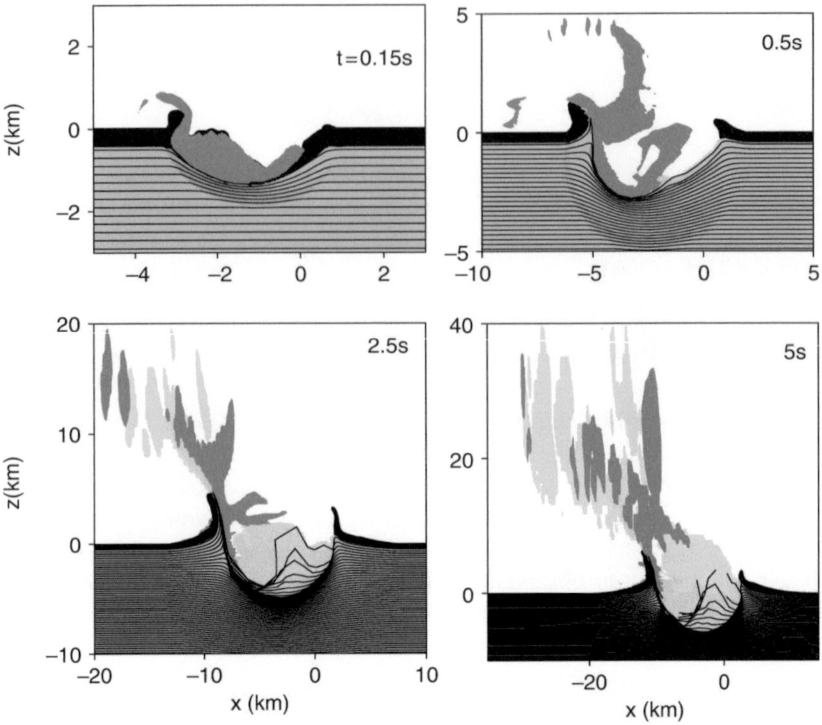

Fig. 7.1 Initial stages of cratering flow after a 45° oblique impact into a 400 m deep sea (in the impact plane). *Black shading* denotes water, *dark gray* projectile material, and *light gray* (the darker the denser) solid target material. Impact direction is from *right* to *left*

melting and brecciation of target rocks along with vaporization of water. Shock compression and succeeding decompression generate a cratering flow resulting in a growing transient cavity. The projectile is decelerated towards the left (downrange) wall of the cavity and gradually moves upwards. Five seconds after impact most of the projectile material is ejected from the crater.

In the first phase the growing transient crater is strongly asymmetrical (Fig. 7.2) with considerably elongated shape. However, already 5 s after the impact the cavity becomes near symmetrical although its center is offset downrange for a distance of about 3 km from the point of initial contact (Figs. 7.1 and 7.2). Note a strong asymmetry of the initial ejecta curtain (e.g., fast ejecta) (Figs. 7.1 and 7.2).

It takes the shock wave about 3 s to expand for a distance exceeding the size of the final crater, i.e., ~40 km. All major shock effects, such as melting and shock metamorphism, occur during these early stages. Figure 7.3 illustrates the shock wave propagation through the target, and shows how rocks and water are compressed during the impact. The shock wave is strongly asymmetrical near the impact point. In particular, the melting zone is asymmetrical and is shifted downrange for approximately 2–3 km from the point of initial contact. At large distances from the impact point, the shock wave develops towards more symmetrical appearance. The incipient

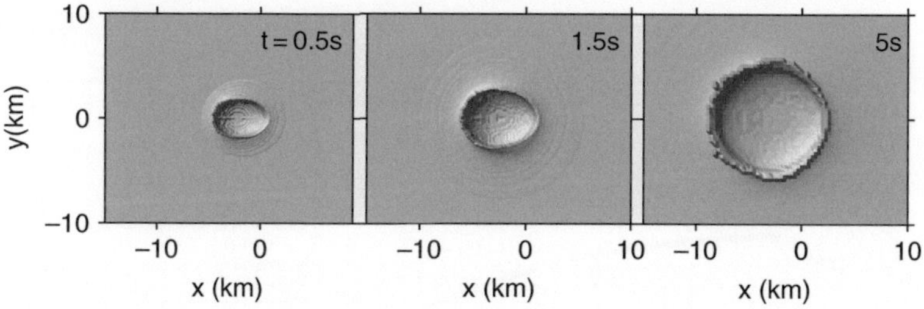

Fig. 7.2 The growing crater geometry as seen directly from above. An iso-surface with bulk density 0.5 g/cm^3 is shown. Within the crater the surface consits of condensed target material, while outside the crater the surface is formed by the sea water. The ejecta curtain consists of both water, impactor and target rocks

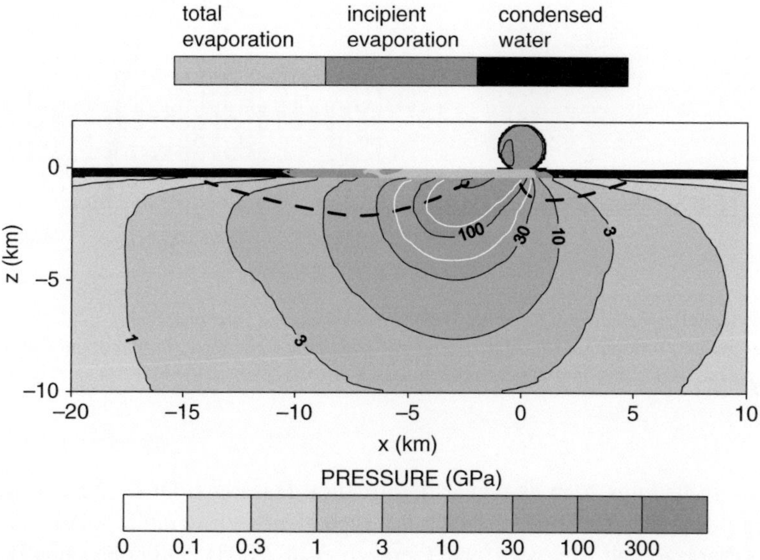

Fig. 7.3 Shock compression of target and projectile material in relation to the initial pre-impact position (at vertical plain, y = 0 passing through impactor trajectory). The intensity of *gray shadings* shows pressure in solid target and projectile (the darker the more compressed), while in the water layer different shadings show phase transformation during impact. The *dashed black line* denotes excavation crater (initial position of ejected rocks). *Thin white lines* in the solid target outlines zones of melting (shocked above 50 GPa) and vaporized (shocked above 140 GPa) material

vaporization of sea water occurs in an area about 10 km wide, while within a 6-km-diameter area the sea water is totally evaporated. The area of vaporization is also shifted downrange for approximately 2–4 km.

Figure 7.4 shows the calculated distribution of target rocks and projectile material with respect to maximum shock pressure. About 70 V ($V = 287$ km^3 is a projectile volume) of the target rocks experience shock compression above 4 GPa, which illustrates that about this amount of target material suffered shock metamorphism

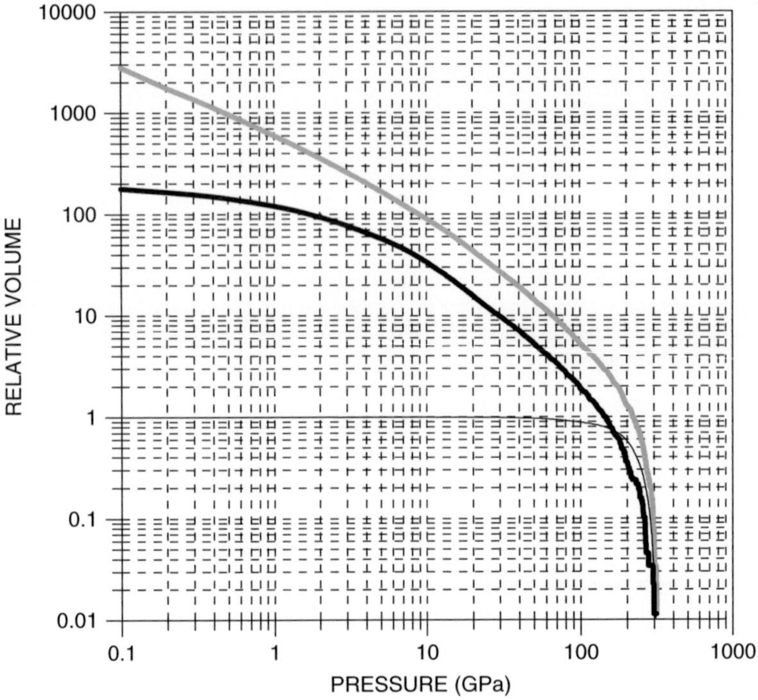

Fig. 7.4 Cumulative distribution of variously shocked ejecta (*thick black line*), total (both ejected and non-ejected) solid target (*thick gray line*) and projectile (*thin black line*) material that is a ratio of the volume of material that experiences pressure above a given value to the projectile volume

resulting in features such as planar deformation features (PDFs). We consider the melting pressure to be 50 GPa, which is a typical value for granite, calcite, and wet basalt (Pierazzo and Melosh 1999; Pierazzo et al. 2005), and consequently about 5 V (i.e., 20 km^3) of target rock is melted (Fig. 7.4). Most of the projectile material experienced shock compression above 46 GPa and even above 140 GPa, which are the melting and vaporization pressures for granite, respectively (Pierazzo and Melosh 1999).

Figure 7.5 shows a sequence of time steps illustrating the late (excavation and modification) stages of a 45° oblique Mjølnir impact. Ten seconds after the impact, the crater reaches its maximum depth of about 5 km (approximately 6 km for a vertical impact case, and 4 km for a 30° oblique impact). At this time most of the projectile material has been ejected from the cavity. The central high is visible from approximately 50 s after impact and is strongly offset in the downrange direction. The peak of the central high moves downrange and by about 70 s after impact it becomes near symmetrical. The central high reaches its maximum height (approximately 1 km) at 90 s, then it slightly descends, and its top portion (highly shocked material lifted from deep layers) spreads along the crater floor. In the final

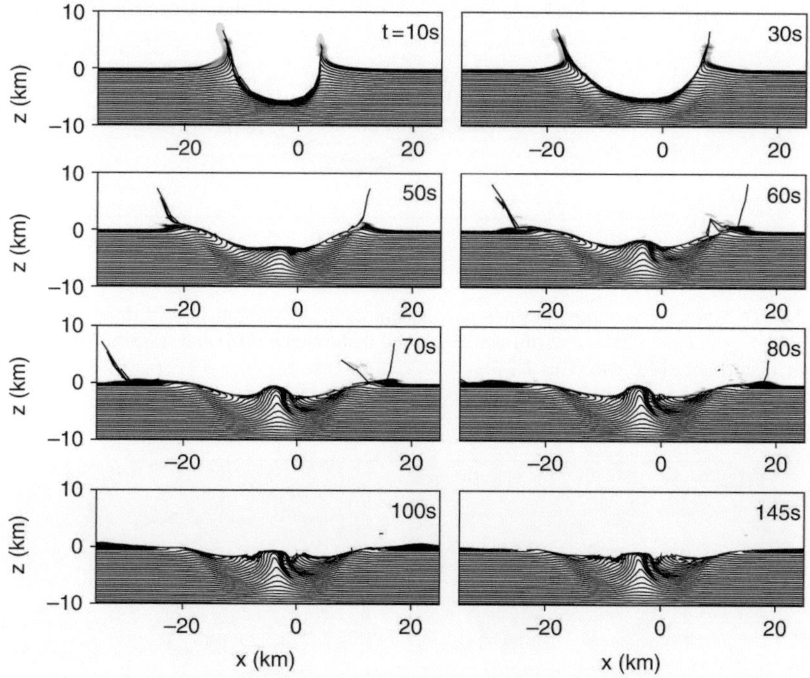

Fig. 7.5 Evolution of the transient crater of a 45° oblique Mjølnir impact. Time steps 10–150 s are shown. *Black lines* show impact induced displacement of the original horizontal platform layers. *Black shading* marks sea water. Note that central peak position is skewed 5 km downrange

crater the rocks below a depth of 5.5-km are only slightly disturbed, while target material from 4.5 km depth rises to the surface in the process of central high growth.

The evolution of the transient cavity for the 45° oblique Mjølnir impact is shown in Fig. 7.6. Envelopes of cavity profiles correlate well with the boundaries of intense and weak seismic disturbance regions, decompacted in order to be representative for the time of impact, utilizing established porosity-depth functions for the region (Tsikalas 1992; Tsikalas et al. 1998b). A point of the rising central high is initially offset in the downrange direction, where the crater reaches its maximum depth. The downrange displacement of the central high occurs due to: (1) a downrange motion and displacement of central high material; and (2) an increased late rise of the crater floor in the downrange part of the crater. Then, the downrange motion practically ends, with the central high finally even moving slightly up range. These quasi-oscillations can result in an almost random final position of the central high (Ekholm and Melosh 2001). In a bird's-eye view, the central high is very close to the geometrical center compared to the one for a circular crater (Fig. 7.7).

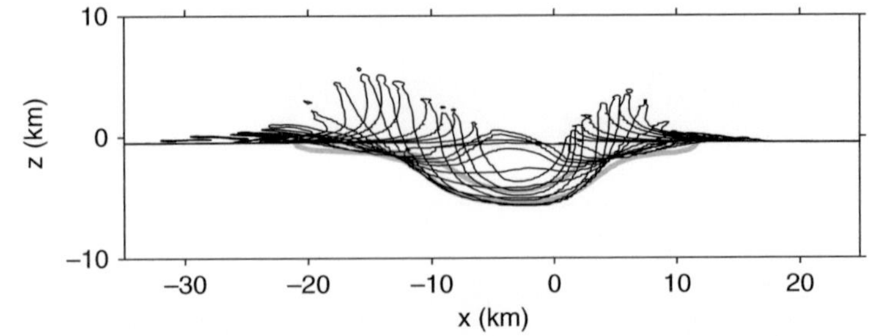

Fig. 7.6 *Thin black lines* show profiles of transient cavity development at different moments of time. *Thick gray lines* show zone of intense seismic disturbance (DZ) and transitional area of less seismic disturbance (TZ) (see Chap. 3)

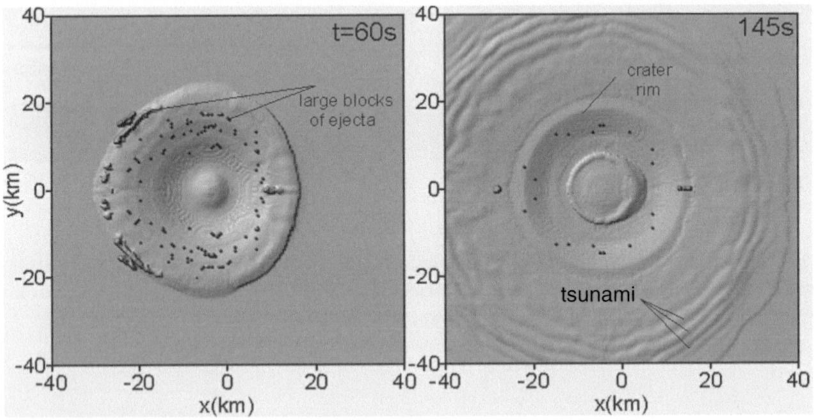

Fig. 7.7 Tsunami waves and the modified crater configuration as seen directly from above. An iso-surface with 0.5 g/cm^3 bulk density is shown. The iso-surface represents a surface of condensed target material within the crater, while outside the crater it denotes the sea-water surface

7.4 Ejecta Formation and Distribution

To describe the Mjølnir ejecta, the influence of water depth and impact angle on the ejecta formation and expansion needs to be accurately considered. In this case, we used a simplified model of ejecta expansion (Shuvalov and Dypvik 2004). At the first step we run 2D or 3D version of the SOVA code to model an initial stage of the impact and to obtain initial ejecta parameters, i.e., mass, angle, and velocity distributions. At the second step, we considered ejecta motion in the atmosphere using a simple ballistic approximation. This approach is not very exact because both laboratory experiments (Schultz and Gault 1992) and numerical simulations (Shuvalov 2002a, b) show that atmospheric drag and interaction with impact-produced vapor significantly modify the ballistic paths of ejecta, which are smaller than some critical ejecta fragment size. Consequently, this modification leads to separation of ejecta fragments by size within the ejecta cone. However, most parts of the ejecta curtain move approximately ballistically, because its bulk density considerably exceeds the ambient air density. Moreover, most of the distal ejecta trajectory paths occur at high altitudes, where atmospheric drag is small. Given these factors the ballistic approximation allows us to describe the ejecta curtain evolution qualitatively and to estimate its distribution on the Earth's surface quantitatively.

It should be noted that the ballistic approximation is rough for both the initial and final portions of the ejecta. The initial (and fastest) ejecta mainly consisting of vapor, has very low bulk density, and, therefore, experiences great atmospheric drag even at high altitudes. In contrast, the final (low velocity) ejecta forming the crater rim has a great bulk density. In that case, the ejecta fragments strongly interact by collision with each other and their motion can not be described as ballistic flight of separated particles. Test simulations (Shuvalov and Dypvik 2004) show that this approach for subaerial craters gives distributions very close to experimental data, as summarized by McGetchin et al. (1973) and Melosh (1989).

The total ejecta volume is about 200 V (820 km^3) as depicted in Fig. 7.3 by the initial asymmetric excavation crater, and thus asymmetrical position of ejected rocks, for the 45° oblique Mjølnir impact case. Furthermore, Fig. 7.4 shows the calculated distribution of ejected rocks with respect to the maximum shock pressure. About 65% of the ejected rocks experience shock compression below 4 GPa, which is the lowest limit of shock pressures in which PDFs form (Stöffler and Langenhorst 1994). Only about 15% of the ejected rocks are compressed above 10 GPa. The volume of ejected impact melt (i.e., shocked above 50 GPa level) is about 20 km^3 (2.5% of the ejected rocks and about 30% of the total impact produced melt). The rest of impact melts cover the transient cavity surface, which is oscillating up and down in the course of crater collapse.

Although the 400 m deep sea only slightly influenced the Mjølnir cratering process (Figs. 7.1–7.7), it strongly influenced the ejecta velocities. The fastest and, consequently, the farthest reaching ejecta was ejected from the uppermost target layers (Shuvalov and Dypvik 2004). In marine impacts, the upper target layer consists of water. Hence, the water ejecta has the highest velocities, and solid ejecta being ejected from deeper target layer is characterized by lower velocities and consequently travel shorter than in the case of similar subaerial/onland impacts.

Figure 7.8 shows how the size of the solid ejecta blanket depends on water depth for a vertical impact of the Mjølnir projectile (Shuvalov and Dypvik 2004) in simulations with high spatial resolution ($h_{r^{\max}} = 50$ m, $h_{z^{\max}} = 50$ m, where $h_{r^{\max}}$ and $h_{z^{\max}}$ are cell sizes at the end of the SOVA calculations). As expected, the area of ejecta deposition strongly decreases as water depth increases. The water depth, however, does not influence the deposition of close ejecta, since for all modeled depths the curves coinside at a distance below approximately 100 km. The shallower the depth of water, the more extensive the region of coincidence. Note that in the case of deeper water, when water depth is comparable to projectile diameter, the expansion of solid ejecta is restricted by the walls of the water transient cavity. No distal ejecta, defined as exceeding distances of 1.5–2 crater radii, occur at all (Lindström et al. 2005).

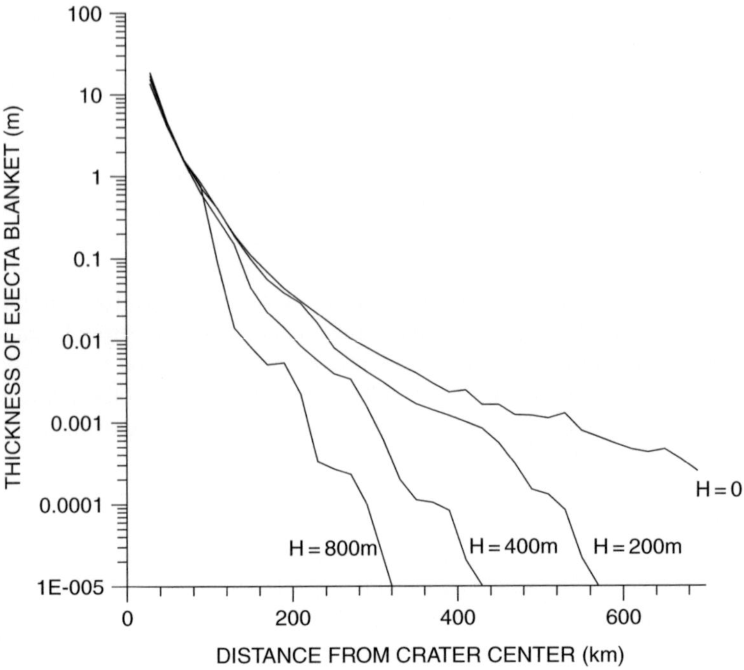

Fig. 7.8 Thickness of solid ejecta blanket versus distance from crater center obtained in calculations with different water depth *H*. Values for a vertical Mjølnir impact with a projectile diameter of 1.6 km

Some cases of marine-target impacts, in contrast to subaerial ones, may produce solid ejecta with a higher maximum velocity. Such increase in ejecta velocity can be due to the possible difference in composition of various solid targets. Volatile-rich sediments that are typical for sea-covered targets may provide a more extensive expansion of shock compressed material than in the drier, subaerial successions (O'Keefe et al. 2001b). This effect, however, dominates only at very shallow water, where the water depth is much less (10–100 times) than the projectile size, i.e., as in the case of Chicxulub Crater (O'Keefe et al. 2001b).

Although most craters, even those resulting from oblique impacts, are nearly circular, the impact angle may strongly influence the ejecta deposition (Pierazzo and Melosh 2000, and references therein). This influence for the case of subaerial impacts was first demonstrated in experiments by Gault and Wedekind (1978). They found that the ejecta deposits remained near circular for impact angles down to 45°, while deposits were slightly offset downrange. As the impact angle decreases below 45°, ejecta deposits became strongly asymmetrical, and the so-called "forbidden" azimuthal zones first appear uprange and then downrange of the evolving crater. Recent experiments by Schultz (1999) show that high-velocity ejecta moves preferably in the downrange direction, whereas the low-velocity ejecta is distributed more evenly around the crater. All these effects have been observed on planetary surfaces that are shaped by craters with features characteristic of oblique impacts (Pierazzo and Melosh 2000). Since the work of Gault and Wedekind (1978), "oblique shapes" of the ejecta deposits have been considered as the most prominent indicator of oblique impacts.

To calculate ejecta deposition caused by an oblique impact, a 3D version of the model was used. First, we carried out numerical simulations of vertical and 45° oblique impacts without water, to test the performance of the code and to compare the obtained results with observational and experimental data. The same 800-m-radius spherical projectile striking the surface at 20 km/s was considered, using a more crude computational grid, i.e., $h_{r^{\max}} = 100$ m, $h_{z^{\max}} = 100$ m.

The distribution of ejecta deposits for the vertical Mjølnir impact is in a good agreement with results obtained in the 2D simulations (Fig. 7.9), but strongly differs from the distribution for the 45° oblique impact. Primarily, the ejecta blanket area strongly increases as impact angle decreases from vertical down to 45°. This increase of ejecta blanket is the result of a significant increase in the downrange ejecta velocity, as first predicted in numerical simulations by O'Keefe and Ahrens (1986). The central part of the ejecta deposits at a distance of a few crater radii looks very similar to experimental data (Gault and Wedekind 1978). "Forbidden" azimuthal zones do not appear either uprange or downrange, but a wedge-shaped region of thin ejecta deposits can be clearly seen downrange at distances above 1,000 km (i.e., 25 crater radii) (Fig. 7.9c). The decrease in impact angle increases the area of ejecta deposition, making it strongly asymmetrical in the more distal locations (Fig. 7.9a, c).

The ejecta deposits of impacts into 400 m of water at different impact angles (Fig. 7.9) substantiate both tendencies discussed above: a decrease of ejecta depositional area in the presence of water layer and an increase of ejecta deposit area with a decrease of impact angle (Figs. 7.9b–e). In the vertical impact case of a Mjølnir-like projectile into 400 m of water, the extent of deposits is restricted to within approximately 600 km of the crater. A decrease of impact angle leads to an increase (only in downrange direction) of this extent up to 3,000–4,000 km, exceeding the modeled ejecta extent after the vertical impact even without water, but being reduced compared to the case with oblique impact without water (Fig. 7.9c). The area of ejecta deposits becomes even more anisotropic than without water. The exact structure of this downrange zone of deposits probably depends on both water depth and projectile shape. These particular distributions are indicative of where the search for ejecta deposits must be concentrated, namely in the downrange direction and in the azimuthal zone subtending 60° at a distance less than 3,000–4,000 km (Fig. 7.9). Note that the spherical shape of the Earth was intentionally neglected, as it does not affect considerably the calculations at distances less than Earth's radius 6,300 km.

The projectile ejecta of a 45° oblique impact into the sea (Fig. 7.9f) have a larger velocity and are even more asymmetrical (move in a more narrow angle) than the target ejecta. Due to high velocity, the projectile ejecta deposits have a local maximum reaching as far as 4,000 km, a distance where almost no target ejecta arrive.

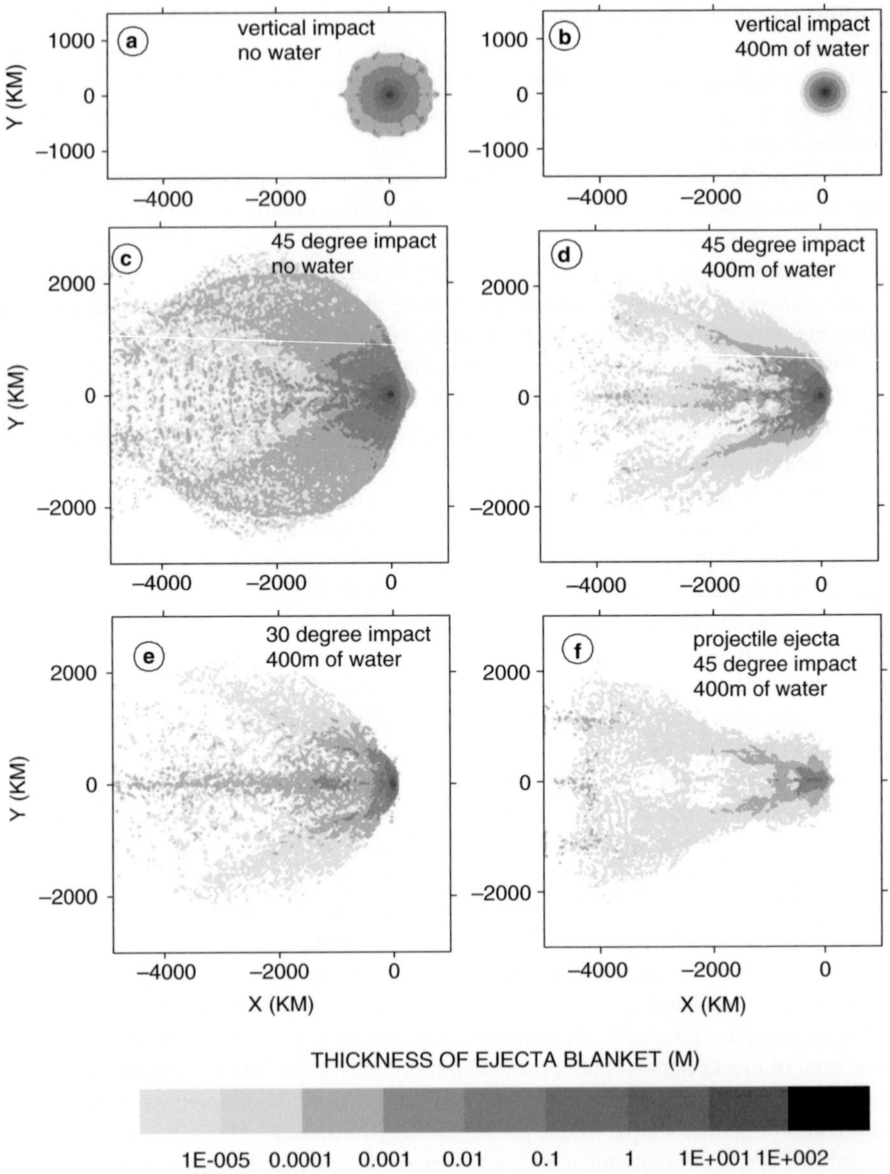

Fig. 7.9 Distribution of the Mjølnir ejecta (from target rock) deposits after a vertical impact without water (**a**); after a vertical impact into 400-m-deep sea (**b**); after a 45° oblique impact without water (**c**); and after 45° (**d**) and 30° (**e**) oblique impacts into 400-m-deep sea, respectively. Plate (**f**) shows the distribution of projectile material after the 45° oblique impact into a 400-m-deep sea. Impact direction is from *right* to *left*

7.5 Resurge Flow and Tsunami Generation

A major consequence of marine target impacts is the generation of tsunamis (Chap. 10). These waves are considered as one of the main hazard-risk factors in the case of asteroid collision with the Earth (Adushkin and Nemchinov 1994). Numerical simulations of tsunamis induced by impacts require much more computer time than the numerical modeling of the crater itself, as the tsunamis form later in the cratering process sequence and expand over a long period of time. Therefore, most results concerning tsunamis and resurge flow were obtained for a vertical impact (using a 2D model), and only initial stages were considered for the oblique impact case.

A mechanism of the Mjølnir tsunami generation is shown in Fig. 7.1 and in Fig. 7.5. The growing crater rim and ejecta curtain pushed the sea water out and upwards, thereby forming a water surge. The water surge grew and then broke up forming several waves. An interaction between these waves resulted in generation of the first leading tsunami. For a vertical impact, a wave ∼200 m in height was formed 300 s after the impact at a distance of about 60 km from the crater center. This amplitude is considerably higher than what was estimated by Tsikalas et al. (1998c), using the relationships valid for shallow water depth (Glasstone and Dolan 1977). These relationships, in turn, are based on the result of the "BAKER test" at Bikini Islands, where nuclear energy of about 20 kilotons was released in a 60 m deep lagoon (Glasstone and Dolan 1977).

For the energy of 250–500 Gtons released in the Mjølnir event the equivalent water depths, resulting from a gas-dynamic similarity, should exceed the water depth in the "BAKER test" by a factor of ∼200 and, therefore, should be ∼12 km. From the impact viewpoint, the 12-km-depth water should be considered as a deep water case for the Mjølnir-size projectile. In such deep-water impact, similar to the Eltanin impact (Gersonde et al. 1997), no crater would have been formed at the seafloor (Artemieva and Shuvalov 2002), and the tsunami should have resulted from the development and collapse of the transient water cavity. In particular, numerically obtained tsunami amplitudes for the Eltanin impact (Shuvalov 2003b) correlate well with the relations of Glasstone and Dolan (1977). In the Mjølnir case, however, there is a different mechanism of tsunami generation and the waves are formed due to motions of solid material due to rim expansion and structural uplift (Chap. 10). In some sense, this mechanism is comparable to earthquake induced tsunamis where water also begins to move due to the motion of solid surfaces.

In conclusion, the higher tsunami-amplitudes calculated for the Mjølnir impact are the results of tsunami formation mechanism in very shallow water. Relationships similar to Glasstone and Dolan (1977) are good approximations for deep water impacts from the viewpoint of cratering processes, but strongly underestimate tsunami amplitude calculations for most marine target impacts.

The returning sea water behind the first (leading) tsunami accelerates to the crater center due to the level gradient and gradually fills the crater as resurge flow. The return takes about 10–20 min and is strongly dependant on both the crater depth and rim height. Numerical simulations do not allow calculation of these values with a high degree of accuracy, due to the low spatial resolution. Consequently, both the real time of infilling and various tsunami parameters obtained in numerical simulations may only be considered as first approximations. Reflection of the resurge flow from the crater center gives rise to the second generation waves characterized by lower amplitudes of ~30 m at a distance of 100 km and longer periods up to 60 min (Fig. 7.10). The longer periods mean that the wave is less subjected to dissipation and disruption due to the Van Dorn effect (Melosh 2003), and, consequently, more hazardous for coastal regions. The late evolution of tsunamis is considered in more detail in Chap. 10.

The initial velocity of the resurge flow directed to the crater center reaches 50–70 m/s and remains at a level of about 20 m/s for at least 30 min (Shuvalov et al. 2002). This high-speed and temporally persistent flow can considerably affect the coarse near-field ejecta displacement and redeposition, as well as the displacement/deposition of fine ejecta which continue to fall into the sea. As a result, the resurge flow can considerably change the final crater topography.

As mentioned above, most tsunami effects were analyzed with a 2D numerical model for the vertical case. A major consequence of obliquity could be an asymmetry of the resulting wave. In Figs. 7.1 and 7.2 it is demonstrated that only the very initial stage of an oblique impact is strongly asymmetrical, and that even the transient cavity becomes circular already near the end of the excavation stage. As we

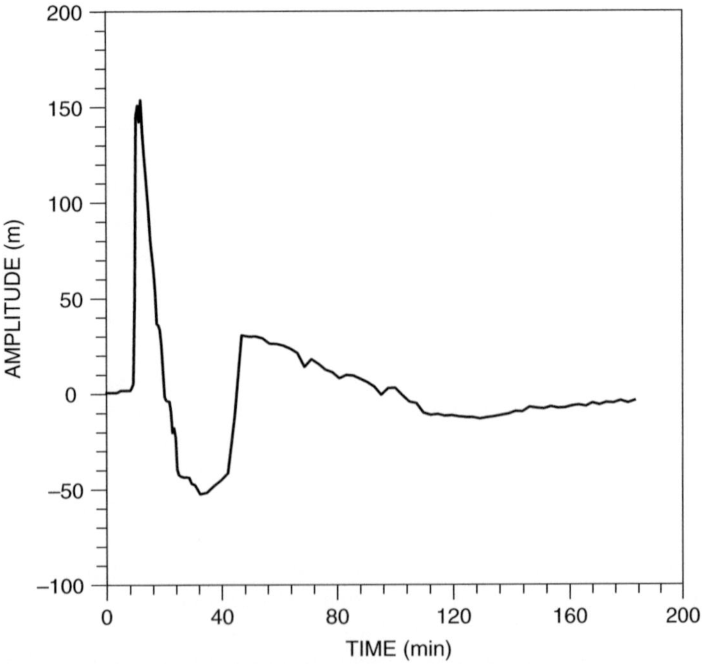

Fig. 7.10 Sea level oscillations at a distance 100 km from the impact point after a 45° oblique impact into a 400 m deep sea (the same as shown in Fig. 7.7)

have seen, the tsunamis form later and a strong asymmetry of the waves is not to be expected. Numerical simulations for deep-water impacts (Shuvalov 2003a) confirm this conclusion (Fig. 7.7). In general, the water waves are axially symmetrical, however local inhomogeneities can appear due to the interaction of water with ejected target rocks.

7.6 Conclusions

The numerical simulations show that a crater of Mjølnir size can be produced by a vertical impact of a 800 m radius stony asteroid or by a 45° oblique impact of a 1,000 m radius stony asteroid. Based on the detailed geological and geophysical analyses (Tsikalas et al. 1998a–c) the Mjølnir impact is considered to be oblique (Shuvalov and Dypvik 2004; Tsikalas 2005). An exceptionally extensive gravitational collapse, as seen in the geophysical observations, can be reproduced in numerical simulations with a composite target strength structure. This is composed of very low strength for the upper 3 km of sedimentary rocks, a gradual increase from 3 to 6 km depth before getting strength values typical for granite rocks.

The impact cratering processes can be roughly divided into three main stages, namely compression/penetration, excavation, and modification (Melosh 1989). In the case of the Mjølnir oblique impact, the compression/penetration stage is strongly asymmetrical, occurring with dimensions highly dependable on projectile size (Figs. 7.1 and 7.2). The modification stage, in contrast, is fairly symmetrical (Figs. 7.5 and 7.7) as seen in the shape of the final crater (i.e., characteristic size of modification process). The latter is considerably larger than the projectile size, and initial asymmetry attenuates at distance. This scenario is comparable to high-energy explosion, where shock-wave propagation and cratering depend solely on the energy release. The excavation is an intermediate stage with the early ejecta phase being strongly asymmetrical and late ejection, being responsible for the crater rim formation, only slightly asymmetrical.

The presence of a 400-m-deep water layer above the sediment target only slightly (to almost not at all) has influenced the Mjølnir cratering process and the size of the final crater. Resurge flow, however, resulting from the collapse of the water-cavity has considerably reworked the crater topography, both due to erosion and ejecta redistribution.

The numerical simulations show that even a 400 m deep sea can strongly influence the formation, expansion and final distribution of distal ejecta. This is to a large degree controlled by the impact angle.

Tsunami waves produced by the Mjølnir impact are considerably greater than the ones predicted by the formula of Glasstone and Dolan (1977), which was derived from nuclear tests. This is mainly due to the different mechanisms of tsunami generation in deep and shallow waters. In the Mjølnir case, the tsunamis are produced by the motion of solid target rocks due to rim expansion and structural uplift. The oblique Mjølnir impact produces axially symmetrical tsunami waves.

Chapter 8
Structural Analysis of Deformed Central Peak Sediments

Roy H. Gabrielsen, Henning Dypvik, and Valery Shuvalov

8.1 Structural Position of the Mjølnir Impact Crater

The buried Mjølnir crater in the Barents Sea (Figs. 1.8 and 1.10) classifies as a complex impact structure with a central peak and an initially subtle peak ring (Gudlaugsson 1993; Dypvik et al. 1996, 2004b; Tsikalas et al. 1999). The Mjølnir bolide hit the paleo-Barents Sea (~400–500 m water-depth at the time of impact) at an impact angle of 45° from a SW-SSW direction (Tsikalas 2005). The crater later underwent collapse, leveling by erosion and redepostion and burial (Smelror et al. 2001a; Tsikalas et al. 2002b). Due to post-impact Cenozoic uplift and erosion the central peak is presently found a few tens of meters below the seafloor.

The Mjølnir crater is situated within the Bjarmeland Platform (Gabrielsen et al. 1990) in the central Barents Sea. This is a tectonically stable structural element surrounded by mechanically weak fault zones with great potential for reactivation (Gabrielsen 1984; Gabrielsen et al. 1992a, 1997). Such fault zones have the ability to absorb strain under renewed stress situations, and could accordingly prevent the transfer of stress into the interior structural elements they delineate (Pascal and Gabrielsen 2001). Hence, the sediments of the inner part of the Bjarmeland Platform were situated in a tectonically stable environment and were probably undeformed at the time of the Mjølnir impact. Although halokinetic structures are known from the Bjarmeland Platform (Gabrielsen et al. 1990), these are not situated close enough to the site of the Mjølnir impact to have influenced the pre-impact local stress field or the general structural development. Therefore, the intense disturbance observed in seismic reflection profiles (Fig. 8.1; Tsikalas et al. 1998a, b, c, 2002a) is totally attributed to impact-related processes. In this context, sedimentological studies of drillhole 7329/03-U-01 at the central peak vicinity (cf. detailed presentation in Chap. 5) have shown that the impact-related sedimentary sequences identified around the central peak were redeposited during the collapse of the central high

R.H. Gabrielsen (✉)
Department of Geoscience, University of Oslo, NO-0316 Oslo, Norway
e-mail: roy.gabrielsen@geo.uio.no

H. Dypvik et al. (eds.), *The Mjølnir Impact Event and its Consequences*, Impact Studies, DOI 10.1007/978-3-540-88260-2_8,

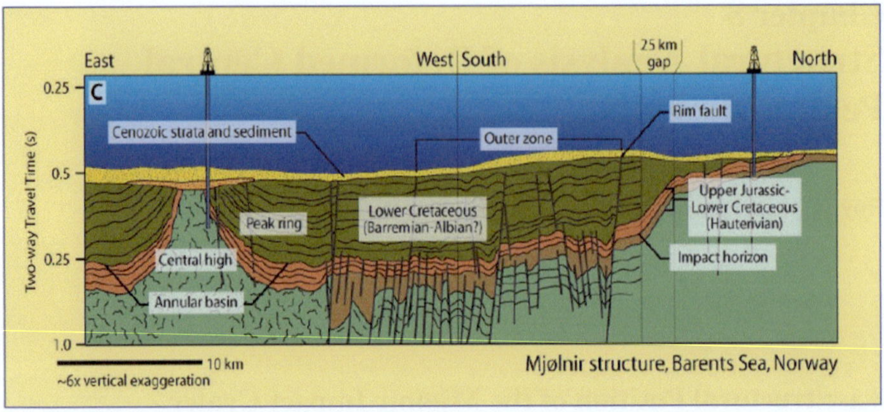

Fig. 8.1 Reflection seismic section displaying the deep structural configuration of the Mjølnir impact structure and the position of scientific well 7329/03-U-01. Modified from Tsikalas et al. (1998a) by K. Evans

itself (Dypvik et al. 2004b, c). The same core, 7329/03-U-01 is the subject of the present detailed structural analysis.

Core 7329/03-U-01 was drilled in 1998 at the edge of the central peak of the Mjølnir crater at water depth of 350 m. The core covers the Lower Cretaceous, post-impact Klippfisk and Hekkingen formations and the strongly disturbed Ragnarok Formation (Figs. 8.2 and 8.3), which was deposited during the turbulent and chaotic conditions following the impact (Dypvik et al. 2004b). The lowermost part of the recovered interval of the Ragnarok Formation was cored between levels 74.05 and 171.0 m below the present seafloor and consists of two units; Unit I and II. The *uppermost part* (Unit II) contains mainly poorly sorted diamictic conglomerates and pebbly mudstones, with a thin basal siderite bed (Fig. 5.2). The unit is believed mainly to represent mass- and gravity-flows (Chap. 5) (Dypvik et al. 2004b, c). Unit I is dominated by chaotic slump-sediments and is strongly folded, fractured, disrupted and sheared (Fig. 8.3), sometimes beyond recognition of the original sediments

8.2 Structural Geological Analysis

Dypvik et al. (2004b, c) reported a number of features related to the deformation of the sediments of the Ragnarok Formation. This includes entities with continuous bedding, breccia and folded and vertically oriented strata as preserved in Units I and II (Fig. 8.5). Furthermore, brittle fractures (faults and joints) and soft-sedimentary deformation features indicative of water-escape processes are commonly seen. These features have been investigated in more detail by macroscopic core analysis and in thin section by use of optical microscope. The studies demonstrated that it is possible to identify and classify several distinct types of deformation. The deformation types seem to be associated with distinct differences in mechanical properties

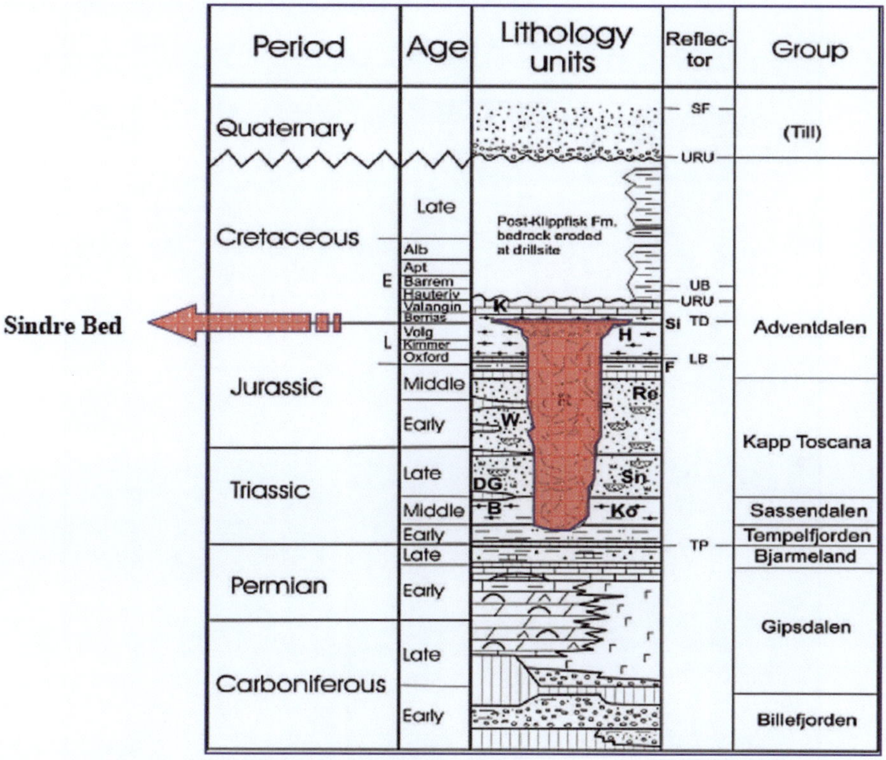

Fig. 8.2 Stratigraphic position of the Ragnarok Formation in the Barents Sea. The unit correlates to the Sindre bed of the Adventdalen Group in Spitsbergen. K = Klippfisk Fm., H = Hekkingen Fm., Re = Realgrunnen Subgroup, W = Wilhelmøya Subgroup, Sn = Snadd Fm., DG = De Geerdalen Fm., B = Botneheia Fm, Ko = Kobbe Fm

of the deforming sediments. To some extent relative dating of the different types of deformation can be established by the use of routine methods for structural investigation of drill-cores (Arthur et al. 1980; Carson et al. 1982). Thus, crosscutting relations give important information in some cases. In addition to direct observations, the relative positions of the strata and already established knowledge about the general development inside the crater can be used in determining the relation between the different types of structures.

8.2.1 Type A Structures: Early Extensional Micro-faults and Fissures

Type A structures are only recognized in strata which survived later stages of deformation and which are preserved in isolated large clasts (5–20 cm) preserved within the chaotic, strongly disturbed and liquefied sediments that generally dominate the

Fig. 8.3 Core photo of core 7329/03-U-01, core depth interval interval 142.86–147.58 m

sections studied (Fig. 8.11). No structures of Type A are found in the matrix of the liquefied units and two subtypes are recognized.

Description: Type A1-structures are single or conjugate mesoscale faults and microfaults with normal-sense displacement, transecting both single layers and packages of two-three layers (Fig. 8.4). Offset of bedding planes as observed in

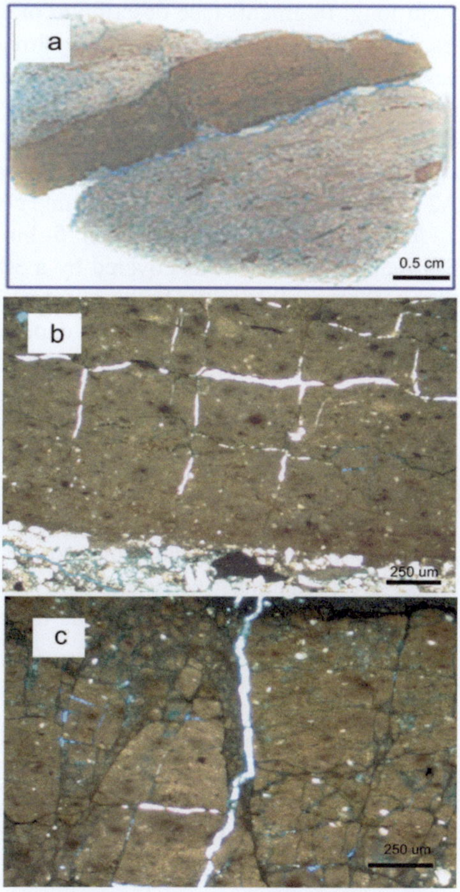

Fig. 8.4 (**a**) and (**b**) Conjugate extensional small-scale- and microfaults Type A1 structures; in siderite bed, well depth 147.30 m. The width of the thin section in (**a**) is 4 cm. (**c**) Orthogonal, ankerite-filled tensile microfractures (Type A2 in siderite bed, 147.17 m description of Type B-structures below)

the cored intervals demonstrates a vertical separation of up to 4–5 mm. As seen on the micro-scale, most of the structures are clay-filled and contain isolated, small clasts of rounded quartz grains. The fractures are most commonly planar to irregular and steeply dipping (dip 60–70°), but more shallowly dipping and even horizontal fractures occur. Subhorizontal fractures seem to represent faults that are oriented subparallel to the surface of the thin-sections. Together with the steep and occasional low-angle faults, this demonstrates that there are more than one distinct fracture set that belong to the type A1 fracture system. Where Type A1 and type A2-fractures (see below) interfere, Type A1-fractures are transected by Type A2-structures, and their internal age relation is established thereby.

Type A2 structures are mineral-filled fissures seen on both micro- and macro-scale in the units that have preserved their cohesion and that consist of interbedded, fine-grained layers of carbonate (siderite) and mud-rich layers, varying in thickness from 4 to 15 mm (Fig. 8.6). These strata are commonly broken into elongated, up to

5 cm long, fragments (see description of Type B-structures below).The individual fractures are characterized by slightly irregular planar to irregular geometry. They have a transverse separation of up to 1 mm and are filled by a clear to whitish ankerite cement. The Type A2 fractures are frequently seen to constitute a system of two orthogonal fracture sets, one oriented parallel to and one transverse to the layers. This fracture system is strictly intra-formational. Fractures of type A1 and A2 are frequently found inside the same clast.

Interpretation: The more than 5 km thick sedimentary succession of the Bjarmeland Platform was covered by a 300–500 m deep epicontinental sea at the time of impact close to the Volgian/Ryazanian transition (Smelror et al. 2001a). The sediments were situated at a burial depth of a few hundred meters before impact, implying that some consolidation had taken place.

In the pre-impact burial process, the Triassic-Jurassic sediments suffered at least two early stages of brittle deformation prior to impact. The first stage (Type A1-structures) was characterized by syn-sedimentary or early post-sedimentary normal (micro-) faulting due to layer-parallel extension. It seems that several sets of extensional faults may have been developed simultaneously and with equal frequency during the pre-impact burial stage, possibly an indication of a homogeneous horizontal stress situation. Such a state of stress would be consistent with a Triassic-Late Jurassic tectonic position well inside the tectonically relatively stable Bjarmeland Platform (Gabrielsen et al. 1990).

The Type A1-microfaulting was succeeded by further compaction and pressure build-up that caused brittle fracturing and carbonate precipitation inside the mode I fractures (Type A2 structures), a process that might have been facilitated by uplift. The Type A2 structures that encompass two sets of orthogonally oriented, carbonate-filled fractures (layer-parallel and at right angle to layers; see above), particularly affect carbonaceous siltstones and marls. The vertical set of Type A2-fractures is indicative of either horizontal, bedding-transverse extension, repeated build-up and release of water-pressure, contraction due to volume loss, or a combination hereof (Henriet et al. 1991; Cartwright and Lonergan 1996; Clausen et al. 2003). Due to being preserved only inside clasts, it has not been possible to establish whether or not the fractures have a preferred orientation, information that could be used to separate between these options. The horizontal set of carbonate-filled fractures, however, suggests that a water-pressure build-up has occurred so that during derformation:

$$P_{H_2O} > \sigma_v = \sigma_1 = \rho g z \tag{8.1}$$

A similar type of fracture in shales of the Hekkingen Formation, inside the realm of the Asterias Fault Complex that delineates the Loppa High to the south, has been reported by Gabrielsen and Kløvjan (1997). They concluded that the fractures were generated either during exhumation and pressure release or during fluid pressure build-up from other causes.

It is evident that the rocks hosting the A1 and A2-type structures had reached some degree of burial and consolidation at the stage of deformation and that a certain contrast in mechanical competence existed between the layers. Hence, in contrast to the bulk of the rocks in the Ragnarok Formation, these relatively competent strata sustained later liquefaction.

This is indicative of early consolidation and differential compaction associated with high water-pressure at a shallow level of burial.

Based on the contrasting styles of deformation and cross-cutting relations between them it is concluded that A1- and A2-types structures are due to uniform extension during an early stage of burial and that type A1 preceded A2. The latter was initiated at a stage when diagenesis and consolidation had lithified some layers. Both structure Type A1 and A2-structures may have been affiliated with the effects of water-escape.

8.2.2 Type B-Structures: Fragmentation of Semi-consolidated or Consolidated Beds

Type B-deformation is characterized by break-up of already consolidated strata, and the deformation products are particularly preserved in units, that encompass the mechanically most competent layers (Figs. 8.5 and 8.11). The rock classifies as an intraformational, monomictic breccia with a light gray, very fine-grained silty matrix.

Description: The clasts of the the breccia affiliated with the Type B structuring are mainly calcite-cemented siltstones of different colors, ranging from light pinkish gray to brick red and there is a spatial mixture of different types of fragments. This demonstrates that they were derived from different strata and transported for sufficient distance and with sufficient energy for the clasts to become mixed. The longest axes of the clasts are from a couple of cm to 10 cm and most clasts are subhorizontally length-oriented. The clasts commonly are subrounded to rounded. The shape of the edges of some of the clasts may reflect that extension took place in concert with compactional boudinage.

Interpretation: The structural style that characterizes the Type B-deformation shows that the slightly consolidated rocks became suddenly disturbed. The shape of the clasts and the general characteristics of the sediments affiliated with Type B-deformation suggest several stages of development. Geometries that are compatible of boudinage in some clasts, indicates that strata started to break up during loading. This is a process that may be seen as a continuation of the development of Type A-structures (see above) and there are examples of Type A1-structures in some of the clasts. However, the strata in which Type A1 and A2-structures are most widespread and consist of silty beds with typical thickness of less than 2 cm. The strata affected by Type B-deformation became ripped up, transported and redeposited as clasts in gravity flows. The monomictic sedimentary breccias seem to have been preserved as larger entities that survived the gravitational collapse as coherent blocks (mega-clasts). It is, therefore, assumed that the breccias that contain Type B-structures were generated prior to the destabilization affiliated by the post-impact deformation as described below.

Fig. 8.5 Intra-formational, monomictic sedimentary breccias with elongated, *gray*, *pink* and *brick-red*, up to 10 cm long clasts (Type B-structures). Note that clasts are sub-rounded to rounded and that some carry signs of semi-ductile deformation, indicating the layers were semi-consolidated by the time of rip-up and redeposition. This unit is the *lowermost* in the cored section (core depth at low end of core cut: 162.00 m). The width of the core is 5 cm

8.2.3 Type C-Structures: Liquefaction and Shearing

Large parts of the Ragnarok Formation are characterized by chaotic structures, where the texture grades from less deformed to penetratively sheared sediments (Figs. 8.6 and 8.11). Two main textural types are seen; Type C1 and Type C2.

Description: The Type C1-structures are zones dominated by isolated, lensoid and sometimes irregularly shaped fragments of light gray siltstone and very fine-grained sandstone (long axis up to 2–3 cm) within a darker gray, very fine-grained sandy to silty matrix. Embedded in this matrix are also round, ball-like clasts that display a faint inner concentric stratification of mineral grains, indicating that the fragments were accumulating when rolling in a fluid "soup" of sediments of low viscosity ("armored balls"). Clay enrichments are common along the margins of the zones dominated by Type C1-structures. These may represent remnants of clay intrusions or fine-grained material accumulated along the margins during fluid flow.

Fig. 8.6 Type C-structures: **(a)** Core section 152.70–152.90 m showing liquefied units with rounded fragments (1), flow banding (2) and clay accumulation (3). The vertical orientation of the layers is believed to reflect rotation that occurred after liquefaction. The width of the core is 5 cm. **(b)** Thin section. Sub-rounded clasts in the liquefied layers. Note that although grain-size is less in the matrix surrounding the larger clasts, there is no sign of grain-crushing

The units that are dominated by Type C1-structures are separated by linings of fine-grained matrix with a pronounced flowbanding (Type C2-structures). The Type C2-deformation zones are typically light gray of color, and are characterized by homogeneous, parallel to wavy fluxion structure. The matrix of Type C2-deformation zones contains isolated sand-sized grains of quartz. In most cases, the fine-grained matrix displays shearbanding that is visible at both meso- and micro-scale.

Interpretation: The Type C-structure influenced units have the characteristic appearance of highly liquefied sediments. It seems that liquefaction occurred abruptly by introduction of large amounts of water resulting in partial to complete loss of cohesion of sediments involved. The fine-grained matrix was subjected to violent shear that most probably occurred during as well as after the deposition. For the Type C1-zones, this process was accompanied by the development of clasts at different stages of rounding and separated by highly sheared matrix. Some fine-grained, ball-like fragments grew by rolling and accumulation of new fine-grained particles along their outer margins, like snowballs. This type of clasts appears in a fine-grained matrix, which is less sheared than that typical for the matrix of Type C2-zones. The distribution and orientation of the Type C2-shear-zones indicate that shearing took place during gravitational collapse and water-escape. The development of Type C-structures is believed to represent the first preserved deformational stages following the impact. The sediments that were exposed due to abrupt uplift of the central peak were catastrophically flooded by water rushing back into the crater. They became completely liquefied, gravitationally destabilized and transported in suspension away from the flanks of the central peak and towards the crater floor. The stratigraphical units dominated by Type C1-structures represent the most liquefied sediments, whereas Type C2-structures probably were generated during flow by decreasing water-saturation, resulting in focused shearing in zones of high pore pressure. In these zones, the typical Type C1-structures became destroyed by shearing, and Type C2-structures developed at their expense.

8.2.4 Type D-Structures: Folds, Rotated Strata and Shear Bands

Rotated strata observed as steeply dipping layers and associated folds are commonly seen throughout the Ragnarok Formation (e.g. core intervals 96.09, 129.09, 134.66, 151.11 and 156.26 m, and Figs. 5.3, 8.7 and 8.11). This structuring is typical for the highly liquefied units, in which Type A, B and C-structures rarely are preserved.

Description: In the cored interval 134.66 m, a tight syncline with a subvertical axial plane and a faintly developed, fan-shaped axial plane foliation is seen (Fig. 8.7). The fold has thinned limbs, but it is not possible to determine whether this is due to primary thickness variations (e.g. channel structures) or shearing associated with the folding itself. Another fold is seen in interval 151.11 m. Here, the axial plane is oriented horizontally and parallel to a package of highly sheared strata below. Also in this case, the core of the fold is affected by axial plane foliation, but the foliation is irregular and penetrative to a varying degree. On the macro-scale, the axial plane foliation is diffuse and sometimes associated with flame-like soft-sediment intrusion structures at the borders between the layers. This suggests that the strata were liquefied at the time of deformation. On the micro-scale the foliation is seen as a system of subparallel shear bands and zones of parallel mineral grains.

Interpretation: The folds that characterize the Type D-deformation developed in sediments that were either less affected by liquefaction or at a later stage in sediments that regained some mechanical strength as compared to those characterized by Type A–C deformation. The mix of folds with contrasting fold styles and the apparent chaotic orientation of fold axes on one hand, and subhorizontal or shallowly dipping orientation of associated shear planes and axial plane foliation on the other, suggest that the folding was gravity-driven. Continued deformation involved folding and rotation of the strata.

Although the folding seems to have been chaotic, consistently oriented, low-angle (to the horizontal of the core) shear zones may be taken as an indication that the transport direction was consistent and perhaps gravity-driven down-slope and that the shearing continued after the folded sequence had settled. High concentration

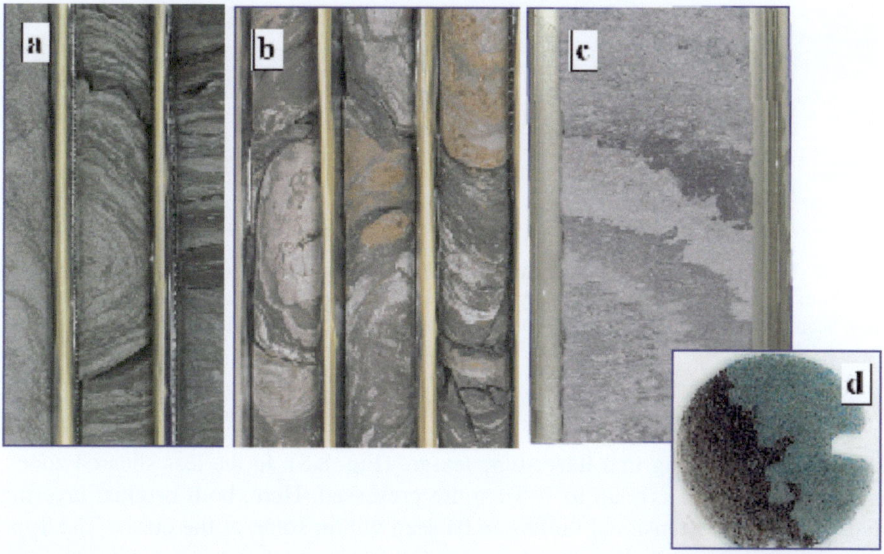

Fig. 8.7 Type D-structures: **(a)** and **(b)** folded strata of the Ragnarok Formation. Most fold axes are horizontal to sub-horizontal. Borehole 7329/03-U-01, 157.00–157.50 and 136.00–137.00 m. The width of the cores is 5 cm. **(c)** Axial-plane parallel shear-planes and **(d)** thin section of flame-shaped intrusions indicate that the strata were water-saturated and mechanically weak during deformation. Borehole 7329/03-U-01, 151.20 m

of strain in some zones indicates that strain weakening occurred, most likely due to preservation of a high water content in some zones, such as along fold axial-planes. Type D-deformation typically is found in rocks of higher mechanical strength than that typical for Type C. This is consistent with Type D-deformation occurring at a later stage, associated with decreasing water pressure. Alternatively, the different deformation styles may be correlated to sediments derived from different depths of burial, brought in contact during the chaotic collapse and redeposition. Finally, the sediments with different deformation style may have originated from different parts of the central peak.

8.2.5 Type E-Structures: Intensely Sheared Sequences

Intensely sheared units occur in particular zones and are frequently seen in association with recumbent folds of the Type D-structures (Figs. 8.8 and 8.11).

Description: Some high-strain zones display characteristics different from the Type D shear-zones. These are classified as deformation Type E, and are associated with units where light gray silt fragments in different stages of shearing and flattening, are found in a dark gray, more fine-grained matrix. Hence, zones of varying degrees of shearing are stacked together. The most strongly deformed units are characterized by lozenge-shaped structures separated by fine-grained, non-equigranular

Fig. 8.8 Type E-structures: Intense shearing in siltstone unit of the Ragnarok Formation at 147.70–148.00 m core depth. Note that primary layering is preserved in some clasts (*1*), whereas others have been sheared beyond recognition of the primary layers (*2*). The width of the core is 5 cm

shear bands, resulting in a flazer-type texture (Fig. 8.8). In the less sheared zones, clasts with longest axis up to 7–8 cm are preserved. Here, both original layering and small-scale extensional faults can be seen within some of the clasts. The Type E-structures affect all the other types of fine-grained siliciclastic sediments and, where observed, cross-cut the soft-sediment liquefaction structures (Type C). The relation between the type E-shear-structures and the folds is less certain, and examples are found in which the late shear-bands themselves are folded (interval 134.66 m). Still, it seems to be more common that the late shear-bands cut the fold limbs. This may suggest that there are at least two generations of such structures, implying that shearing and folding were initiated simultaneously and that shearing continued after folding terminated.

Interpretation: Type E-deformation probably represents a continuation of structural Type D-development, occurring during ongoing dewatering and increasing consolidation of the strata. The Type E-structures display great variation in shear intensity from one layer to another, occasionally also within the same host rock. In some cases a:c-ratio of the clasts is up to 1:20, indicating strong flattening in addition to shear. There is a strong parallelism within the units characterized by Type E-deformation, indicating a uniform transport direction. It is therefore likely that although there are great contrasts in shear intensity in the Type E-structural units, the orientation of the principal stress axes remained constant during the deformation. This pattern could be explained by gravity-driven down-slope slump transport away from the central uplift.

8.2.6 Type F-Structures: Late Brittle Fractures and Microfaults

Type F1-structures are extensional microfaults that are found in several intervals of core 7329/03-U-01 (Figs. 8.9 and 8.11). Typical throws are in the order of a few mm to 1 cm and the faults occur in swarms of equally spaced, planar and parallel structures.

Description: Two types of late fracture sets are found in distinct zones at various places in the studied core. As seen in the microscope, they have several characteristics in common with the Type A-structures (see description above), but are different in that they affect larger parts of the sedimentary column and appear in concentrated swarms. The widths of the Type F fractures are 0.1–0.3 mm and are seen in the microscope as shear bands or deformation bands where isolated quartz grains (0.01–0.1 mm) occur in a denser, clay-rich matrix. In some cases, a faint internal flow-banding can be observed. The Type F1-faults are developed as swarms of parallel to subparallel (max deviation angle of 25–30°) structures separating 0.5–2 cm wide lamellas. The faults typically dip 60–80° relative to the layering and the strata within these lamellae are commonly systematically rotated, with typically inclination of 5–20°. In some cases the fractures merge at depth with high-angle, sometimes listric, more chaotic floor-faults. The F1 fractures also appear as swarms of anastomosing, curved and interlinking structures. However, in the latter cases there is a good parallelism between the structures, as seen on the short cut of the cores (Fig. 8.9).

Type F2-structures are seen as brittle calcite-filled fissures with planar to planar-irregular geometries. They are subvertical to the layering, cross-cut several layers and have been encountered only in a few cases, but consequently in association with Type F1-structures.

Fig. 8.9 **(a)** Arrays of rotated extensional fault lamellae (Type F-structures) with dip-slip displacements in the order of centimeters, are seen in restricted zones, where they off-set Type D-structures. The width of the core is 5 cm. **(b)** The fault planes are clay-filled and normal drag is sometimes seen on the micro-scale. **(c)** Fault planes are planar to irregular and up to 1 mm wide

The brittle fractures of Type F seem to postdate all the other structures described above. Furthermore, the generation and development of this type of structures require a certain mechanical strength of the sediments, suggesting that the fracturing took place after the major phases of dewatering/liquefaction (Types C–E).

Interpretation: The Type F-structures are characterized by brittle faulting and display cross-cutting relations to all other (Types B-E) structures. The Type F-deformation style suggests a higher grade of consolidation and hence higher mechanical strength compared to that of Types B–E. The Type F-structures were generated in situation of layer-parallel, relative uniform extension, although the steep orientation of the faults allows for only moderate bulk displacement. It is also noteworthy that Type F1-structures, in contrast to, e.g., Type A-structures, are not intraformational, but affect entire sedimentary units. The orientation and parallelism of the Type F1-structures suggest that they were initiated in a stress situation where the two principal horizontal stresses were distinctly different, in contrast to what seems to have been the case for the Type A1- structures. In the few cases where Type F2-structures have been encountered, they clearly transect the Type F1-structures. It is reasonable to associate the Type F2-structures with late uplift and water-escape.

8.3 Deformation History of the Ragnarok Formation

In the formation of complex submarine impact craters, characterized by generation of a central peak and peak rings formation, three main stages of development are commonly identified; (1) compression stage (including crushing, melting, and evaporization of the rocks of the impact site), (2) excavation stage (formation of ejecta and transient crater) and (3) modification of the transient crater by slides, slumps, avalanches, tsunami and gravity flows (Melosh 1989; French 1998). Study of seismic reflection data (Tsikalas et al. 1998 a, b, c) and the Mjølnir drillcore 7329/03-U-01 with its blend of microfossils of different ages (late Early Triassic to Late Jurassic; Bremer et al. 2004; Dypvik et al. 2004b, c, 2006, in combination with numerical modeling, Shuvalov et al. 2002) show that the Mjølnir impactites have undergone all these stages. The numerical modeling suggests that it takes the shock wave about 3 s to expand for a distance exceeding the size of the final crater (compression stage, Fig. 8.10). A more than 6 km thick sedimentary sequence was disturbed by the Mjølnir impact during the first 10 s (excavation stage). In the succeeding modification stage the crater rim migrated outwards due to gravitational collapse of crater walls, reaching a maximum diameter of 40 km. This was accompanied by simultaneous elastic uplift and rotation of the strata of the central crater peak, which probably started collapsing immediately. At the impact site, the 400 m deep water column was forced aside, the seafloor was set on fire and violent tsunamis flooded the crater (tens of minutes after the impact). During these processes, the central peak quickly became submerged, gravitationally destabilized and its rims collapsed (Shuvalov et al. 2002).

The post-impact erosion of the Bjarmeland Platform has been estimated to between 1,000 and 2,000 m (Nyland et al. 1992; Skagen 1992). The shallow burial depth at which the crater top-level is presently found (50–150 m below seabed),

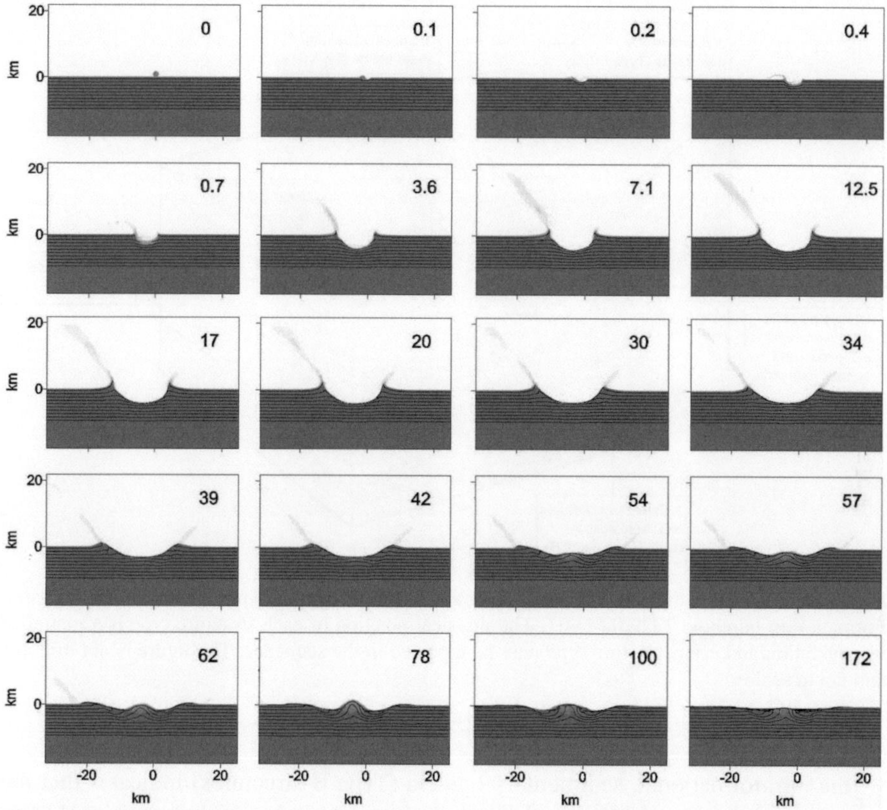

Fig. 8.10 Modeled development of the Mjølnir crater after the 45° oblique impact (after Shuvalov and Dypvik 2004). Time steps 0–172 s are shown. *Black lines* show impact induced distorsion and displacement of the original horizontal platform layers. Impact direction is from *right* to *left*

substantiates that only the uppermost, collapsed part of the central peak was penetrated in core 7329/03-U-01, and hence that no syn-impact rocks have been available for study. Thus, Dypvik et al. (2004c) suggested the lower boundary of the Ragnarok Formation to represent post-impact slump and massflows of regionally reworked fall-out/fall-back units. The base of the Ragnarok Formation is correlated to a mappable reflection at ca. 1.3 km below the present seabed, showing it to be an intra-crater depositional unit of regional extent.

Based on this interpretation, and supplemented with data from the present structural geological study, the following evolution is suggested from pre-impact burial through impact and subsequent the post-impact collapse of the central peak (Fig. 8.11):

1. The Type A-structures are results of a normal phase of burial under tectonically quiet conditions on the Bjarmeland Platform during Triassic-Late Jurassic times. The lack of preferred orientation of the normal, soft-sedimentary faults suggests that a homogeneous stress situation prevailed and hence that the principal horizontal stresses were not distinctly different. From the generation of

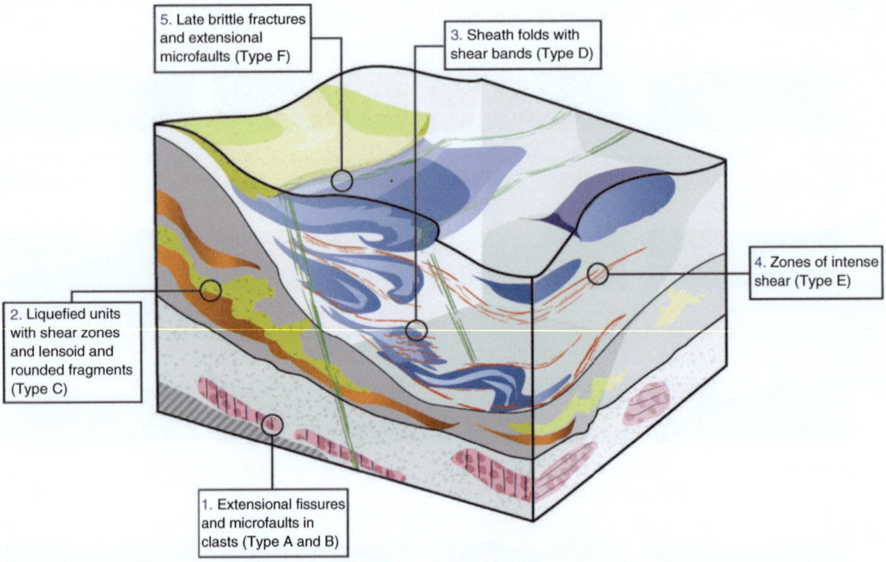

Fig. 8.11 Schematic diagram showing the main structural types (Types **A–E**) of the Ragnarok Formation in borehole 7329/03-U-01. The physical position of each structure type is a matter of interpretation and each structure type may be repeated in the sequence. The figure is not oriented and not to scale

the intraformational, sedimentary breccia (Type B-structures) indicates that the sediments became destabilized in a shallow marine or brackish depositional environment after they had reached a certain level of consolidation. The energy of this process was moderate, but still sufficient for the layers to become ripped up. Fragments were transported far enough to become rounded (which is likely not to be very far; the clasts were not mechanically strong) and mixed with neighboring layers before they finally became re-deposited. Accordingly, it is suggested that the structures from the two first stages of development (Type A- and B-structures) are of pre-impact origin and only preserved in mega-clasts that remained cohesive during the impact and the following collapse of the central peak.

2. The sediments carrying Type C-structures display all signs of extreme liquefaction and quite fast deposition when still oversaturated by water. The ball-like fragments probably stayed in suspension and some grew by rolling and are interpreted as "armored balls". The sediments settled as water escaped, causing stabilization of the unit. However, the sequence continued to move, probably in one overall preferred direction. This is manifested by the development of zones with internal shear structures. In total, these observations suggest that deposition possibly occurred on the upper slope of the central peak. It is suggested (by both observations and modeling, Fig. 8.10) that the crater was invaded by several large tsunamis and that the central peak drowned several times, each flooding

being separated by events of subaerial exposure. It is reasonable to correlate the development of the Type C-structures to these tsunami-triggered events (a few minutes after the impact).

3. Type D-structures are documentation of violent deformation, rotation of strata to become steeply dipping, vertical and even overturned. This process also included chaotic folding, which was followed by shearing in the later stage of development (Fig. 8.11). For such deformation to occur, the affected sediments must have preserved or regained some mechanical strength, implying that these units did not become liquefied or that this deformation took place when the liquefied sediments had become stabilized. Nevertheless, these structures show that the sedimentary pile was still unstable, and probably located at a relatively steep slope. The Type E-structures may be interpreted to be generated simultaneously with Type D-structures, or, alternatively, as a continuation of the process of down-slope transport along the margin of the central peak.
4. Type F-structures (Fig. 8.11) herald less violent deformation indicative of sedimentary compaction taking part in a relatively stable position. Again two possibilities exist: the sediments containing the Type F1-structures may have been develop in a more stable position and brought into their present position as a megablock, or the structures were developed after the Type D- and E-structures. Since it is difficult to conceive that a stable regime dominated by compaction can have existed among the violently deformed sediments present at a short distance, and since the Type F1-structures effect some of the liquefied sediments, it is considered most likely that F1-structures represent a next step in the structural development succeeding the Types D and E (from tens of minutes to millions of years). This is supported by the association with Type F2-structures, which are assumed to be related to later uplift and dewatering events.

The structures encountered in the sedimentary crater infill of Mjølnir impact drillcore have many characteristics in common with reports on deformed sediments from inside the Upheaval Dome impact crater in Utah (Kenkmann 2002, 2003; Kenkmann et al. 2005, Okubu and Schultz 2007).

These include folds, reverse and normal faults of different scales and zones of fluidization. In the case of the Upheaval Dome impact crater, the present surface of erosion is reported to represent a level of about 2,000 m below the target surface at the time of impact. It is fully exposed and has been studied in three dimensions (Kenkmann et al. 2005).

Folds from the Upheaval Dome impact crater reveal a variety of geometries (Kriens et al. 1999; Kenkmann et al. 2005). Sheath-folds and tight to isoclinal folds with varying axial plunge are common, resembling what is seen in the Mjølnir core. In the Upheaval Dome impact crater, folds are set in connection with different types of mass transfer directly related to the impact and to the subsequent collapse of the crater. Because three-dimensional information is lacking from the Mjølnir core, no further comparison except for the fold style and plunge of fold axes, can be made. However, the relation between the folds and other structural features like faults, fluidized zones and shear-zones suggests that the folding in the Mjølnir impact sediment fill took place at a somewhat later stage (post-impact) and during conditions of very high water pressure.

Fluidized sediments are common in the sediment fill of both the Upheaval Dome and the Mjølnir impact crater. It is interesting to note that the fluidized units of the Upheaval Dome reveal a significantly stronger component of grain-size reduction by grain-crushing (Kenkmann 2003; Kenkmann et al. 2005) compared to that seen for the fluidized parts of the Mjølnir core, where flow banding without significant grain crushing is common. This contrast may be consistent with the different conditions for deformation, where acoustic fluidization dominated for the Upheaval Dome and fluidization affiliated with high water pressure prevailed for the Mjølnir structure.

Finally, micro- to meso-scale, brittle fracture systems are mainly found in crater sediments from both Upheaval Dome and the Mjølnir structure. These are of two types and display strikingly similar geometries. Systems with complex and partly chaotic and anastomosing fracture networks affect sand- and siltstone units both places. In the case of the Mjølnir impact, these are intraformational and stratabound and are believed to represent pre-impact loading, whereas arrays of micro-scale to cm-scale rotated fault-blocks are believed to be affiliated with post-consolidation compaction and destabilization of the slump pile. In the Upheaval Dome, however, both these types of structures represent syn-impact deformation.

The comparison between deformation structures of the two craters emphasizes the need for even more detailed studies on structural geological aspects of impact cratering. For example, fluidization is an important, and perhaps not completely understood agent for mechanical destabilization. From comparison of the Upheaval Dome impact crater and the Mjølnir impact structure it is indicated that similarly looking types of structures may be generated at different stages and by contrasting mechanisms in impact craters. By combining detailed studies on all available scales (field, reflection seismic data, micro-structures), however, it is possible to separate between the mechanisms and conditions of deformation, as illustrated by the comparison of micro-structures in the fluidized units given above.

Chapter 9
Postimpact Deformation Due to Sediment Loading: The Mjølnir Paradigm

Filippos Tsikalas and Jan Inge Faleide

9.1 Postimpact Burial

Although protective at initial stages, extensive burial and associated processes, such as mechanical- and chemical-compaction, and diagenesis, may eventually lead to considerable changes in the original crater structure and morphology. Extensive postimpact modifications may obscure many marine impact craters formed in sedimentary, water-covered targets. The same postimpact processes may result in alterations in typical/expected geophysical signatures at such structures. The postimpact structural and morphological crater modification is, generally, an overlooked process because planetary research of impact structures (where postimpact sediment loading is mostly absent) dominated the impact-related research until recently. In addition, the terrestrial impact record is dominated by crystalline-target impacts on land (e.g., Melosh 1989; Ormö and Lindström 2000; Dypvik et al. 2004a; Turtle et al. 2005). Furthermore, postimpact modifications are difficult to quantify if an extensive and dense geophysical dataset of seismic reflection profiles and potential field data is not available, which is most commonly the case.

During an impact, the propagating shock-pressure waves give rise to extensive insitu fracturing and autochthonous target-rock brecciation (Fig. 9.1). Impact-induced porosity changes have been identified in damage zones below several craters (e.g., Pilkington and Grieve 1992; O'Keefe et al. 2001b) and it is considered that the lateral change in porosity within an impact structure (especially in a sedimentary target) is a key property to understand impact processes, postimpact compaction, and crater burial history (Tsikalas et al. 2002b; Tsikalas and Faleide 2007). Although several impact craters have experienced considerable postimpact burial (Table 9.1), quantification of this process is almost entirely absent from the terrestrial impact record, and, therefore, a great need exists for such studies.

F. Tsikalas (✉)
Eni Norge AS, NO-4064 Stavanger, Norway; Department of Geosciences, University of Oslo, NO-0316 Oslo, Norway
e-mail: filippos.tsikalas@eninorge.com

H. Dypvik et al. (eds.), *The Mjølnir Impact Event and its Consequences*, Impact Studies, DOI 10.1007/978-3-540-88260-2_9,

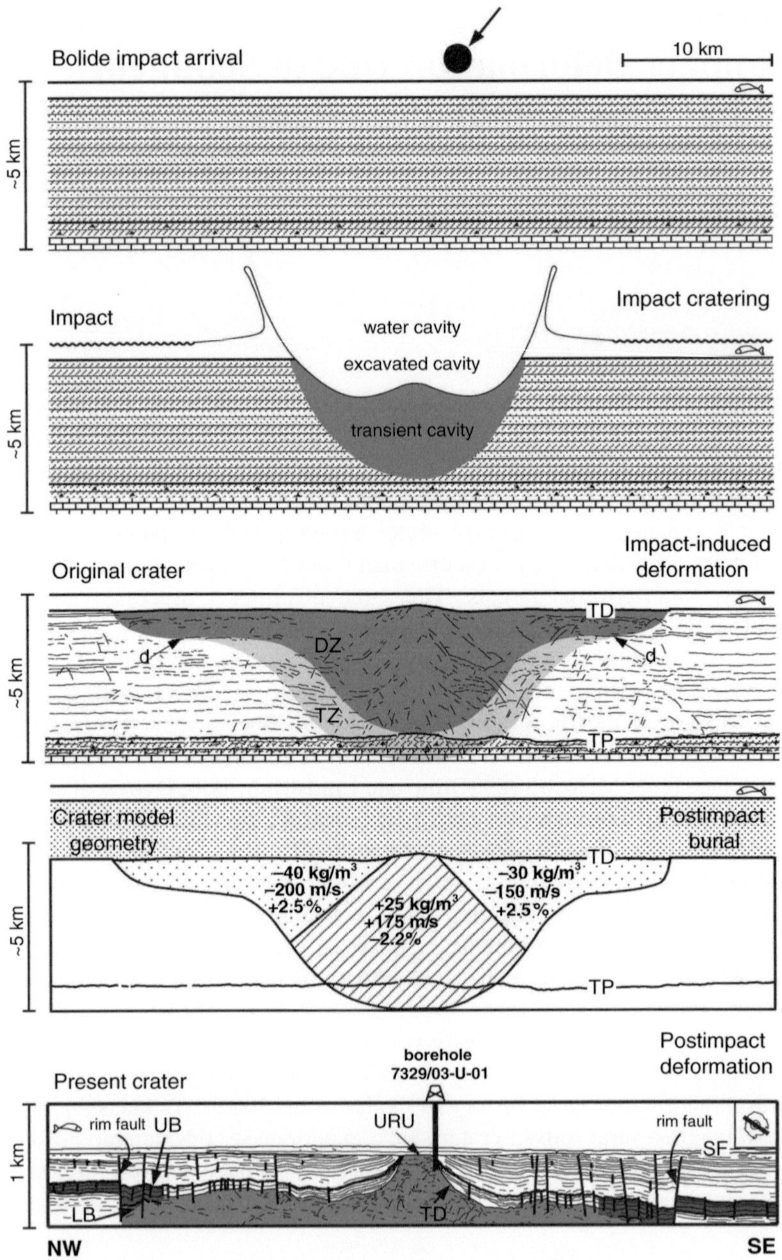

Fig. 9.1 Schematic cross-section showing the Mjølnir impact, resultant radially-varying physical property changes, and deformation types. Reflectors UB and LB bound the time of impact (confer Fig. 9.2). SF, seafloor; URU, Late Cenozoic upper regional unconformity; UB (*upper boundary*), lower Barremian reflector defining the top of a Upper Ryazanian-Lower Barremian limestone unit; TD, impact horizon (*top disturbance*; the first continuous reflector above the seismic disturbance); LB (*lower boundary*), upper Callovian/middle Oxfordian; TP, Top Permian; d,

Table 9.1 Well-preserved and well-studied impact craters that experienced postimpact burial and related modifications (see references in text for the individual craters)

Crater	Locality	Age (Ma)	Diameter (km)	Target water-depth (m)	Maximum postimpact overburden (km)
Chicxulub	Yucatan, Mexico	64.98 ± 0.05	180–200	<50	1–1.5
Chesapeake Bay	Virginia, USA	35.5 ± 0.3	85	200–500	0.5–1
Montagnais	Nova Scotia, Canada	50.5 ± 0.76	45	100–150	1
Mjølnir	Barents Sea, Norway	142 ± 2.6	40	400–600	2–2.5
Bosumtwi	Ghana	1.07	10.5	(Exposed)	0.3

In marine impacts, postimpact infilling starts with the impact-induced water cavity during the modification stage (Fig. 9.1). Collapse of this cavity starts at its base and causes a flow of water and rock debris towards the crater center (e.g., Shuvalov 2002b). These processes at marine-target craters may explain the lack of a raised crater rim at Mjølnir (Tsikalas et al. 1998a,b; Tsikalas and Faleide 2004), Lockne (Sturkell and Lindström 2004), Chesapeake Bay (Poag et al. 1994, 2004) and Chicxulub (e.g., Morgan et al. 2002a) craters, and the bevelled crater rim at Montagnais crater (Jansa et al. 1989; Jansa 1993) (Figs. 9.2 and 9.3). In marine impact craters, the impact event leaves a depression on the sea floor, into which a considerable amount of sediments is subsequently deposited (Fig. 9.1). The discussion in this chapter deals only with imposed effects due to postimpact sediment loading and not other postimpact geological processes, such as long-term tectonism (e.g., Sudbury Crater, Milkereit et al. 1994) or hydrothermal (Naumov 2002) alterations that may cause considerable changes in the impact structure or morphology.

9.2 Mjølnir Crater

The impact event left a crater relief on the seafloor (Fig. 4.2), which acted as substratum for considerable amounts of sediments. The geometry and the structural relations of these sedimentary strata provide information on amplitude, spatial distribution, as well as the mode of the postimpact deformation. The deformation is

Fig. 9.1 low-angle décollement; DZ (disturbed zone), area of intensely disturbed seismic reflections; TZ (transition zone), transitional area of less disturbance. The crater model geometry (second panel from *bottom*) illustrates the modeled impact-produced physical property changes (confer Fig. 9.7): modeled density-contrasts are given in kg/m^3, seismic velocities in m/s, and porosity anomalies in percentage

Fig. 9.2 Mjølnir crater high-resolution single-channel seismic reflection profile examples, and interpretations, across the crater center (**a**), and across the crater rim (**b**). M, marginal fault zone; P, peak-ring; IB1 and IB2, intra Barremian reflectors. Other annotations as in Fig. 9.1

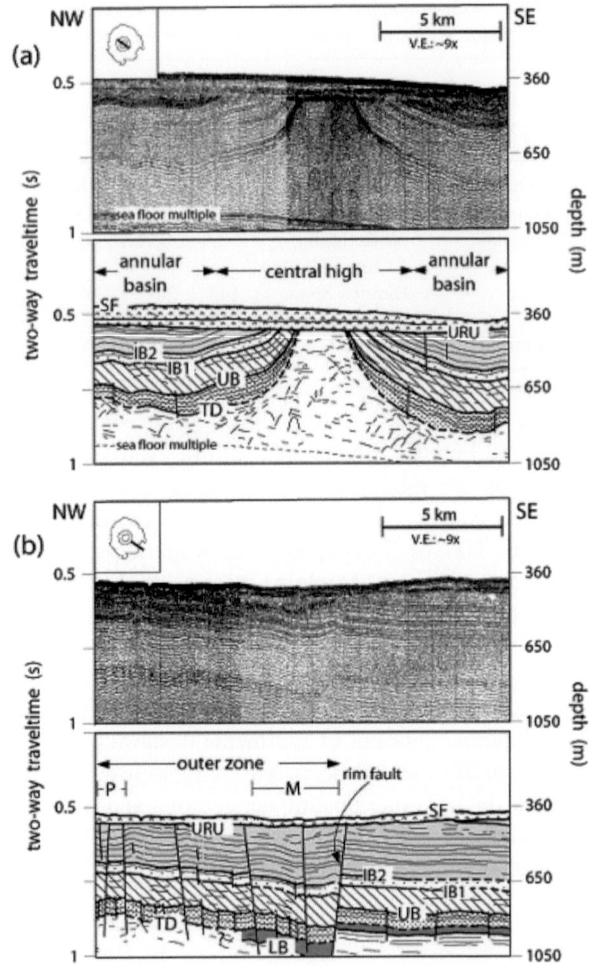

expressed by reactivation of impact-induced faults, initiation of new faults, and differential vertical movements (see Chap. 8). The effects of the extensive deformation are cumulative and enhance the underlying structural morphology, giving rise to the present ring structure (Fig. 3.10).

9.2.1 Postimpact Infilling

At Mjølnir (Table 9.1), continuous sedimentation until Barremian time completely buried the impact structure (Fig. 9.2), and about 1.5–2 km of siliciclastic sediments were subsequently deposited during Cretaceous-Tertiary time (e.g., Dimakis et al. 1998; Tsikalas et al. 1998a). Thereafter, Late Cenozoic uplift and glacial

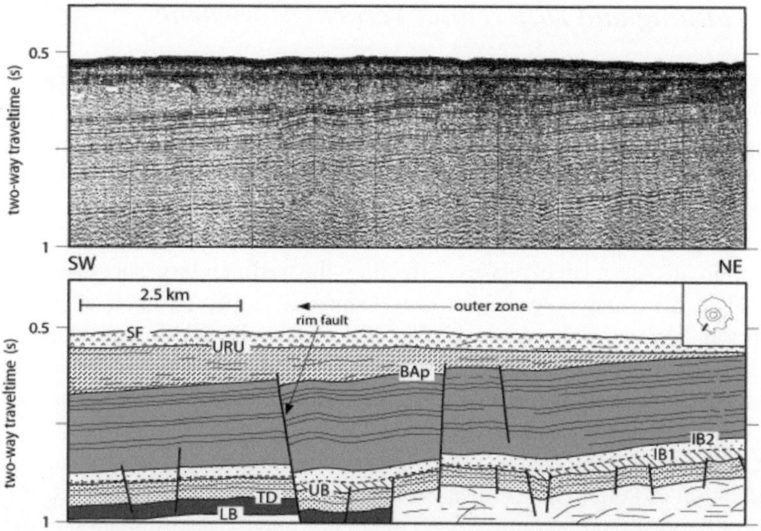

Fig. 9.3 High-resolution single-channel seismic profile, and interpretation, across the crater rim at the southern edge of the Mjølnir structure. BAp, base Aptian. Other annotations as in Fig. 9.1

erosion removed most of the postimpact sediments, and sheared the top of the central high (Fig. 9.2a). The shallow boreholes (7430/10-U-01 and 7329/03-U-01) together with the extensive seismic reflection coverage provide a detailed chronostratigraphic framework for the postimpact strata at Mjølnir (Dypvik et al. 1996; Smelror et al. 2001a; Tsikalas et al. 2002a). The cross-sectional profiles (Fig. 9.2) reveal that sedimentary thickness variations are influenced by the crater. In particular, most of the postimpact units over the marginal fault zone thicken and sag considerably (Fig. 9.2b). The discernible thinning of the unit deposited immediately-after-impact over the peak ring (TD-UB unit of earliest Berriasian-earliest Barremian age, Fig. 9.2; TD: top disturbance, impact horizon; UB: upper boundary, lower Barremian reflector defining the top of a Upper Ryazanian-Lower Barremian limestone unit) is evidence for the early existence of the peak ring as a positive impact feature. Moreover, the depression around the central high exhibits gradual infilling by progressive accumulation of sediments into the original annular basin, which became completely buried prior to the deposition of the intra Barremian unit (IB1-IB2, Fig. 9.2a). The residual depth of the annular basin at the level of the impact horizon (reflector TD) is ~50 ms (70 m), decreasing to ~35 ms (50 m) at reflector UB (lower Barremian), whereas the relief is completely smoothed above reflector IB1 (intra Barremian). The seismic profiles also indicate onlap against the central high for most of the early, pre-IB1 (pre-intra Barremian), postimpact sediments. The increase in reflection coefficients between the strata onlapping the central high (Fig. 9.2a), suggests lateral changes in depositional facies, possibly including turbidite deposits on the steep flanks of the high.

9.2.2 Faulting and Differential Vertical Movements

Analysis of faulting at several craters (Table 9.1) has shown secondary postimpact phases of faulting in addition to the primary impact-induced phase. Postimpact faulting is closely related to reactivation of major faults generated during impact, and initiation of several new faults related to the differential compaction processes (Fig. 9.2).

The marginal zone faults, generated during the impact event, show postimpact reactivation in successive phases, affecting the sedimentary succession beneath URU (Figs. 3.14 and 9.2b; URU: Late Cenozoic upper regional unconformity). In addition, we recognize several other large faults inward of the marginal fault zone that continue into the impact-related strata. Small-offset faults restricted to the postimpact sediments are also observed. Some of the faults are associated with gentle folding (Figs. 3.14 and 9.2b).

We have studied the successive phases of faulting by: (1) fault restoration and quantification of fault displacement for the various stratigraphic intervals, (2) observations of depositional features around the faults, and (3) determination of the radial fault distribution for the various deformation phases. At the Mjølnir crater, detailed structural and stratigraphic analysis show that the faulting is of varying amplitude and spatial extent, and that there are two major postimpact phases of faulting in addition to the impact-induced phase (Tsikalas et al. 1998a; Tsikalas and Faleide 2007). The impact-induced faulting (LB-TD; LB: lower boundary, upper Callovian/middle Oxfordian) is mainly observed within the marginal fault zone and giving rise to rim faults with throws of ~30–70 ms (45–105 m) (Fig. 9.2). It was succeeded by the postimpact intra Barremian (UB-IB1) and post-intra Barremian (IB2-URU) phases, which are separated by a tectonically quiet period during the deposition of unit IB1-IB2 when the surface expression of the crater was buried and disappeared (Figs. 3.14 and 9.2).

In terms of the number of reactivated and new faults, the post-intra Barremian extensional faulting was more intense and complex than the preceding postimpact phase. It affected the entire structure and, to a larger degree than earlier phases, the surrounding platform (Figs. 3.14 and 9.2b). We clearly recognize faulting affecting the top of the large clinoform (reflector BAp: base Aptian) (Fig. 9.3). The faults do not penetrate or are hidden within the above-lying shaley unit. Note that this unit is only preserved at the southern edge of the structure (Fig. 9.3). The same or a younger event of faulting gave rise to faults truncated by URU (Figs. 3.14 and 9.2b). The unit geometries and the progressive infilling of the annular depression (Fig. 9.2a) show that the surface expression of the crater disappeared prior to the deposition of unit IB1-IB2, which comprises a quiet period (Fig. 3.14).

We relate both the UB-IB1 and the IB2-URU phases of faulting to differential vertical movements caused by the development of the thick overburden above the structure. An important element of the postimpact burial history of the Mjølnir crater is the Early Barremian regional shelf-progradation toward the south. We speculate that a paleo-shelf edge was located just to the north of the structure and that it later may have been removed during the northern hemisphere glaciations (Fig. 9.3). A prograding depositional apron exceeding a critical threshold weight induced differential compaction, which initiated differential vertical movements. This process may explain the evolution of a prominent central high standing above the rim level, the enhancement of the originally subtle peak ring and marginal grabens, and the structural sag of the entire postimpact succession within the structure. In such a model, most of the postimpact faults could be interpreted as compaction faults, as in the case of Chesapeake Bay crater (Poag 1996; Poag et al. 2004). A significant observation is the absence of faulting at the central high (Figs. 3.14 and 9.2a), which shows that the differential motion was flexural in character. Coeval regional phases of faulting may have affected the faulting at Mjølnir. However in contrast to what would be expected in such a case, neither is there any clear evidence for preferred fault directions, nor is the deformation concentrated in any specific sector of the crater. During and/or after the post-intra Barremian deformation, southward tilting of the entire Bjarmeland Platform took place (Fig. 1.7). The regional tilting of the platform is linked to extensive Late Cretaceous-Tertiary uplift and to isostatic response to glacial erosion, which preferentially affected the northern Barents Sea (Gabrielsen et al. 1990; Nyland et al. 1992; Faleide et al. 1993, 2008).

9.3 Other Craters: Chesapeake Bay, Chicxulub, Bosumtwi, and Montagnais

Numerous shallow boreholes in the near vicinity of the Chesapeake Bay crater, USA, (Table 9.1; Fig. 9.4a) (Poag et al. 2004) and the new USGS-NASA Langley (Horton et al. 2005a, b) and ICDP (International Continental Drilling Program)-USGS (e.g., Gohn et al. 2006, 2008) (Eyreville) coreholes provide a detailed chronostratigraphic scheme for the siliciclastic postimpact units. There are no precise estimates of the maximum amount of postimpact sediments deposited over the structure. From available seismic reflection profiles (Poag et al. 2004) and utilizing uniform interval velocities of 2–2.5 km/s for the entire postimpact sequence the maximum postimpact succession is estimated to be ~800 m (range of 700–950 m). The impact-related Exmore breccia at Chesapeake Bay crater is overlain by the Upper Eocene Chickahominy Formation (Fig. 9.4a), which is a silty clay unit that exhibits a 60–100 m thickness and distinct thickness variations within the crater, and diminishes in thickness beyond the crater rim (Poag 1996). In addition, sagging and differential subsidence make the Chickahominy Formation mimic the geometry of the underlying Exmore breccia surface (Fig. 9.4a) (Poag et al. 2004). A similar trend is observed at Mjølnir where the TD-UB limestone bed mimics the original crater relief (Fig. 9.2) (Tsikalas et al. 1998a; 2002b). Furthermore, all postimpact units exhibit discernible depositional and structural thickening and sagging above the Chesapeake Bay structure (Fig. 9.4a). Similar to the Chickahominy Formation, several Oligocene and Lower Miocene postimpact units are considerably thin or absent beyond the crater's rim (Poag 1996; Poag et al. 2004), providing evidence for extensive persistence of the impact-induced morphological relief for

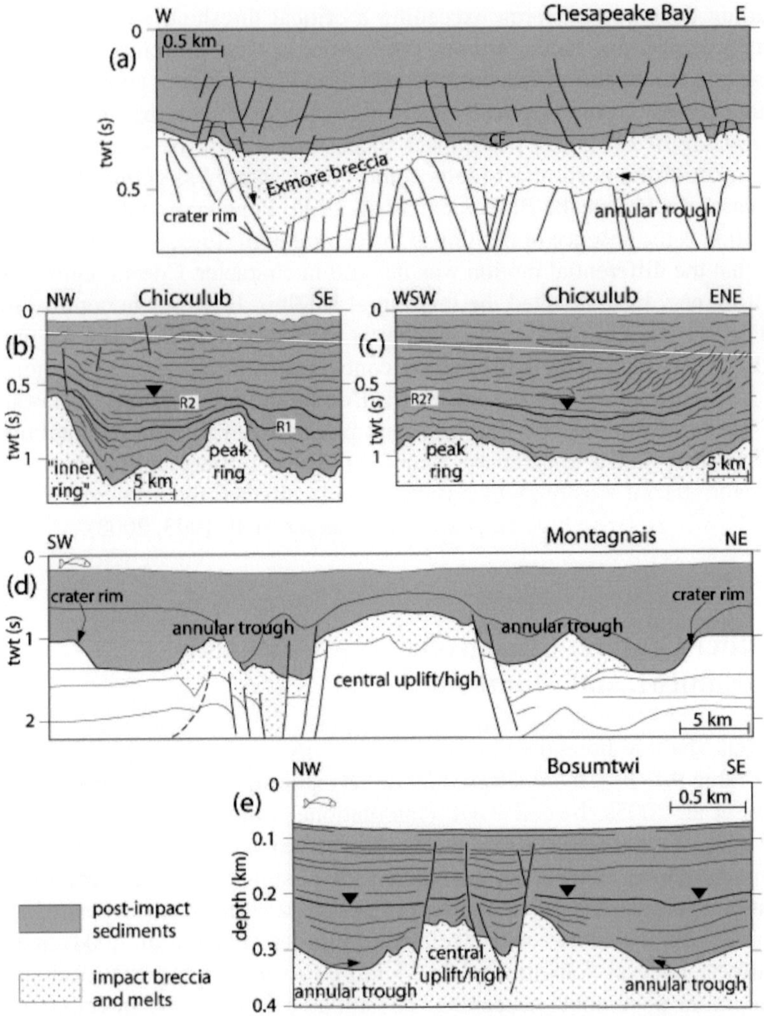

Fig. 9.4 Interpreted seismic reflection profiles across well-preserved impact craters that experienced postimpact burial and related modifications. (**a**) Chesapeake Bay crater (profile crossing the NW side of crater rim, modified from Poag 1996). CF, Chickahominy Formation; (**b**) Chicxulub crater (re-interpreted part of profile Chicx-B across the NW side of the crater rim; Snyder and Hobbs 1999b). For reflectors R1 and R2 see discussion in the text; (**c**) Chicxulub crater (re-interpreted part of profile Chicx-A1 across the NW side of the crater rim; Snyder and Hobbs 1999b); (**d**) Montagnais crater (profile across the entire structure; modified from Jansa et al. 1989); and (**e**) Bosumtwi crater (re-interpreted part of profile B2000-1; Scholz et al. 2002; see also Scholz et al. 2007)

long time after impact. At Chesapeake Bay crater, differential compaction of the impact-related Exmore breccia is responsible for an extensive number of normal-offset growth faults and fault systems within the postimpact sedimentary succession (Fig. 9.4a) (e.g., Poag 1996; Poag et al. 2004). The faults mainly cluster in concentric orientations throughout the postimpact strata. The majority of these compaction faults cut through most of the postimpact succession and their throws decrease up-section, indicating that they are growth faults. Simultaneously long-term continuous or intermittent movements took place (Fig. 9.4a) (e.g., Poag et al. 2004).

The Chicxulub structure (Yucatan, Mexico) (Table 9.1) is considered a relatively intact and pristine impact crater covered by a 1 km thick postimpact Paleogene and Neogene successions. In the central crater region thicknesses of ~1–1.2 s two-way traveltime of marine sediments appear (Figs. 9.4b, c). Several onshore wells and first of all the ~1.5-km-deep ICDP Yaxcopoil-1 bore hole provide detailed information for the postimpact Cenozoic stratigraphy above the structure (e.g., Urrutia et al. 2001, 2004; Popov et al. 2004).

Utilizing seismic velocity information from the seismic reflection and refraction experiments (Brittan et al. 1999; Christeson et al. 1999; Mackenzie et al. 2001; Morgan et al. 2002a, b) a uniform velocity of ~2.5 km/s can be picked as a representative average for the entire postimpact succession at Chicxulub. This number has been used for depth-conversion in our analysis. Variable classification schemes have been used over the years to describe the discernible impact-induced structural ring features within and in the near vicinity of Chicxulub (e.g., Morgan and Warner 1999; Snyder and Hobbs 1999a). However, it seems that a consensus recently has been reached defining a prominent peak-ring within the crater, which is bounded by the crater rim, with an outer-ring and a weak exterior-ring (e.g., Morgan et al. 2002a). Figure 9.4b comprises part of the Chicx-B seismic profile (Snyder and Hobbs 1999b) that does not exhibit exactly the typical dimensions for the structural elements at Chicxulub. The feature in the SE on Fig. 9.4b is the peak-ring, while the one in the NW can be part of a more composite peak-ring structure, or can be considered as an "inner-ring" (Morgan and Warner 1999, their Figs. 4 and 6; Poag et al. 2004, their Fig. 10.25) that is closely related to the crater rim as it exhibits the largest morphological relief. We chose the latter terminology ("inner-ring") as more convenient to differentiate the two prominent elements (Fig. 9.4b) in our further analysis. Furthermore, the depressions around the prominent peak-ring and "inner-ring" appear to have filled gradually by sediments (Fig. 9.4b, c). In particular, the residual depth of the depression between the two elements at the level of the impact horizon (top of impact breccia; K/T boundary at the base of the impactite) is 550–600 ms (~700 m), decreasing to 200–250 ms (~300 m) at reflector R1, and 100–125 ms (~150 m) at reflector R2, whereas the relief is almost smoothed stratigraphically farther above (black triangles, Fig. 9.4b, c).

The seismic profiles also indicate onlap against the peak-ring for most of the pre-R1 postimpact sediments (Fig. 9.4b, c). Reflector R1 represents the first continuous reflector to have surpassed and covered the impact-induced relief, whereas reflector R2 represents the stratigraphic level above which the inherited original impact structure is minimal. Based on interpolation of the onshore Yaxcopoil-1 borehole stratigraphy and seismic profile ties, a preliminary age of ~40 Ma (Middle Eocene) was assigned close to reflector R2 (Bell et al. 2004). At Chicxulub, several small-throw faults restricted to the Cenozoic postimpact succession are observed, and mainly concentrated above the "inner-ring", the peak ring (Fig. 9.4b) and the crater rim (e.g., Bell et al. 2004). In detail, the internal postimpact depositional patterns reveal discrete fluxes in relative postimpact vertical movements, with sedimentation initially reducing the impact-generated relief. The subsequent differential compaction created additional relief, which was filled (black triangles, Fig. 9.4b, c). The latter is possibly accommodated by faulting on top of the prominent structural elements (Fig. 9.4b).

At the Montagnais crater offshore Canada, (Table 9.1), continuous reflections from postimpact strata of Eocene to Holocene age are observed along seismic reflection profiles tied to the deep-well drilled on the central uplift/high (Jansa et al. 1989; Pilkington et al. 1995), and there is clear evidence of draping of these sequences on the flanks of the central uplift (Fig. 9.4d). Several periods of submarine current erosion at postimpact times are also observed, particularly recognizable by the deep channel that cuts into the deposits southwest of the Montagnais central uplift (Fig. 9.4d) (Jansa et al. 1989). It is believed that the central high/uplift of the structure was a positive relief feature above the seafloor and an obstacle to ocean-bottom currents during post-Early Eocene times (e.g., Jansa 1993). This is inferred because the depth of erosional channeling diminishes northward from the center of the structure (Fig. 9.4d). The central part of the structure is covered by ~0.5 km of postimpact sediments that reach ~1 km thicknesses in the annular depressions. The variable thickness of impact breccia is closely connected to both varying degrees of postimpact erosion and thickness variations of the postimpact strata. This indicates prolonged influence of the impact-induced relief and differential sediment loading, especially in the annular depressions surrounding the central high/uplift of the structure (Fig. 9.4d). At Montagnais crater, postimpact reactivation of impact-related faults is observed, mainly at the faults bounding the thick breccia bodies surrounding the central high/uplift (Fig. 9.4d).

At the Bosumtwi crater, Ghana (Table 9.1), the postimpact sediments covering the crater structure are 150–300 m thick (Fig. 9.4e) (Karp et al. 2002; Scholz et al. 2002, 2007) and the recent ICDP coring resolved the postimpact stratigraphy (e.g., Coney et al. 2007a, b; Koeberl et al. 2007a, b). The present-day central uplift/high, with a diameter of 1.9 km and maximum height of 130 m, was a positive feature immediately after impact, as progressive onlap and distinct thickness variations are present within the postimpact sediments (Fig. 9.4e). Following impact and subsequent sediment accumulation, the impact-induced morphologic relief diminished (black triangles, Fig. 9.4e) and progressive accumulation of lacustrine sediments led to postimpact deformation. This is evidenced by the slight sagging of the sequences deposited after complete burial of the original impact relief, and by fault reactivation on the central high. Most probably these faults were initiated during collapse of the transient crater (Scholz et al. 2002; Artemieva et al. 2004) and later reactivated due to differential sediment loading and compaction (Fig. 9.4e). At the Bosumtwi crater, postimpact fault displacements within the lacustrine section is in the order of 1–3 m, with a maximum offset of 15 m just above the central high that was prone to reactivation (Fig. 9.4e) (Scholz et al. 2002).

Several other impact structures are known to have been buried by considerable amounts of postimpact sediments, which potentially could have triggered modifications, but data for comprehensive analysis do not exist. Such structures include (among others) the Manson (USA) (e.g., Hartung and Anderson 1996), Ames (USA) (e.g., Carpenter and Carlson 1997), and Lockne (Sweden) (von Dalwigk and Ormö 2001) craters. The offshore Silverpit structure (Stewart and Allen 2002, 2005) is not considered here as it has not yet been shown to be a genuine impact crater.

Postimpact infilling can complicate seismic interpretation and it may be difficult to differentiate impact-related and postimpact sequences. Seismic profiles with good resolution (Figs. 9.2 and 9.4) may help reduce these complications. The depositional influence of the impact-induced relief can take millions of years to reduce depending on the local sedimentation rate. It may be followed by complete burial of the crater-related morphology. In particular, at Mjølnir crater complete burial of the impact-induced morphological relief was completed during Barremian time, i.e., ~15 million years after the impact (Jurassic-Cretaceous boundary, Smelror et al. 2001a). Continued deposition subsequently created a substantial overburden. For the Chicxulub crater, Bell et al. (2004) postulated a spatially progressive infilling. In particular, the western and northwestern parts of the Chicxulub postimpact Cenozoic basin were filled ~25 million years after the impact, whereas during a major marine regression a shelf progradation took place in the east ~45 million years after the impact (in Early Miocene). Similarly, at Chesapeake Bay complete burial of the structural crater relief may have required ~15 million years, as lower Miocene units are considerably thin or absent beyond the crater's rim (Poag 1996; Poag et al. 2004).

9.4 Original Crater Relief Reconstruction

Well-preserved impact structures (Table 9.1; Figs. 9.2 and 9.4) have shown that the distribution and character of postimpact successions have been influenced initially by the crater structure and morphology. The long-term subsidence of a thick overburden, differential between the unconsolidated impact breccia (autochthonous and allochthonous) inside the crater and the semi-consolidated or unconsolidated preimpact sedimentary column outside the crater had additional effects (Fig. 9.5). An exponential reduction function is considered to best approximate the porosity-depth relationship for a compacted sedimentary section (e.g., Wyllie et al. 1956, 1958; Gardner et al. 1974; Sclater and Christie 1980). By introducing an impact-induced porosity change (Fig. 9.5), the new porosity-depth function, $\text{Ø}'$, which defines the compaction in the crater, becomes:

$$\text{Ø}' = \text{Ø}_0 \mathrm{e}^{-cz} + \Delta\text{Ø} \qquad (9.1)$$

where Ø_0 is the initial porosity, c is the compaction constant, $\Delta\text{Ø}$ is the porosity anomaly induced by the impact, and z is the total, present and eroded, overburden.

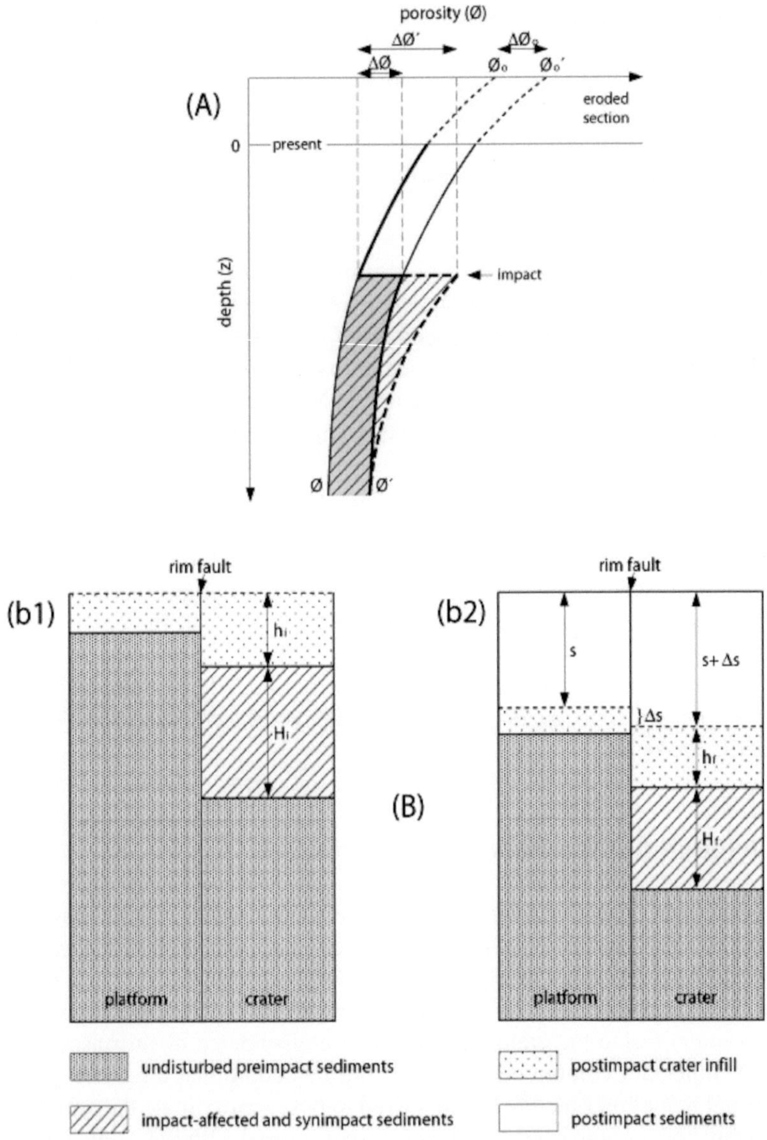

Fig. 9.5 (**a**) Generalized porosity-depth functions for sedimentary rocks incorporating also the impact-produced porosity change. *Shading* denotes a constant porosity anomaly function, whereas *hatching* denotes an exponentially decreasing function. The various porosities, Ø, are described in the text; $Ø_0'$ is the expected initial/surface porosity of the impact-induced porosity-depth function $Ø'$, and $\Delta Ø_0$ is the porosity anomaly between the initial/surface porosities $Ø_0$ and $Ø_0'$ (modified from Tsikalas et al. 2002b). (**b**) Schematic diagram of postimpact differential compaction across the crater rim: *(b1)* immediately after the postimpact infilling of the crater, and *(b2)* at present. H_i and H_f, initial and present (after postimpact burial) thickness of impact-affected and synimpact sediments, respectively; h_i and h_f, initial and present (after postimpact burial) thickness of postimpact crater infill, respectively; s, the present and eroded overburden above the infilled crater; Δs, observed differential compaction (modified from Tsikalas et al. 2002b)

The impact-induced porosity anomaly ($\Delta Ø$) can be considered either as constant (retained until present) or as exponentially decreasing during burial (see Eq. 9.2 and Fig. 9.5).

$$\Delta Ø = \Delta Ø' e^{-cz} \quad (9.2)$$

The reconstruction of the immediately-after-impact crater relief is consequently of great importance to elucidate the postimpact evolution, and to provide an original crater relief for volume balance (cratering scaling laws) calculations and comparison with the terrestrial impact record. The approach employed in this study is reconstruction and backstripping of all postimpact sequences in which the deposition has been influenced by the underlying crater relief. Sediment decompaction and fault restoration at time steps, corresponding to postimpact seismo-stratigraphic boundaries, were carried out using the Basin Modelling Toolbox (Fjeldskaar et al. 2004) software. These calculations were done for the Mjølnir, Chicxulub, and Bosumtwi craters, which can be considered end-members in the postimpact burial spectrum, receiving an overburden of 2–2.5, 1–1.5, and 0.3 km, respectively (Table 9.1). For Mjølnir, we update the results based on the analysis of Tsikalas et al. (1998a), whereas for the Chicxulub and Bosumtwi craters, we have conducted an analysis that represents a new aspect of their existing studies.

9.4.1 Mjølnir

At Mjølnir (Fig. 9.6), the depth-conversion was based on the velocity-depth function of Tsikalas et al. (1998b). A well-log derived porosity-depth relationship for siliciclastic sediments ($Ø_0$, ~55%; c, 0.42 km^{-1}) developed for the southwestern Barents Sea (Tsikalas 1992) was used to decompact the postimpact sequences. In addition, a uniform Early Cretaceous paleowater-depth of 500 m was incorporated, consistent with the shallow-marine depositional environment (Dypvik et al. 1996; Smelror et al. 2001a). The Mjølnir crater currently lies beneath ~50–800 m of postimpact sediments (Fig. 9.6f) (e.g., Tsikalas et al. 1999). In the reconstruction analysis it is of major importance to realize that 1.5–2 km of siliciclastic sediments were deposited during Cretaceous-Tertiary times (Fig. 9.6d) (e.g., Dimakis et al. 1998) and were later removed (Fig. 9.6e), during the Late Cenozoic glacial erosion. Reconstruction suggests that the impact resulted in a very shallow structure with an average crater depth of only ~30 m (Fig. 9.6a), taken as the depth relative to the pre-impact top surface of the allochthonous breccia (reflector TD) (Tsikalas et al. 1998a). Furthermore, the reconstruction dramatically changes the width and height of the central high, as the original crater exhibited a narrower-than-present central high, 4.5–5 km in diameter, that stood ~80 m above the surrounding platform level, in comparison with the present 8-km-diameter width and 250-m-height (Fig. 9.6a, f). On the other hand, the annular basin was quite prominent with a maximum depth of ~70 m, while the peak ring was not well developed in the original crater (Fig. 9.6a, f).

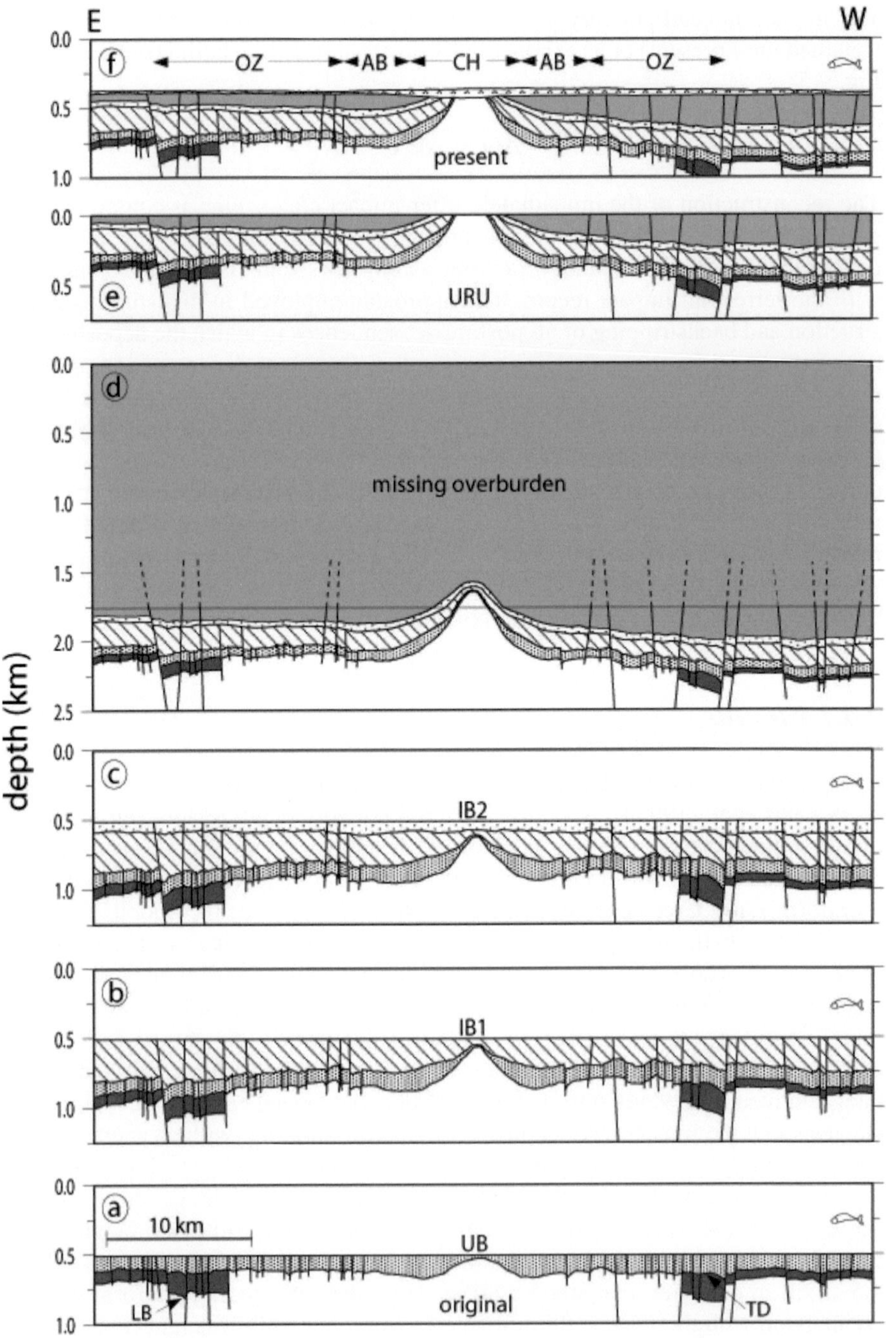

Fig. 9.6 Reconstruction of the Mjølnir original crater relief along an E-W trending profile by decompaction and fault restoration of the entire postimpact sedimentary succession. Time steps (**a**–**f**) correspond to the main unit boundaries. CH, central high; AB, annular basin; OZ, outer zone. Other annotations as in Figs. 9.1 and 9.2 (modified from Tsikalas et al. 1998a)

Furthermore, the reconstruction suggests that the original thickness of the allochthonous breccia in the marginal fault zone was ~125–200 m and reduced by compaction to 75–115 m. Similarly, the original throw on the rim faults, when corrected for reactivation and compaction was only 75–105 m, as compared with the cumulative throw of 120–150 m (Tsikalas et al. 1998a). In summary, the decompaction and fault restoration approach at Mjølnir (Fig. 9.6) demonstrates that the deformation associated with the extensive postimpact overburden considerably enhanced the structural expression of an originally subtle crater and that the present distinct crater expression is largely a postimpact burial phenomenon (Fig. 9.6) (Tsikalas et al. 1998a, 2002b; Tsikalas and Faleide 2007). At Mjølnir, a quantitative model for the porosity change caused by the impact has been developed using density and seismic travel time/velocity distributions, and postimpact sediment deformation (Fig. 9.7) (Tsikalas et al. 1998c, 2002a). This model uses well-known equations describing the inter-relation between porosity, density and seismic velocity (Wyllie et al. 1956, 1958). The model integrates the impact-induced porosity anomaly ($\Delta Ø$) and the corresponding new porosity-depth function ($Ø'$) (Fig. 9.5a), as well as postimpact differential compaction effects (Fig. 9.5b). The integrated geophysical modeling at Mjølnir demonstrates a close correspondence of geophysical anomalies to the radially-varying distribution of structural and morphological units, and to the physical-properties distribution. Specifically, the impact resulted in an extensive, 850–1,400 km^3 (Tsikalas et al. 1998b), seismically disturbed volume at the impact-site (Fig. 9.7). These formations exhibit relatively increased porosities, and decreased densities and velocities in the highly fractured, brecciated and collapsed porosities and increased densities and velocities, where crater floor rebound and structural uplift processes prevailed (Fig. 9.7).

This setting resulted in differential subsidence and compaction faulting within the structure, reactivated older faults (especially rim faults) and initiated new ones (Figs. 9.2, 9.6 and 9.7). The extensive postimpact burial (Fig. 9.6d) led the brecciated periphery to further compaction than the denser central crater and consequently a central Mjølnir high standing taller than the surrounding platform beds (Figs. 9.2, 9.6, and 9.7). In comparison to the dramatic impact cratering processes, postimpact burial produced more subtle, long-term alterations which reduced the porosity (and thus density) contrasts between the crater periphery and the platform. According to the compilation of gravity anomalies of impact craters in sedimentary targets (Pilkington and Grieve 1992), a structure of the size of Mjølnir should produce an annular low of about –7 mGal. However, the observed value for Mjølnir is only –1.5 mGal (Fig. 9.7), though this is still within the wide range of –1 to –11 mGal determined from a number of craters. Although we recognize that the Mjølnir impact in relatively soft marine sediments may have resulted in less pervasive brecciation than in crystalline settings, we conclude that the reduced annular gravity low is closely connected to the extensive postimpact burial. Therefore, interaction of impact-induced and postimpact processes accounts for the lower-than-expected gravity (and seismic velocity) values, and the corresponding porosity distributions and levels (Fig. 9.7).

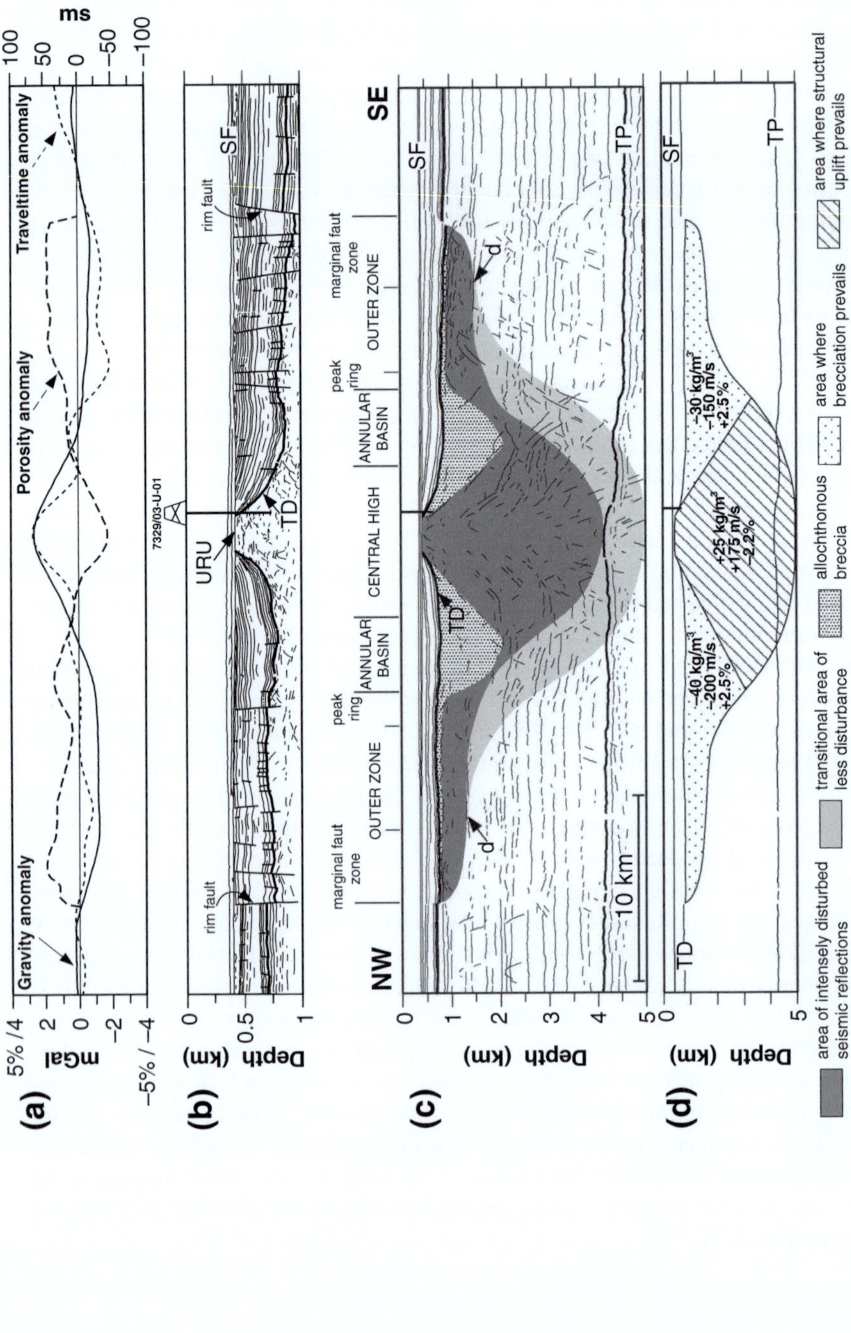
(a)
(b)
(c)
(d)
mGal
ms
Depth (km)
Gravity anomaly
Porosity anomaly
Traveltime anomaly
rim fault
7329/03-U-01
URU
TD
SF
TP
NW
SE
marginal faut zone
OUTER ZONE
peak ring
ANNULAR BASIN
CENTRAL HIGH
10 km
d
−40 kg/m³ −200 m/s +2.5%
+25 kg/m³ +175 m/s −2.2%
−30 kg/m³ −150 m/s +2.5%
area of intensely disturbed seismic reflections
transitional area of less disturbance
allochthonous breccia
area where brecciation prevails
area where structural uplift prevails

A second reconstruction approach consisted of isopach mapping of the postimpact strata that show thickness variations correlating with the crater relief. In order to estimate the original crater relief, we initially mapped units TD-UB and TD-IB2. The TD-UB unit represents deposition immediately after impact and closely mirrors the relief of the impact horizon. This thin unit was deposited in relative deep water and probably did not completely fill the structural anomaly created by the impact. As a result, crater depth estimates based on the TD-UB isopach are minimum estimates. On the other hand, since the impact structure was completely buried at the level of reflector IB1, an estimate of crater depth based on thickness variations in the TD-IB1 isopach will not be influenced to the same degree by uncertainty in paleo-depth. The possibility of sagging under load makes this depth estimate a maximum one. Because reflector IB2 can be mapped with a greater confidence, the maximum depth estimate was instead based on the TD-IB2 isopach. This approach was possible because of the uniform thickness of the IB1-IB2 unit.

We averaged the two isopach maps and removed a second-order polynomial surface from the data corresponding to a uniform southward thinning trend. The resulting residual crater-depth values were gridded into 0.5×0.5 km cells and manually re-contoured. The original crater relief in Fig. 9.8 exhibits structural features quite similar to the present crater morphology (Fig. 3.10), however with reduced amplitudes. The original crater becomes very shallow with an average depth of only ~10–20 ms (25–40 m). To facilitate comparison with the corresponding results obtained by backstripping, the thickness values were decompacted using similar parameters. The ~5 km diameter central high is less prominent than at present (Fig. 3.10). It is surrounded by a ~30–50 ms (60–90 m) deep and 4.5–5.5 km wide annular basin (Fig. 9.8). Furthermore, the peak ring and the marginal grabens are less well-developed than at present.

In summary, both reconstruction approaches (Figs. 9.6 and 9.8) clearly demonstrate that the deformation associated with the extensive postimpact overburden considerably enhanced the structural expression of an originally rather subtle crater. Consequently, the present very distinct expression of the Mjølnir crater is largely a postimpact burial phenomenon.

9.4.2 Chicxulub

For the Chicxulub impact structure, the depth-converted section illustrating the peak-ring and "inner-ring" features was used in the compaction reconstruction

Fig. 9.7 Geophysical type section along a NW-SE trending profile at the Mjølnir Crater. (**a**) observed free-air gravity and seismic travel time anomalies, and modeled porosity anomaly, (**b**) interpreted high-resolution single-channel profile, (**c**) interpreted multi-channel profile (**a–c**, modified from Tsikalas et al. 1998a–c, 2002b), and (**d**) impact crater model with calculated physical property distribution. The crater model geometry in (**d**) is corrected for regional tilt, and the modeled density contrasts are given in kg/m^3, the seismic velocities in m/s, and the porosity anomalies in percentage. The type section is representative for the structural and morphological radial zonation of the Mjølnir Crater. Annotations as in Fig. 9.1

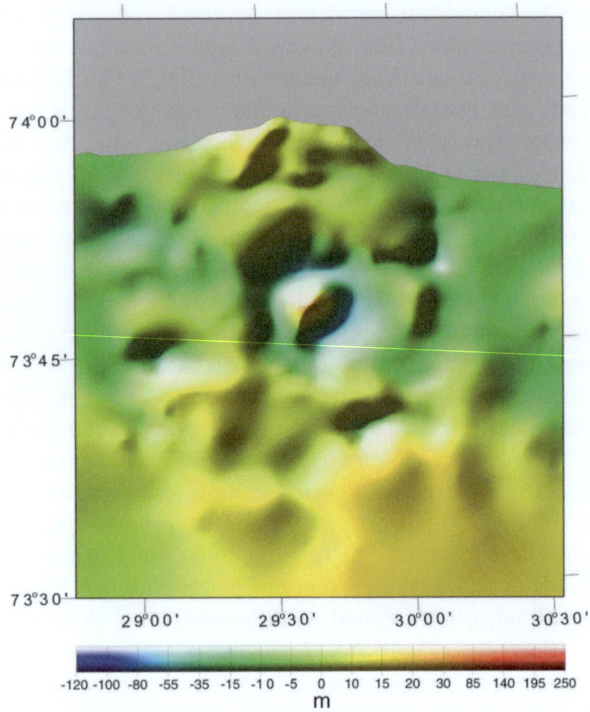

Fig. 9.8 Illuminated perspective image of the reconstructed original Mjølnir crater relief (expressed as depth residuals relative to a substracted regional southward thinning trend; see text and confer Fig. 3.10) from isopach mapping of the postimpact strata that exhibit crater-influenced sedimentation. The view is directly from above; light sources are at azimuths 30°, 290°, and 340°. The *grey area* in the north shows where intra Barremian reflector IB2 is truncated by erosion. Vertical exaggeration ~20×

(Fig. 9.9a). The peak-ring and "inner-ring" with average radius of 40 and 60 km from the crater center appear irregular and rugged (Figs. 9.4b, c and 9.9a).

The seismic reflection profiles at Chicxulub display typically a narrow and prominent peak-ring, ~550 m above the crater floor in Fig. 9.9a (400–600 m elsewhere), in the west and northwest. It is broader and less prominent (200–300 m above crater floor) to the east and northeast (Fig. 9.4c) (e.g., Morgan et al. 1997). Backstripping reconstruction at Chicxulub was employed at two time-steps defined by the sequences bounded by reflectors R2 and R1 in which the deposition was influenced by the underlying crater relief (Fig. 9.9). The Tertiary sediments of the Yaxcopoil-1 borehole are dominated by carbonaceous siltstone to limestone deposits (e.g., Arz et al. 2004; Kenkmann et al. 2004; Popov et al. 2004; Vermeesch and Morgan 2004). Nevertheless, it was argued that the thick offshore postimpact strata may contain several additional siliciclastic deposits, as revealed from the seismo-stratigraphic depositional patterns including, prominent prograding clinoform build-ups (Bell et al. 2004). In order to include various alternative possibilities, numerous porosity-depth relationships ($\emptyset_0$ and c combinations) ranging between siliciclastic (sand- and shale-dominated alternatives and mixtures), carbonaceous and dolomitic strata were used to decompact the postimpact succession at Chicxulub (Fig. 9.9, models 1–3). In addition, two different paleodepths were used. Initially, a uniform paleodepth of 60 m (that approximates the current water depth) was used

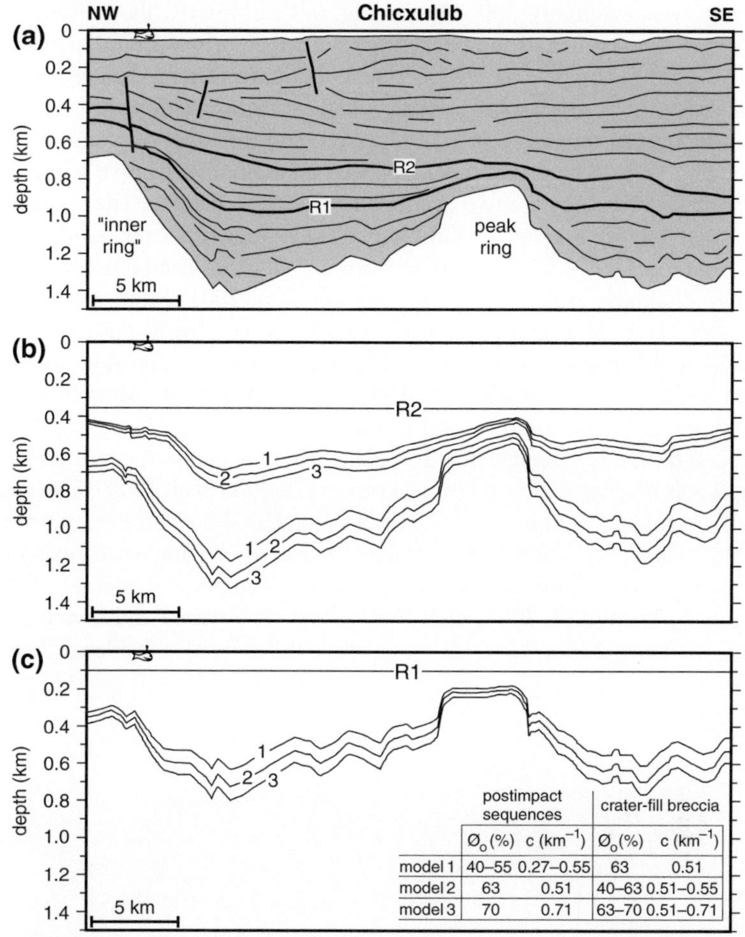

Fig. 9.9 Reconstruction of the Chicxulub original crater along the depth-converted (**a**) profile in Fig. 9.4b (re-interpreted part of profile Chicx-B across the NW side of the crater rim; Snyder and Hobbs 1999b) by backstripping and decompaction utilizing paleowater depths of 350 and 100 m for reflectors R2 and R1, respectively. Reconstruction time-steps at reflectors R2 (**b**), and R1 (**c**). A full spectrum of porosity-depth compaction relations of impact and postimpact rocks is incorporated (models 1–3). $Ø_0$, initial/surface porosity; *c*, compaction constant. See text for more comprehensive discussion on the reconstructions

for the backstripped time-steps. Finally, a differentiated paleodepth of 350 and 100 m was used for the R2 and R1 time-steps, respectively (Fig. 9.9b, c), based on the estimates derived from the progradating clinoform geometries (Bell et al. 2004).

Reconstruction at Chicxulub shows that the restored peak-ring was not only considerably subdued by comparison with the present configuration, but that the two nearby structural elements; peak-ring and "inner-ring" underwent a differentiated postimpact evolution (Fig. 9.9). In particular, the current ~700-m-relief of the "inner-ring" was originally 300–450 m (Fig. 9.9). Similarly, the current ~550 m (range 535–575 m) relief of the peak-ring above the surrounding depressions was originally ~500 m (range 420–570 m). Note that currently the "inner-ring" stands ~150 m above the peak-ring (Fig. 9.9a), whereas reconstruction shows that immediately after impact this relation was reversed, and the peak-ring stood higher by ~100 m (Fig. 9.9c). Therefore, there has been a cumulative relative vertical movement of ~250 m between the two prominent structural features. The reconstruction suggests that the postimpact evolution is closely related to the structurally disturbed volume at Chicxulub, containing zones of weakness and a brecciated region of variable thickness at depth (e.g., Christeson et al. 2001; Morgan et al. 2002b). Furthermore, the peak-ring lies directly over the inner edge of the collapsed transient cavity and appears to have no consistent gravity or seismic velocity signature (Pilkington et al. 1994; Sharpton et al. 1996; Brittan et al. 1999; Morgan et al. 2000; Christeson et al. 2001). The position of the peak-ring above the inner edge of the last slumped blocks is evidence that the peak-ring is not vertically uplifted basement material that is emergent through impact breccia (Brittan et al. 1999). The peak-ring in Figs. 9.4b and 9.9a is underlain by a thick volume of crater-fill breccia in comparison to the "inner-ring", which is underlain by a smaller volume of impact breccia on top of the inward-slumped blocks (Fig. 9.10) (Morgan and Warner 1999, their Figs. 4 and 6; Poag et al. 2004, their Fig. 10.25). The nature and location of both structural elements governed their postimpact differential subsidence behaviour,

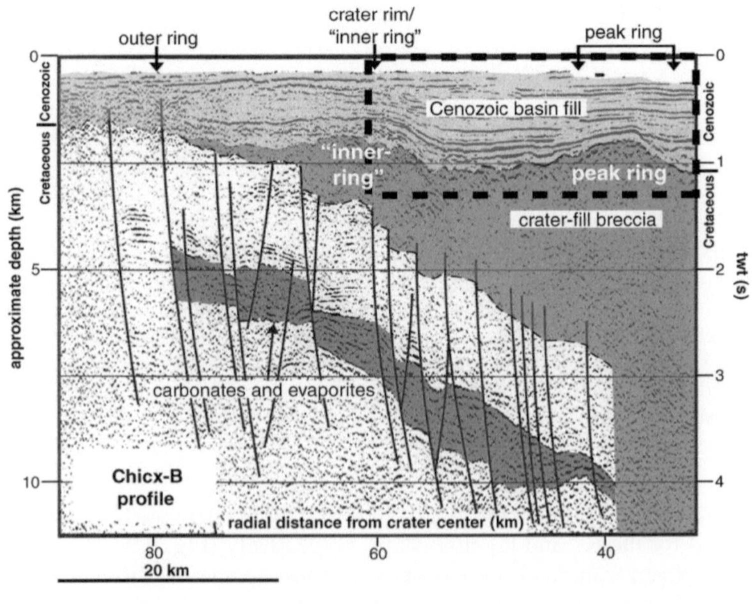

Fig. 9.10 Part of Chicx-B seismic profile at the Chicxulub crater periphery (Snyder and Hobbs 1999b) with overlaying interpretation (modified after Morgan and Warner 1999; Poag et al. 2004). The profile part that was analyzed and reconstructed (Figs. 9.4b and 9.9) is indicated by the *thick dashed rectangle*

with the peak-ring and its surrounding depressions being prone to greater subsidence in comparison to the "inner-ring" (Figs. 9.9 and 9.10). Based on isostatic balancing assumptions utilizing the elastic thickness of the crust and time constants for crustal relaxation, Bell et al. (2004) provided subsidence estimates that reveal an original impact-horizon topographic relief in the range of 450–700 m between the crater and the surrounding platform. In Figs. 9.4b and 9.9a (part of profile Chicx-B; Snyder and Hobbs 1999b) this specific relief corresponds to the relief between the "inner-ring"/crater-rim and the depression bounding it from the peak-ring. Our analysis employing detailed backstripping and decompaction demonstrates that the model of Bell et al. (2004) underestimates the postimpact load effect. The conducted reconstruction (Fig. 9.9) provides a robust model for the postimpact structure evolution as it accounts for a broad spectrum of porosity-depth compaction relations between impact-related and postimpact rocks and fits with the impact-induced structures and geometries.

9.4.3 Bosumtwi

For the Bosumtwi crater, we used the depth-converted section in Fig. 9.4e in the reconstruction analysis. Backstripping and fault restoration was performed at the postimpact stratigraphic level where the impact-induced morphologic relief was diminished (black triangles in Figs. 9.4e and 9.11).

As at Chicxulub, we used a broad-spectrum of porosity-depth relationships ($Ø_0$ and c combinations) to decompact the postimpact succession at Bosumtwi crater (Fig. 9.11). We used two different uniform paleowater depths, one of 75 m that approximates the current water depth (Fig. 9.11a) and another of 200 m as the maximum possible paleowater depth level (Fig. 9.11b).

Reconstruction at Bosumtwi shows a discernible postimpact evolution for the central uplift/high and the two depressions/annular troughs (western and eastern) surrounding it (Fig. 9.11). In particular, the current 110 m height (on this particular profile) of the central uplift/high, measured from the base of the western depression, was originally 95 m (range 85–105 m), whereas the current 101 m height of the central uplift/high, measured from the base of the eastern depression, was originally 103 m (range 95–110 m). On the other hand, the current 46 m depth of the western depression (measured from its base up to the relief on its right side) was originally 55 m (range 43–68 m), whereas the current 50 m depth of the eastern depression (measured from its base up to the small terrace on its left side) was originally 52 m (range 49–55 m) (Fig. 9.11). Recent integrated analysis of gravity and borehole petrophysical data and modeling has shown laterally-varying physical properties at Bosumtwi, which are related to impact cratering processes (Ugalde et al. 2005, 2007).

Our reconstruction results (Fig. 9.11) appear to be in accordance with the latter gravity and petrophysics analysis. Similar to the Mjølnir and Chicxulub craters, the postimpact evolution of Bosumtwi crater is closely related to the impact disturbed

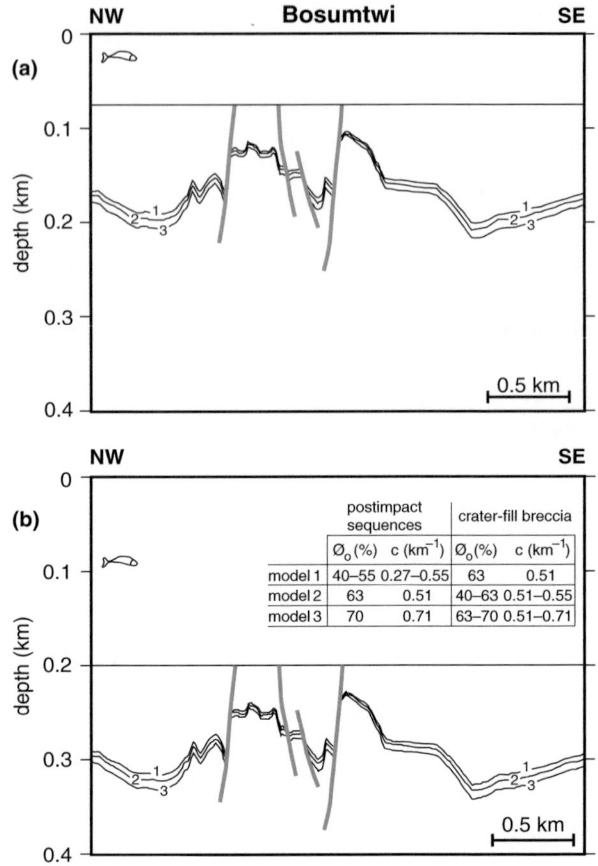

Fig. 9.11 Reconstruction of the Bosumtwi impact crater along the depth-converted profile in Fig. 9.4e (Scholz et al. 2002) by backstripping and fault restoration at the postimpact stratigraphic level where the impact-induced morphologic relief was diminished (*black triangles* in Fig. 9.4e). (**a**) A uniform paleowater depth of 75 m was utilized; (**b**) A uniform paleowater depth of 200 m was utilized. A full spectrum of porosity-depth compaction relations of impact and postimpact rocks is incorporated (models 1–3). Ø_0, initial/surface porosity; *c*, compaction constant. See text for more comprehensive discussion on the reconstructions

target-rock volume and a brecciated region of variable thickness and physical properties at depth (Fig. 9.12). The western part of the reconstructed profile (Fig. 9.11) compacted relatively more, as it was underlain by a more porous impact breccia unit. The central high and the eastern depression subsided relatively less, as they were underlain by less porous, melt-rich breccia related to structural uplift (Fig. 9.12) (Ugalde et al. 2005, 2007).

9.4.4 Chesapeake Bay

Based on the methodology developed for the Mjølnir crater (Tsikalas et al. 2002a), we have produced a preliminary quantitative model for the porosity changes in the Chesapeake Bay crater (Fig. 9.13) using the density distributions of Poag et al. (2004; their Fig. 4.37b). Compared to the surrounding sediments, the porosity increased immediately after impact up to 8.5% in the collapsed and brecciated

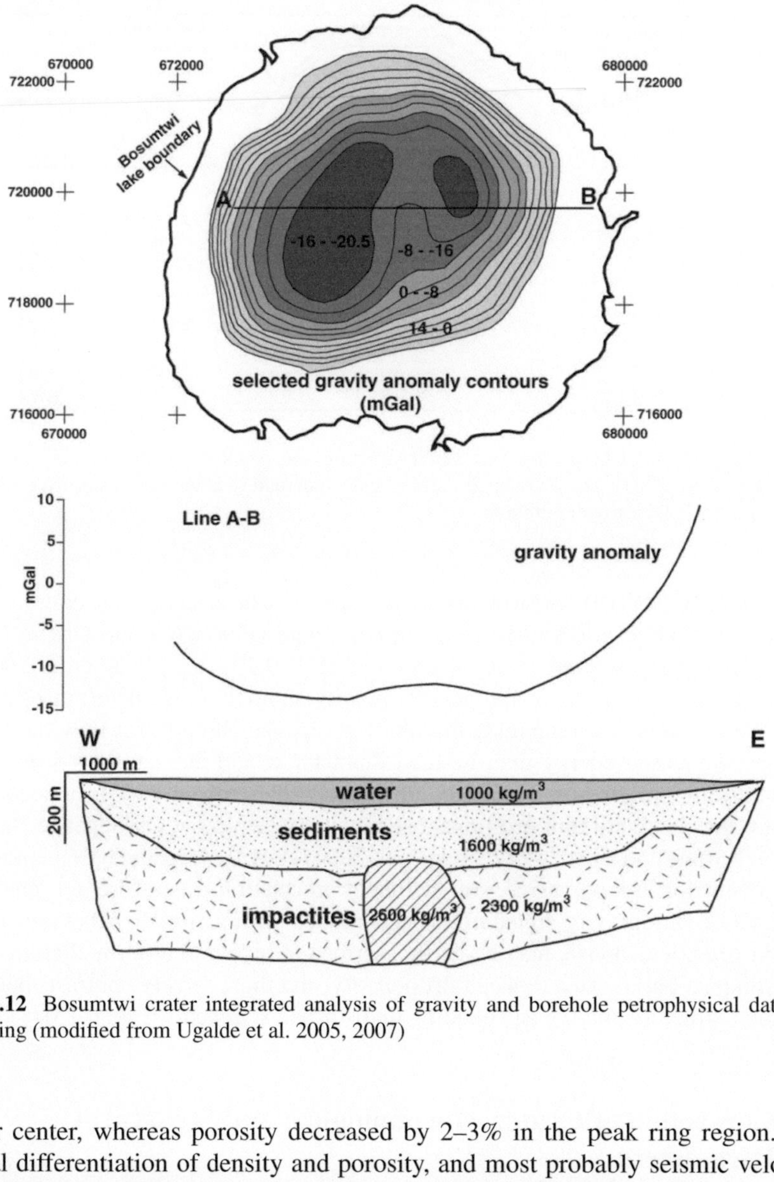

Fig. 9.12 Bosumtwi crater integrated analysis of gravity and borehole petrophysical data and modeling (modified from Ugalde et al. 2005, 2007)

crater center, whereas porosity decreased by 2–3% in the peak ring region. The lateral differentiation of density and porosity, and most probably seismic velocity, is attributed (as at Mjølnir) to the interaction and local spatial prevalence of cratering processes. These include brecciation, gravitational collapse, and structural uplift and were responsible for the large seismically disturbed rock volume at the impact-site (Figs. 9.4a and 9.13). Following impact, the crater was progressively buried by a ~800 m (range of 700–950 m) thick overburden (Table 9.1), which caused differential compaction of the extensively brecciated central part of the crater fill. This compaction significantly affected the postimpact crater evolution, and decreased the porosity anomaly to values of +6% in the center and –3 to –4.5% in the peak

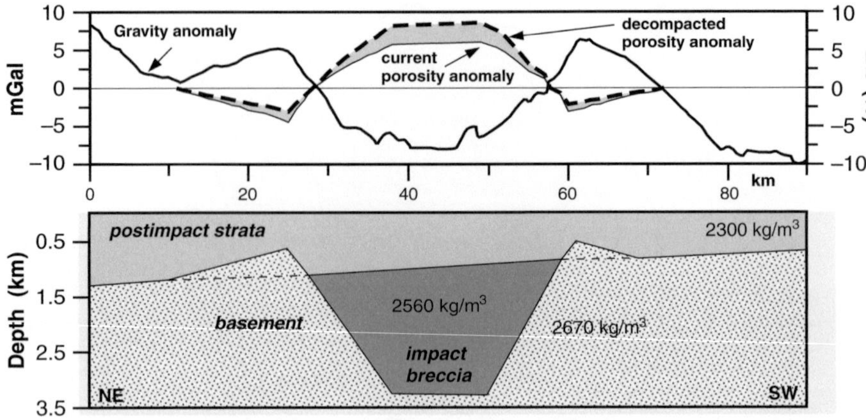

Fig. 9.13 Simplified Chesapeake Bay crater structure and residual gravity anomaly (modified from Poag et al. 2004), and calculated current and immediately-after-impact porosity anomaly using the modeled density distribution

ring region (Fig. 9.13). A structure of the size of Chesapeake Bay crater should produce an annular gravity low of about –15 mGal (Pilkington and Grieve 1992). Nonetheless, the observed value of −8 mGal (Fig. 9.13) (e.g., Poag et al. 2004) is still within the −7 to −20 mGal annular gravity anomaly range determined from a number of craters. We recognize that the Chesapeake Bay impact may have been smaller in released energy than its final diameter would indicate based on recent insights from numerical modeling (Collins and Wünnermann 2005). In contrast to Mjølnir, the relatively thin pre-impact sedimentary section at Chesapeake Bay (1–1.5 km; Powars and Bruce 1999; Poag et al. 2004) above crystalline basement would not account for less pervasive, soft sediment brecciation. Therefore, the moderate Chesapeake Bay gravity signature, in accordance with a similar behaviour of the Mjølnir gravity response, may be partly ascribed to lesser, long-term alteration due to postimpact burial. This reduced the porosity, and thus, density contrasts between the crater structure and the surrounding undisturbed sediments (Fig. 9.13).

9.5 Correction of Crater Morphological and Structural Parameters

9.5.1 Parameters Prone to Postimpact Burial Modification

The quantitative reconstructions for the Mjølnir, Chicxulub and Bosumtwi craters represent the relict transient crater rim uplift, at the location where the transient crater diameter reached its maximum possible extent (Fig. 9.14a). However, there is no general agreement for the formation mechanisms of rings in multi ring basin-scraters, these reconstructions together with studies of other craters (Table 9.1;

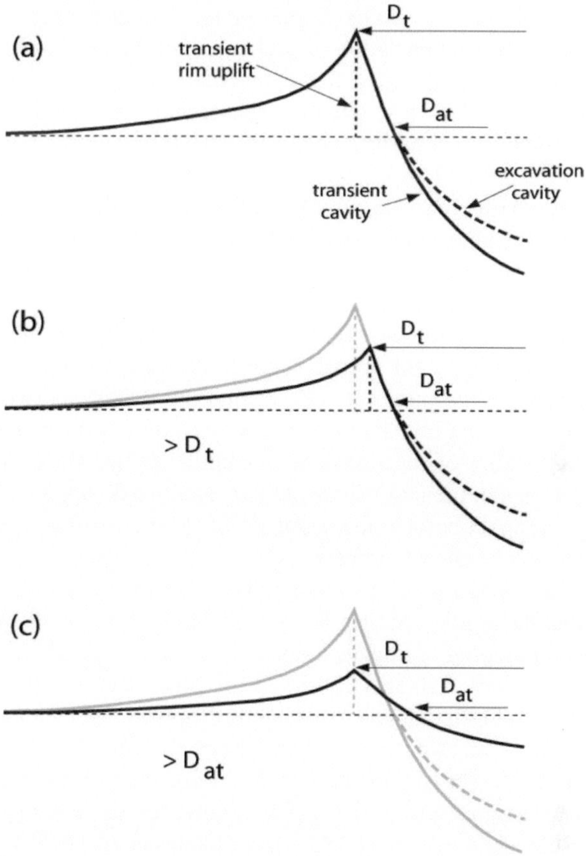

Fig. 9.14 Schematic diagram of the transient cavity, transient rim uplift, and excavation cavity reconstructions without (**a**) and with (**b**), (**c**) considerations of postimpact burial enhancement and relief exaggeration. D_t, true transient crater diameter; D_{at}, apparent transient crater diameter (referred also as the excavated crater diameter). See text for more detailed description

Figs. 9.2 and 9.4) demonstrate the great importance of long-term deformation processes operating after impact. As we have shown, the effects of extensive deformation are cumulative and may enhance or subdue the underlying crater morphology (Figs. 9.6, 9.9, and 9.11), and change the geophysical properties (Figs. 9.7 and 9.13).

Peak-ring and inner-ring features are prominent morphological elements in complex impact craters (e.g., Melosh 1989; Turtle et al. 2005). The mechanism of rings in multi-ring basins and a variety of models have been proposed (e.g., Merrill and Schultz 1981; Melosh 1989; O'Keefe and Ahrens 1999; Grieve and Therriault 2000; Turtle et al. 2005). At Chicxulub, the peak-ring formation (considered crucial in understanding the impact cratering mechanics) is thought to have formed at the region of interaction between upward and outward mass displacement through structural uplift and inwards collapse of the transient cavity (e.g., Morgan et al. 2000). Peak-ring locations in complex craters are used to differentiate the true transient crater diameter (D_t), which differs from the apparent transient crater diameter (D_{at}, referred also as the excavated crater diameter) (Fig. 9.14a). Utilizing the present-day peak-ring elevation (~550 m height, Fig. 9.9a), Morgan et al. (1997) estimated the true and apparent transient crater diameters for Chicxulub crater to be ~118 and ~100 km, respectively. These, and other similar estimates, have been incorporated into cratering mechanics analyses, volume balance (cratering scaling laws) estimates, and impact-energy calculations for Chicxulub (e.g., Sharpton et al. 1993, 1996; Hildebrand et al. 1998; Morgan et al. 2000). Our detailed decompaction and backstripping analysis for Chicxulub crater (Fig. 9.9) indicates considerably less prominent peak-ring and "inner-ring"/crater-rim features in the original crater morphology. The presence of an originally less prominent peak-ring introduces some interesting geometrical relations that may reduce the estimated extent of both the true and apparent transient crater diameters (Fig. 9.14). In particular, the fit and height of the paraboloid of resolution, which approximates the form and vertical extent of the transient crater cavity, is quite different when a less prominent peak-ring (relict of the transient rim uplift) is introduced (Figs. 9.14b, c). Figure 9.14 shows that incautious use of the current peak-ring height without any correction for postimpact burial enhancement may lead to overestimation of the true and apparent transient crater diameters.

Impact craters in sedimentary rather than crystalline targets have the advantage that the regular, pre-impact stratification provides reference horizons for correlation and mapping of impact-induced structures (e.g., Morgan and Warner 1999; Tsikalas et al. 1999; Tsikalas 2005). At the Mjølnir crater, the seismic profiles provide evidence of reflector segments bending upward beneath the central high and the annular basin, indicating elevation of deep strata to shallower levels (Fig. 9.7) (Tsikalas et al. 1998b, c; Shuvalov et al. 2002). Based on such geometries, structural uplift was estimated to be 1.0–1.5 km, however, these values are low in comparison with established empirical relations (Tsikalas et al. 1998b). This is because the impacted sediments at Mjølnir have compacted significantly under the load of a maximum postimpact overburden of 2–2.5 km (Fig. 9.6) (Tsikalas et al. 1998a, 2002b). When decompacted, the structural uplift estimates for Mjølnir become ~1.5–2.3 km. These values fit with the predicted structural uplift based on the dimensions of the Mjølnir crater and the well-established empirical relations of Pilkington and Grieve (1992) and Cintala and Grieve (1994), with best estimates of 2.5 and 2 km, respectively. For the Chicxulub crater, seismic velocity models used to refine the deep crustal structure indicate a lower limit on the vertical extent of structural uplift of 9 km in the crater center (Christeson et al. 2001). It is reasonable to consider that the ~1–1.5 km postimpact overburden and consequent compaction at Chicxulub considerably reduced any impact-induced porosity excess in the central crater and that this is also reflected in increased densities and seismic velocities in the same area. Therefore, we believe that structural uplift values based on velocity models for Chicxulub are overestimated, as they do not account for the influence of postimpact burial. The present-day enhanced peak-ring morphology and structural uplift estimates for Chicxulub have provided direct geophysical values to calibrate numerical modeling of cratering mechanics and released impact energy. This information has been used to infer the anticipated level of environmental perturbation (Schultz and D'Hondt 1996; Pierazzo et al. 1998; Pierazzo and Melosh 1999; Morgan et al. 2000).

9.5.2 Postimpact Modification Correction Factor

Several studies have used crater structural and morphological parameters (e.g., estimates of structural uplift, true and apparent transient crater diameters, peak-ring height, rim-height, and annular trough/basin depth) to reveal the cratering processes operating during an impact (Croft 1985; Melosh 1989, Pilkington and Grieve 1992; Cintala and Grieve 1994; Grieve and Pesonen 1996; Turtle et al. 2005; Hayden et al. 2008). The resulting scaling laws are then utilized to average and calibrate other parameters, such as the excavated volume, melt production, and released-energy spectrum. Therefore, overestimation of morphological and structural features that have been prone to postimpact burial modification, may lead to erroneous conclusions regarding impact-related consequences on a variety of operational scales.

We propose that the establishment of a "postimpact modification correction factor" is prerequisite for several craters. The factor is an estimate of the postimpact morphological and structural changes discussed above, geophysical signature changes, and subdued transient cavity dimensions and structural uplift estimates. Application of the "postimpact modification correction factor" should lead to more accurate estimation of the impact-released energy and, therefore, of related impact consequences. We have made a preliminary attempt to quantify and average the postimpact burial deformation at the five craters discussed herein (Table 9.2). Based on integrated analysis and original crater reconstructions, the postimpact modification correction factors are on the order of 0.35–0.65, 0.25–0.55, 0.25–0.35, 0.10–0.35, and 0.05–0.15 for the Mjølnir, Chicxulub, Montagnais, Cheseapeake Bay, and Bosumtwi craters, respectively (Table 9.2). The correction factors (p_f) are indicative of the degree of postimpact burial deformation, and provide a qualitative relation between modification response and overburden thickness, and a quantitative

Table 9.2 Postimpact modification correction factors (p_f) for impact craters that suffered postimpact burial. See text for discussion and proper application

Crater	Max. postimpact overburden (km)	Postimpact modification correction factor (p_f)
Mjølnir	2–2.5	0.35–0.65
Chicxulub	1–1.5	0.25–0.55
Montagnais	1	0.25–0.35
Chesapeake Bay	0.5–1	0.10–0.35
Bosumtwi	0.3	0.05–0.15

correction of crater morphological and structural parameters utilized in cratering mechanics.

These parameters can be corrected as follows:

$$\begin{pmatrix} \textit{original crater} \\ \textit{morphological/structural} \\ \textit{parameter} \end{pmatrix} = \begin{pmatrix} \textit{observed} \\ \textit{present-day} \\ \textit{paramemter} \end{pmatrix} \times \left[1 - p_f\right] \qquad (9.3)$$

Existing and future borehole petrophysical data regarding postimpact deposits at Bosumtwi, Chesapeake, and Chicxulub will further constrain the postimpact modification correction factors and their further application to other impact craters.

Chapter 10
The Mjølnir Tsunami

Sylfest Glimsdal, Geir K. Pedersen, Hans P. Langtangen, Valery Shuvalov, and Henning Dypvik

10.1 Introduction

Propagation characteristics of impact-generated tsunamis are different from most tsunami originating from other sources in that both nonlinearity and dispersion remain important for a long time after generation. This is particularly true for bolides with diameters that are comparable to, or larger than, the ocean depth. Submarine earthquakes and mass gravity flows on the other hand generally produce waves with amplitudes of only a few meters. Such tsunamis are linear during generation as well as propagation, while nonlinear effects become significant only close to the shore. Tsunamis of yet other origins, such as airborne slides, huge rock falls, or exploding/collapsing volcanoes, may locally display features reminiscent to impact tsunamis, but the far-field propagation is again linear. Oceanic impacts of asteroids and comets, however, may produce huge waves in mid ocean that stay strongly nonlinear during propagation over hundreds and thousands of km.

Impact tsunamis may be categorized into two extreme types. First, small asteroids, with a diameter much less than the water depth (e.g., with ratio diameter to depth less than 0.1) will produce a surface cavity in the sea with rim elevations (Gault and Sonett 1982; Artemieva and Shuvalov 2002). The waves evolving from this kind of source will inherit much energy on wavelengths that are short compared to the depth and are thus highly dispersive. For the Eltanin impact the disintegration of the impactor, ejecta formation, wave generation, and the early phase of tsunami propagation were modelled by Shuvalov (2003a). During impact, the ocean surface suffered violent vertical excursion of several km, while the amplitude of the tsunami was of about 1 km at a distance 20 km from the impact center. Gisler et al. (2004) studied tsunami generation by impactors of varying compositions and size, ranging from 250 m to 1 km, into water depths of 5,000 m. In the impact region interpretations of huge surface excursions as wave height are doubtful; splash-up may be a

S. Glimsdal (✉)
Norwegian Geotechnical Institute, NO-0806 Oslo, Norway; International Centre Geohazards (ICG) at Norwegian Geotechnical Institute, Oslo, Norway
e-mail: sylfest.glimsdal@ngi.no

H. Dypvik et al. (eds.), *The Mjølnir Impact Event and its Consequences*, 257
Impact Studies, DOI 10.1007/978-3-540-88260-2_10, © Springer-Verlag Berlin Heidelberg 2010

better term. Still, at a distance of 20 km from the impact center the tsunami amplitudes were well defined and above 1 km for the most energetic bolides. Even though the first phases of tsunami propagation are strongly affected by nonlinearity, dispersion combined with radial spreading will eventually reduce the amplitudes. The far field propagaton may then be efficiently modelled by Fourier transforms, combined with optical approximations, to describe the far-field tsunami propagation (Ward and Asphaug 2000, 2002).

The second extreme class includes large objects that lead to crater formation on the sea floor and a temporarily dry sea bed (see, e.g., simulations of the 10 km large bolide which formed the Chicxulub crater in the Gulf of Mexico 65 million years ago (Matsui et al. 2002)). In this extreme class the tsunami generation is characterized by intense wave breaking and resurge into the crater. Given that the crater radius is large compared to the water depth; long waves with large amplitudes are eventually formed. Subsequently, these waves will behave very differently both from seismic tsunamis and waves from deep water impacts. This kind of huge tsunamis crossing the oceans may lead to strong mixing and sediment transport that may change the environment with drastic consequences for marine life (see distribution of tsunami deposits at KT boundary in Claeys et al. 2002).The Mjølnir impact belongs to this second category.

10.2 Tsunami Models

A variety of different hydrodynamic models are in use for tsunami modelling. Generally, the actual choice of method for a given task is a compromise with regard to computational efficiency, accuracy, inherent physical effects, and simplicity of implementation. A single case study will generally require a combination of methods. This is indeed true for the Mjølnir tsunami.

The most general group of computational models discretize primitive equations for conservation of mass and momentum, combined with additional relations from thermodynamics, rheology, etc. The multi-material SOVA code, which has been used for the Mjølnir impact, is one example. Other models solve, for instance, the Navier-Stokes equations with free surface and turbulence. The computational domain is resolved in a large number of grid boxes, each with its own set of discrete unknowns. However, three dimensional applications of such methods involve extremely time-consuming computations. In some of the impact simulations, the computational requirements are reduced through the assumption of radial symmetry, which reduces the dimension from three to two.

For the impact and tsunami generation, only the models based on primitive equations will do. However, in propagation of surface gravity waves the fluid flow may often be regarded as non-rotational (no vortical water motion) except in thin boundary layers at the bottom and at the free surface. This is naturally not the case for the region close to the impact where wave breaking and splash-up give rise to a substantial vorticity in the water. Still, the vorticity is advected by the particle velocity only and is rapidly left behind by the outgoing waves (see Fig. 10.3 and the related discussion in Sect. 10.3). On the other hand, long time wave propagation will be affected by the Coriolis force that will induce vertical vorticity. In mid-range the assumption of non-rotational flow holds and potential theory can be employed. Unfortunately, also full potential theory rapidly becomes unmanageable for large scale computations. In the Mjølnir study it has so far been used only for general verification of simpler models and special solutions such as solitary waves or undular bores.

Oceanic propagation of tsunamis from earthquakes and landslides are most frequently described by shallow water theory. This is the simplest of the theories for waves that are long in comparison to the depth. In shallow water theory the vertical configuration of the flow is extremely simple and both the vertical velocity and coordinate may be omitted in the equations. The number of dimensions is then reduced and the structure of the equations becomes simple and well suited for numerical solution. Breaking waves may be included in shallow water theories as "jumps" in the field variables, while other features of wave propagation as dispersion are lost. Unfortunately, shallow water theory is inadequate for the propagation of the Mjølnir tsunami, which generates high and moderately long waves that are genuinely influenced by dispersion and nonlinearity alike.

We then resort to the next level of long wave equations, often named Boussinesq equations. These equations include both nonlinearity and dispersion. Even though they are much more complex than the shallow water equations they may still be formulated as depth integrated equations with the number of spatial dimensions reduced by one. Among the Boussinesq type equations we again face a multitude of choices. The equations used herein are based on potential flow, but is otherwise of a standard type. The waves from the Mjølnir impact are clearly somewhat short for the Boussinesq equations, but comparison with full potential theory indicates errors less than 10% in the wave heights. A simple breaking facility is included in the Boussinesq model as a dissipation term which is activated whenever the amplitude crosses the threshold for breaking. This feature presumably is adequate as long as the breaking is a gently spilling and the wave shape remains fairly close to that of a solitary wave.

On sequential computers, or affordable parallel systems, even Boussinesq types of models are too time-consuming for simulation of the whole Barents Sea with an appropriate resolution. An alternative is then other descriptions, such as ray theory (geometrical and physical optics). Waves of a particular class may propagate in a slowly varying medium and adjust its properties gently without loss of identity or noticeable diffraction. The celerity of the wave is as in constant medium and its energy, or a related quantity like wave action, is preserved. This is the basic idea of the optical theories, which are widely used for sinusoidal waves (Mei 1989; Peregrine 1976), also in the context of impact generated tsunamis (Ward and Asphaug 2000). Herein we adopt a ray theory for solitary waves (Miles 1977; Pedersen 1994). Solitary waves are briefly discussed in Sect. 10.3.1 below. Equations are then formulated in terms of the characteristics and propagation of the wave crests, as opposed to the theories described above where field quantities as particle velocity, pressure and surface elevations are primary unknowns. For surface waves the effective number of spatial dimensions is then reduced to one and the grid resolution in time and space may be coarse, because there is no need to resolve the wave periods. Besides significantly increased computational efficiency, the optical theories also provide simple closed form solutions that render the physics in the wave propagation more transparent.

10.3 Tsunami Generation

The Mjølnir impact event itself and the response of the crust are described in detail elsewhere in this volume (see Chaps. 4 and 7). The crater configuration up to 150 s after impact, including the ocean water, is shown in Fig. 10.1. The water was blown away, and it did not return permanently before 15–20 min. The exposed sea bottom was ignited during this period (Dypvik et al. 2008b; Wolbach et al. 2001). After 3 s the crushed seafloor cavity was deeper than 5 km and the crater radius almost 5 km, while after 30 s the radius of the crater was more than 10 km. Then the short-lived transient crater started to collapse due to the gravitational force, and the central part of the crater rose, forming the central peak.

In Table 10.1 we have compared energies from the Mjølnir event to those of the giant earthquake off Sumatra on December 26, 2004 and the Krakatau tsunami

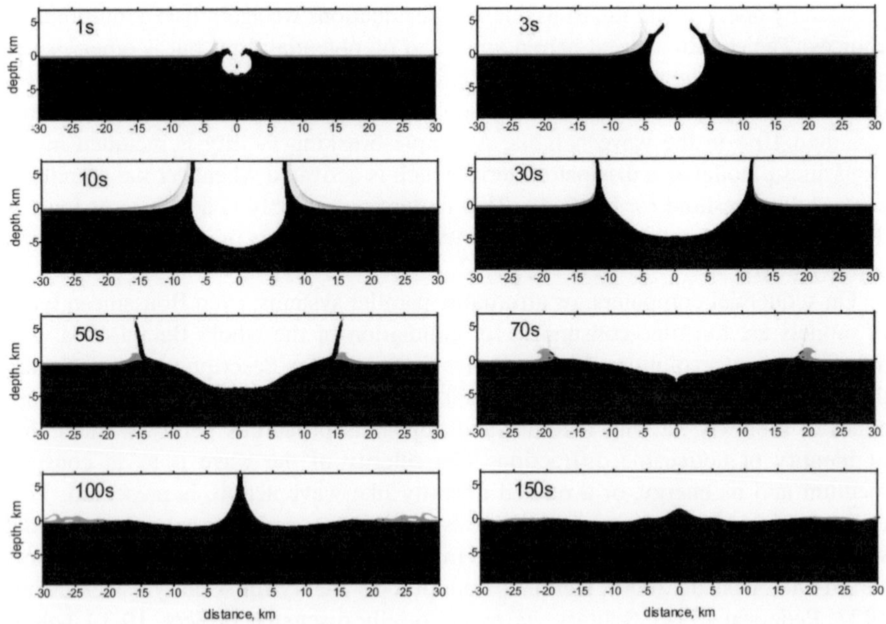

Fig. 10.1 Impact simulation by SOVA at time steps between 1 and 150 s after the impact. The areas colored black indicate sediments and solid rock, while the grey layer above it represents water. The figure is from Shuvalov et al. (2002)

Table 10.1 Energies of the Mjølnir event compared to other large geological and tsunami events. For data collection, see the text

Event	Energy
Kinetic energy of the Mjølnir bolide	1.3×10^{22} J
Earthquake Dec. 26, 2004	4.3×10^{18} J
Mjølnir tsunami	2.1×10^{18} J
Tsunami Dec. 26, 2004	3.5×10^{15} J
Tsunami Krakatau 1883	9.6×10^{15} J

of 1883. We have employed a density of 3,000 kg/m^3, a diameter of 1.6 km and a velocity of 20 km/s for the Mjølnir impactor. The energy for the December 26, 2004 earthquake is taken from the Bilham (2005). For the Indian Ocean tsunami we assume an initial sea surface elevation of 2 m, over a region of 1,200 km times 200 km, which yield a high estimate. The energy of the Krakatau tsunami corresponds to an initial cavity of 50 km^2 times 200 m. The energy of the tsunamis are calculated directly from numerical solutions.

We observe that the total energy of the Mjølnir impact was a factor 600–700 higher than that of the 2004 earthquake. Moreover, integration of the wave energy in the Mjølnir tsunami shows a total of 2.1×10^{18} J, which is comparable to the total energy of the December 26, 2004 earthquake. The energy of the earthquake tsunami is about a factor 1,000 less than that of the Mjølnir tsunami. Naturally, the parameters behind the comparisons are uncertain. Still, it is clear that both the geological consequences and the tsunami of an impact of a large asteroid are orders of magnitude larger than those of even the largest earthquakes recorded.

10.3.1 Near Field Evolution

At $t = 300$ s (Fig. 10.2) wave like features are evident, in particular as high bore propagation outward from the crater. The central peak and its vicinity are still dry and there are several breaking crests on the resurge of water into the crater. The flow is dominated by vorticity and turbulence up to 50 km, say, from the impact center.

After 600 s the water is climbing the central peak. The front is no longer breaking and a marked, smooth peak has evolved, with a slightly reduced height of 250 m. This reduction is due to breaking and radial spread, but is counteracted by the growth of the peak (see discussion below). At this time the ratio amplitude to depth is 0.63, which is somewhat too high for Boussinesq equations.

At 800 s a second peak is visible at the front of the leading wave system. Later, at 1,000 s, yet a third peak is discernible and it is apparent that an undular bore (see below) is in progress. At this time the leading peak is located 116 km from the impact center. Closer than 20 km to the impact center another wave elevation is being produced from the resurge.

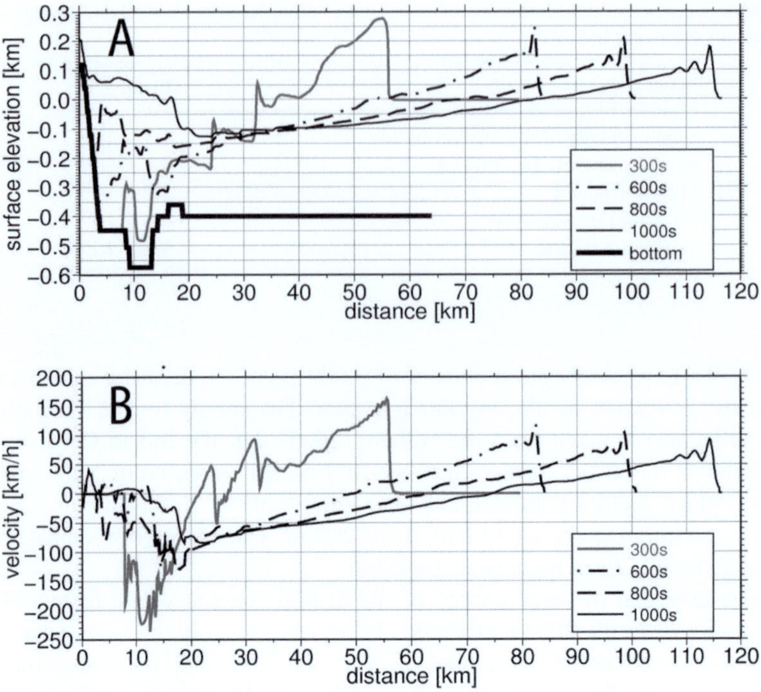

Fig. 10.2 Surface elevation (**A**) and depth averaged horizontal velocities (**B**) after the Mjølnir impact. Simulations based on the SOVA-model. The *lower axis* shows the distance from center of impact, r. The spatial resolutions are $\Delta r = 160$ m and $\Delta z = 40$ m

The development of the flow from $t = 600$ s to $t = 1{,}000$ s is illustrated by means of the vorticity (curl) in Fig. 10.3. Due to coarse grid and numerical errors in its computation the vorticity is probably overestimated in regions where it should have been very small. Still, we may observe that the region within 35 km from the impact center is heavily influenced by rotation. Moreover, for $t = 600$ s we observe a distinct vortex, located close to 70 km from center, which originates from wave breaking. At $t = 1{,}000$ s this feature is still apparent and have moved only a few km. Meanwhile the wave advance is almost 30 km and the vortex is thus left behind by the waves.

After 1,000 s the amplitude to depth ratio for the leading crest has fallen below 0.5 and it seems reasonable to convey the tsunami from SOVA to the radial symmetric Boussinesq model. We observe that the evolution of the leading wave as an undular bore continues, with an increasing number of apparent peaks, see Figs. 10.4 and 10.5. Moreover, in the Boussinesq simulation the amplitude of the leading peaks increase significantly from $t = 1{,}000$ s and culminates with a maximum at $t \sim 1{,}150$ s. This amplification is in agreement with the dynamics of plane undular bores, where the highest peak may reach double the initial amplitude according to long wave theory (Peregrine 1966). When the growth of the individual waves slows down, as they

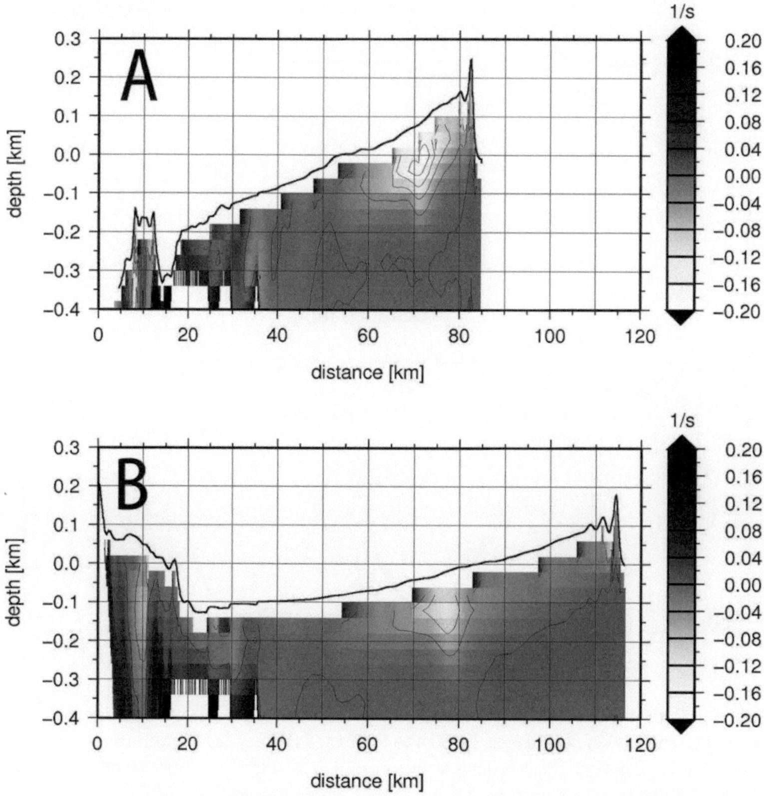

Fig. 10.3 The out of plane component of the curl of the solution after 600 s (**A**) and 1,000 s (**B**). The partially filled cells at the surface is left out and the computational scheme is inaccurate for the adjacent cells

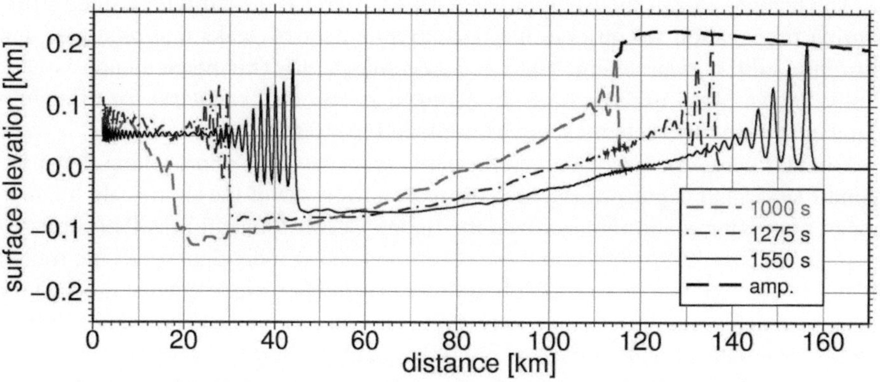

Fig. 10.4 Surfaces from Boussinesq model. The curve with key "amp" is the amplitude tracking of the leading wave. Resolution: $\Delta r = 50$ m, $\Delta t = 0.8$ s

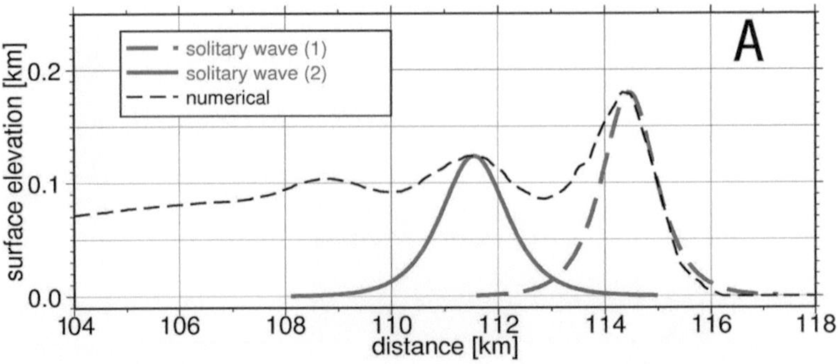

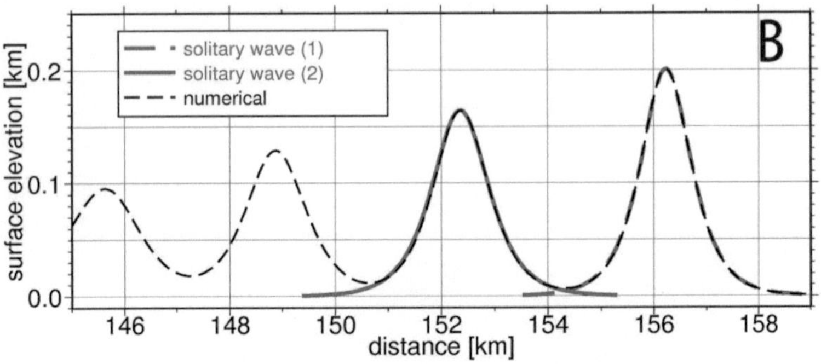

Fig. 10.5 Separation into solitary waves in the evolution of the Mjølnir tsunami. *Upper panel* (**A**) Matching of the SOVA solution at 1,000 s to the exact solitary wave solution by Tanaka (1986). *Lower panel* (**B**) Numerical Boussinesq solution matched with the corresponding solitary wave solution. The solitary wave solutions are calculated from the maxima of the numerical surfaces and the depth. Resolution: $\Delta r = 50$ m, $\Delta t = 0.8$ s

become separated, the opposing attenuation due to radial spread becomes dominant and the amplitudes start to drop as is seen for $t = 1{,}550$ s in Fig. 10.4.

The leading part of the tsunami wave pattern starts as a shelf with a steep front. During propagation we observe fission into a number of peaks that separates and turn into solitary waves (Fig. 10.5) as r becomes large. This phenomenon is generally known as an undular bore. In nature it is observed regularly for tides in a number of rivers and estuaries, see example in Fig. 10.6.

Undular bores are also observed for earthquake tsunamis approaching the shore (Shuto 1985) and is presumably also the reason for some of the short scale features apparent in images and footage from the Indian Ocean Tsunami in 2004 (Glimsdal et al. 2006; Grue et al. 2008). In these cases the leading individual peaks are trailed by the bores, while the fission of solitary waves is rather complete when the Mjølnir tsunami reaches the shore as a long train of solitary waves. This yields different characteristics for the effect of breaking and the nature of coastal inundation. The undular bore has an apparent likeness with periodic waves similar to swells or those generated by ships. However, this similarity is superficial and misleading. The undular bore has quite different properties as well as mechanisms for generation.

Fig. 10.6 Tidal bore of Dordogne river (France) on September 27, 2000. Courtesy of Hubert Chanson

The evolution of undular bores is well established in the literature for plane waves propagating in constant depth. In systematic studies on evolution on undular bores in variable depth we have found that solitary wave like crests do evolve even in bathymetries with bottom gradients that are an order of magnitude higher than those in the Barents Sea. Moreover, tests also indicate that the variable depth influence is small for the first 1,000 s of the wave propagation (3%). We also find that, when first generated, the individual crests retain their identity as solitary waves, until they break in shoaling water.

A solitary wave is a single crested wave of permanent shape and constant celerity, increasing with amplitude that may exist due to the combined action of nonlinearity and dispersion. The concept of solitary waves goes back to the first half of the nineteenth century (Russel 1845) and has retained interest in the context of surface gravity waves since. From the late sixties throughout the seventies the solitary wave theory was fashion in various physical settings and a number of its remarkable properties, manifested through closed form mathematical solutions, came to light. These include the particle like interaction properties for small, but finite, amplitudes, and the genesis of solitary waves from general initial conditions (see review in Miles 1980). While the approximate long wave theories inherit formal solitary wave solutions of any amplitude, the full potential theory yields a maximum height $A = 0.83h$, where A is the amplitude and h is the depth. However, these waves are subject to longitudinal instabilities for $A > 0.78h$ (Tanaka 1986) and transverse instabilities for $A > 0.72h$ (Kataoka and Tsutahara 2004). Wave crests that are close to solitary waves and surpass this limit will develop breaking that may start as smaller or larger plunging jet from the peak. The crests will then evolve into an aerated and turbulent region near the crest that dissipates energy.

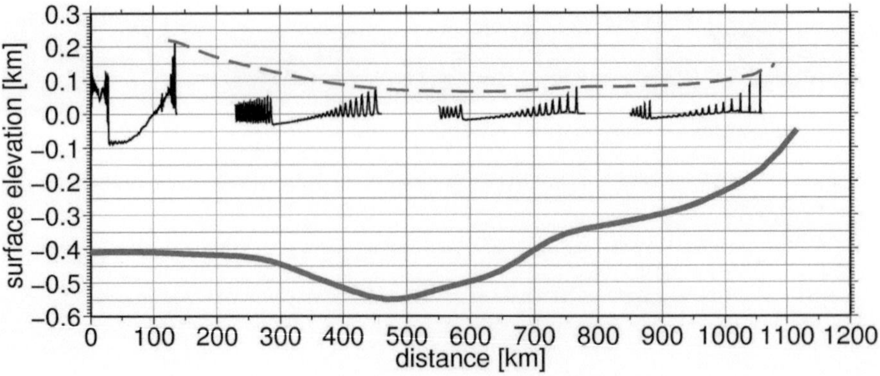

Fig. 10.7 Solutions of wave propagation with depth profile extracted from impact center towards paleo-Greenland (255° to North). The solutions are printed at $t = 23$ min, 1 h 33 min, 2 h 45 min, and 4 h 6 min, respectively. Only the primary wave system and the leading part of the secondary wave system are plotted. The spatial and time increments are $\Delta r = 100$ m and $\Delta t = 1.6$ s, respectively. The bottom is drawn with a thick *grey line*, while the *dashed line* shows the amplitude evolution

According to Fig. 10.4 the waves from the resurge of water into the crater develop into a secondary undular bore. It must be noted that these waves are more nonlinear than the leading system due to the intermediate trough. The highest amplitude nearly reaches 0.7 times the depth in front of the system. Hence, breaking cannot be ruled out. Still, as radial dilution takes its toll on the wave heights it is probable that a smooth undular bore evolves. At later stages (see Fig. 10.7) the secondary system is much weaker than the primary one.

10.3.2 Far Field Propagation

We have reconstructed the bathymetry of the paleo-Arctic Seas based on our own field work and regional compilations as shown in Fig. 10.8, where also the positions of today′s coastlines are indicated. The paleo-Arctic Seas at the time of the impact were dominated by shallow water (200–600 m), with the deepest part between paleo-Norway and paleo-Greenland.

10.3.2.1 Estimates of Far-Field Behaviour

An estimate of the evolution of the amplitude for the solitary crests may be readily obtained from the optical theory. For simplicity we assume radial symmetry around the impact center and employ an approximate expression for the energy. Then we find a simple formula for the amplitude of the solitary waves, namely

$$A = A_0 \frac{h_0}{h} \left(\frac{r_0}{r}\right)^{\frac{2}{3}}, \tag{10.1}$$

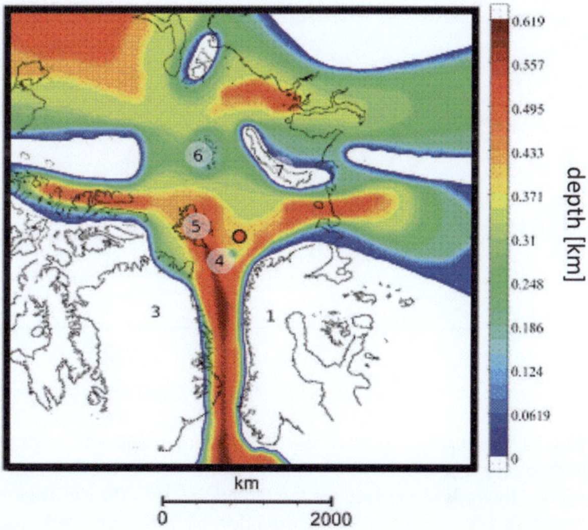

Fig. 10.8 The reconstructed bathymetry of the paleo-Arctic Seas at the actual time (approximately 150 million years ago). The center of impact is marked with a red bullet in the center of the figure. White areas are land. The keys for the numbers are as follow: *1* – Norway/Scandinavia, *3* – Greenland, *4* – pre-Bear Island, *5* – pre-Svalbard, *6* – pre-Franz Josef Land, *7* – Novaya Zemlya. At this time Bear Island (sea mount), Svalbard, and Franz Josef Land were submerged

where the amplitude A_0, the depth h_0, and the distance r_0 define a reference state, see Glimsdal et al. (2007) for further details. It is instructive to compare (see Eq. 10.1) to its counterparts for linear waves. For solitary waves, radial spreading yields an ampliude attenuation $\sim r^{-2/3}$. Correspondingly, for linear, highly dispersive and nearly periodic waves the amplitudes are reduced markedly faster in proportion to r^{-1}, which is the combination of two factors of $r^{-1/2}$ from dispersion and extension of wave crests lengths, respectively. If the initial disturbance involves a net integrated elevation or depression there is a front of the wave train that attenuates as $r^{-2/3}$ (Clarisse et al. 1995), which is the same rate as for solitary waves. Numerical computations of impact and wave generation (Gisler et al. 2004) (see introduction) infer damping rates that are systematically stronger than r^{-1}, with $r^{-2.25}$ up to 100 km from the impact as the most extreme. However, these are the results of extremely complex computations, where it is difficult to assess all relevant particulars. In general, higher damping rates than r^{-1} must be explained by breaking, high numerical damping, or under-resolution. Misinterpretation of splash up, wave group dynamics, or interference with compression waves may also give higher damping rates. In shoaling water the amplitude of the solitary wave increases as h^{-1}, until breaking, while linear long waves amplify as $h^{-1/4}$. In short, the solitary waves attenuate slower than dispersive deep water waves and amplify substantially faster than linear waves in shoaling water.

For the leading solitary crest we have an amplitude $A_0 = 200$ m in depth $h_0 = 400$ m at a distance 160 km from the impact center. Inserting these parameters, together with the critical amplitude $A = 0.72h$ into equation (see Eq. 10.1) we find an estimate of breaking depth as function of distance from impact center. This depth is displayed in Fig. 10.9 together with selected depth profiles from the paleo-bathymetry, taken along different direction through the impact center. At distances

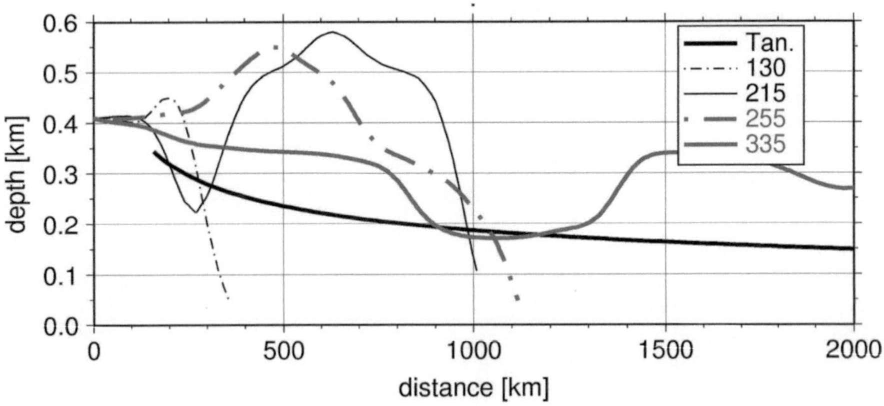

Fig. 10.9 Breaking-depth of the Mjølnir tsunami plotted with depth-profiles taken from the paleo-bathymetry. The labels for the depth-profiles reflect the degrees of the direction out from the impact center. Towards Greenland (in the direction 215°) the pre-Bear Island area is located 280 km from the impact center, while in the direction of 335°, the pre-Franz Josef Land region is located at a distance of 1,100 km

as large as 2,500 km the tsunami still will break at depths close to 140 m. Toward paleo-Norway (130°), we must expect breaking at 300 m depth, approximately 120 km outside the coast. The tsunami will be breaking over the pre-Bear Island and pre-Franz Josef Land regions (in the directions 215 and 335 degrees, respectively). At Greenland (255°) breaking occurs at depth of 185 m about 1,040 km from the impact center. However, it must be noted that these are only rough estimates that are based on a formula for radial symmetry.

10.3.2.2 Computations of Far-Field Behaviour

Figure 10.10 displays summarizing results from three-dimensional simulations. After 1 h and 10 min the leading waves are completely separated into solitary waves, see lower panel. Here, the leading crest is estimated to a height of 144 m and the five first waves are all higher than 50 m. The overall maximum amplitude during the whole simulation is up to 220 m (upper panel). See also the cross sections of the surface elevation in Fig. 10.11. The simple breaking facility in our model has been activated over pre-Bear Island area and toward paleo-Norway at depths less than approximately 310 and 260 m, respectively. These depths are slightly less than the predicted breaking depth due to under-resolution in the Boussinesq model.

The Boussinesq model features a simple breaking facility that dissipates energy in breaking areas. In Fig. 10.11 cross sections of breaking solutions taken along a line from impact center toward Norway are shown. After 1 h there is no breaking, but after 1 h and 10 min the front of the solutions is significantly reduced by breaking (spilling) and the wave crests above the critical value of A/h are damped as they

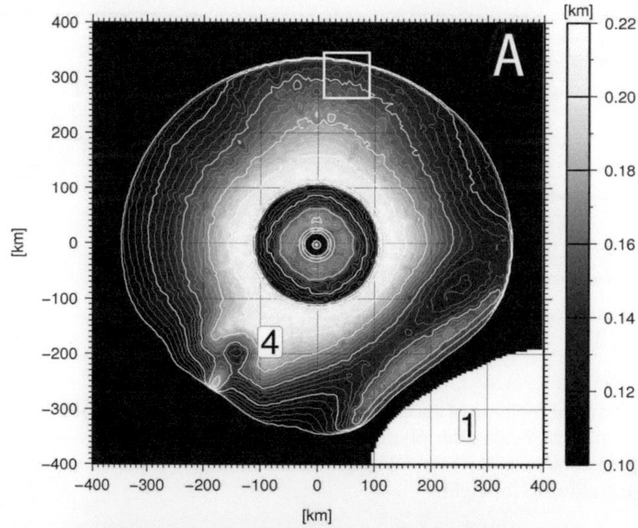

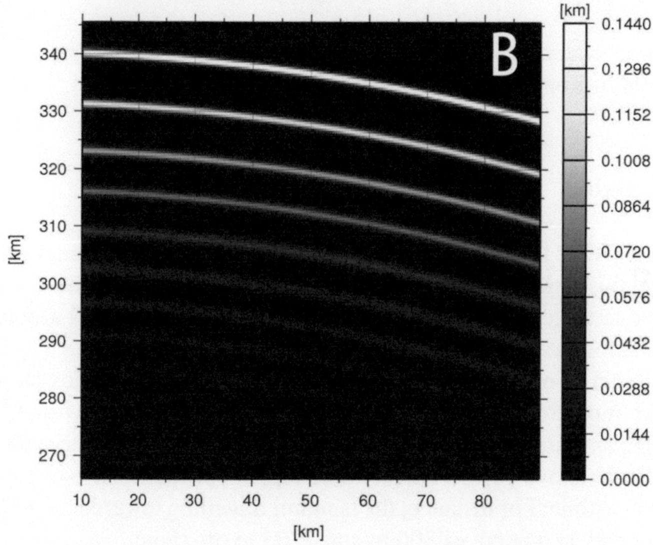

Fig. 10.10 Simulations using the reconstructed palaeography/paleobathymetry and the three-dimensional Boussinesq equations. The surface elevations (see colorbars) are given in kilometers.The *upper panel* (**A**) shows the maximum surface elevation during the tsunami propagation from $t = 17$ min to 1 h 10 min. The *lower panel* (**B**) shows a snapshot of the surface elevation after 1 h and 10 min of a local part of the solution (indicated by a *white rectangle* in *panel* **A**). The resolutions are $\Delta x = \Delta y = 0.4$ km and $\Delta t = 6.3$ s (see Fig. 10.8 for explanation of the key numbers)

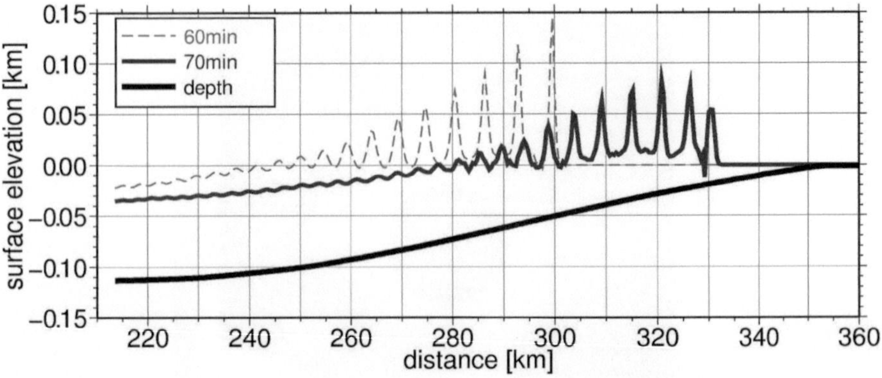

Fig. 10.11 Breaking of the train of solitary waves towards Norway. The profiles are cross sections in the three-dimensional fields after 60 and 70 min. The thick black curve is the sea bottom, h (depth in kilometers shown at left vertical axis). The lower axis is the distance from impact center, and waves are travelling to the right

enter shallower water. Offshore breaking of waves generated by impact tsunamis is also discussed in Korycansky and Lynett (2005).

10.4 Discussion

A few minutes after the impact, the tsunami started to develop as an undular bore. Furthermore, this bore developed into a train of solitary waves. These wave characteristics are due to genuine nonlinear and dispersive features, and are not captured neither by shallow water models nor linear theory.

The total energy of the Mjølnir tsunami was several magnitudes larger than the energy of the monstrous Indian Ocean tsunami. Simulations show that the Mjølnir impact caused amplitudes above 200 m, and we have examples where the amplitude after 4 h still was above 100 m. Due to these high amplitudes, the tsunami will break far from the coastline. For instance, the tsunami travelling toward paleo-Norway will presumably break at a depth of 300 m and 120 km off-shore.

At the time of impact the sedimentation of organic rich muds and clays dominated, and loose, unconsolidated deposits formed the sea floor. Hence, there was no well defined sea bottom, but a thick (turbid) layer with gradually increasing density downward. The tsunamis, causing repeated pulses of currents with particle velocities on the sea-floor in the range 30–90 km/h, must have generated substantial mixing and reworked the sea floor dramatically. A total change of the marine environment throughout the paleo-Barents Sea is thus likely. The direct effect of re-sedimentation is difficult to detect in the sediments, but the prolific bloom of the prasinophycean algae (*Leiosphaerida*), and the reduction of the benthic foraminifera assemblages in the impact-influenced deposit (Smelror et al. 2002; Bremer et al. 2004; Dypvik et al. 2004c; Smelror and Dypvik 2005), surely indicate environmental changes. The unconsolidated sea floor and the suspended fine grained sediments might also have caused a substantial damping of the tsunami. Similar effects may be important for seismic tsunamis or storm surges, flooding tidal flats and entering estuaries and other shallow regions that are rich in sediments. We, however, expect the coastal deposits to carry the most clear-cut tsunami evidence in the paleo-Barents Sea region.

We have not yet computed the run-up of the Mjølnir tsunami. The tsunami reached the shores as series of crests that were both higher and shorter, than the more common earthquake tsunamis. Hence, they will break in deep water and loss of energy will be substantial. This will in turn reduce the inundation lengths. The state of the art modelling of this kind of breaking waves (Kennedy et al. 2000), are based on onset of energy dissipation according to criteria that are adapted to solitary waves. In the present case the length of the period with breaking will yield wave forms that are completely different from solitary waves and the application of the state of the art criteria is doubtful. Anyway, in view of the monstrous amplitudes vast areas of the neighbouring coasts presumably were flooded.

References

Adushkin VV, Nemchinov IV (1994) Consequences of impacts of cosmic bodies on the surface of the Earth. In: Gehrels T (ed) Hazards due to Comets and Asteroids. University Arizona Press, Tucson and London, pp 721–778

Ahrens TJ, O'Keefe JD (1983) Impact of an asteroid or comet in the ocean and the extinction of terrestrial life. Proc Lunar Planet Sci Conf 13th Part 2 J Geophys Res 88:A799–A806

Alexopoulos JS, McKinnon WB (1994) Large impact craters and basins on Venus, with implications for ring mechanics on the terrestrial planets. In: Dressler BO, Grieve RAF, Sharpton VL (eds) Large meteorite impacts and planetary evolution. Geological Society of America Special Paper 293, Boulder, pp 29–50

Alvarez W, Claeys P, Kieffer SW (1995) Emplacement of Cretaceous Tertiary boundary shocked quartz from Chicxulub crater. Science 18:930–935

Amundsen H, Evdokimov A, Dibner VD, Andresen A (1998) Geochemistry and petrogenetic significance of Mesozoic magmatism on Franz Josef Land, Northeastern Barents Sea. Nor Polarinstitutt Medd 151:105–120

Anderson RR, Witzke BJ, Roddy DJ (1996) The drilling of the 1991–1992 geological survey bureau and US Geological Survey Manson impact structure research cores. In: Koeberl C, Anderson RR (eds) The manson impact structure, Iowa: Anatomy of an impact crater. Geological Society of America Special Paper 302, Boulder, pp 45–88

Artemieva NA, Shuvalov VV (2001) Extraterrestrial material deposition after the impacts into continental and oceanic sites. In: Buffetaut E, Koeberl C (eds) Impact studies (Geological and biological effects of impact events). Springer, Berlin-Heidelberg, pp 249–263

Artemieva NA, Shuvalov VV (2002) Shock metamorphism on the ocean floor (numerical simulations). Deep Sea Res Part II Top Stud Oceanogr 49(6):959–968

Artemieva NA, Karp T, Milkereit B (2004) Investigating the Lake Bosumtwi impact structure: insight from numerical modelling. Geochem Geophys Geosyst 5:1–20

Arthur MA, Carson B, van Huene R (1980) Initial tectonic deformation of hemipelagic sediment at the leading edge of the Japan convergent margin. Initial Report Deep Sea Drilling Project, 56, 57, Part 1:569–613

Arz JA, Alegret L, Arenillas I (2004) Foraminiferal biostratigraphy and paleoenvironment reconstruction at Yaxcopoil-1 drill hole, Chicxulub crater, Yucatan Peninsula. Meteorit Planet Sci 39:1009–1111

Asphaug E, Ward S (2002) Impact tsunami, Eltanin. Deep Sea Res Part II 49:1073–1079

Balkwill HR (1978) Evolution of Sverdrup Basin, Arctic Canada. Am Assoc Pet Geol Bull 62:1004–1028

Barton PJ (1986) The relationship between seismic velocity and density in the continental crust - a useful constraint? Geophys J R Astron Soc 87:195–208

Beals CS, Halliday I (1967) Impact craters of the Earth and Moon. J R Astron Soc Canada 59:199–216

H. Dypvik et al. (eds.), *The Mjølnir Impact Event and its Consequences*,
Impact Studies, DOI 10.1007/978-3-540-88260-2,

Bell C, Morgan JV, Hampson GJ, Trudgill B (2004) Stratigraphic and sedimentological observations from seismic data across the Chicxulub impact basin. Meteorit Planet Sci 39:1089–1098

Bergendahl E (1989) Halokinetisk utvikling av Nordkappbassengets sørvestre segment. Cand scient (candidatus scientiarum) Thesis. University of Oslo, Oslo, p 120

Bergh SG, Grogan P (2003) Tertiary structure of the Sørkapp-Hornsund Region, South Spitsbergen, and implications for the offshore southern extension of the foldthrust Belt. Norwegian J Geol 83:43–60

Berglund LT, Augustson J, Færseth R, Gjelberg J, Ramberg-Moe H (1986) The evolution of the Hammerfest Basin. In: Spencer AM et al (ed) Habitat of hydrocarbons on the Norwegian Continental Shelf. Graham and Trotman, London, pp 319–338

Birkenmajer K (1980) Jurassic-Cretaceous succession at Agardhbukta, East Spitsbergen. Stud Geol Pol 66:35–52

Bilham R (2005) A flying start, then a slow slip. Science 308:1126–1127

Bjærke T (1977) Mesozoic palynology of Svalbard II. Palynomorphs from the Mesozoic sequence of Kong Karls Land. Nor Polarinstitutt Årbok 1976:83–99

Bjærke T, Manum SB (1977) Mesozoic palynology of Svalbard I. The Rhaetian of Hopen with a preliminary report on the Rhaetian and Jurassic of Kong Karls Land. Nor Polarinstitutt Skrifter 165:1–48

Brekke H, Kalheim JE, Riis F, Egeland B, Blystad P, Johnsen S, Ragnhildstveit J (1992) Two-way time map of the unconformity at the base of the Upper Jurassic (north of 69°N) and the unconformity at the base of the Cretaceous (south of 69°N), offshore Norway, including the main geological trends onshore, scale 1:2 millions. Norwegian Petroleum Directorate, Continental Shelf Map No 1

Brekke H, Sjulstad HI, Magnus C, Williams RW (2001) Sedimentary environments offshore Norway – an overview. In: Martinsen OJ, Dreyer T (eds) Sedimentary environments offshore Norway – Paleozoic to recent. Norwegian Pet Soc Spec Publ 10:7–37

Breivik AJ, Gudlaugsson ST, Faleide JI (1995) Ottar Basin, SW Barents Sea: a major Upper Paleozoic rift basin containing large volumes of deeply buried salt. Basin Res 7:299–312

Bremer GMA, Smelror M, Nagy J, Vigran JO (2004) Biotic responses to the Mjølnir meteorite impact, Barents Sea: evidence from a core drilled within the crater. In: Dypvik H, Burchell M, Claeys P (eds) Cratering in marine environments and on ice, Springer Series in Impact studies. Springer, Berlin-Heidelberg, pp 21–38

Brittan J, Morgan JV, Warner MR, Marin L (1999) Nearsurface seismic expression of the Chicxulub impact crater. In: Dressler BO, Sharpton VL (eds) Large meteorite impacts and planetary evolution II. Geological Society of America Special Paper 339, Boulder, pp 281–290

Bruhn R, Steel RJ (2003) High-resolution sequence stratigraphy of a clastic foredeep succession (Paleocene, Spitsbergen): An example of peripheral-bulge-controlled depositional architecture. J Sediment Res 73(5):745–755

Bugge T, Elvebakk G, Fanavoll S, Mangerud G, Smelror M, Weiss HM, Gjelberg J, Kristensen SE, Nilsen K (2002) Shallow stratigraphic drilling applied in hydrocarbon exploration of the Nordkapp Basin, Barents Sea. Mar Pet Geol 19:13–37

Burchell MJ, Mackay NG (1998) Crater ellipticity in hypervelocity impact on metals. J Geophys Res 103:22761–22774

Byerly GR, Lowe DR (1994) Spinel from Archean impact spherules. Geochim Cosmochim Acta 58:3469–3486

Carpenter BN, Carlson R (1997) The Ames meteorite-impact crater. Oklahoma Geol Surv Circ 100:104–119

Carson B, van Huene R, Arthur M (1982) Small-scale deformation structures and physical properties related to convergence in Japan. Trench slope sediments. Tectonics 1:277–302

Cartwright JA, Lonergan L (1996) Volumetric contraction during compaction of mudrocks: a mechanism for the development of regional-scale polygonal fault systems. Basin Res 8:183–193

Channell JET, Smelror M, Jansen E, Higgins S, Lehman B, Eidvin T, Solheim A (1999) Age models for glacial fan deposits off East Greenland and Svalbard (Sites 986 and Site 987). Proc Ocean Drill Prog Sci Results 162:149–166

Christeson GL, Buffler RT, Nakamura Y (1999) Upper crustal structure of the Chicxulub impact crater from wide-angle ocean bottom seismograph data. In: Dressler BO, Sharpton VL (eds) Large meteorite impacts and planetary evolution II. Geological Society of America Spesial Paper 339, Boulder, pp 291–298

Christeson GL, Nakamura Y, Buffler RT, Morgan JV, Warner M (2001) Deep crustal structure of the Chicxulub impact crater. J Geophys Res 106:21751–21769

Christie JM, Raleigh CB (1959) The origin of deformation lamellae in quartz. Am J Sci 257:385–407

Cintala MJ, Grieve RAF (1994) The effects of differential scaling of impact melt and crater dimensions on lunar and terrestrial craters: Some brief examples. In: Dressler BO, Grieve RAF, Sharpton VL (eds) Large meteorite impacts and planetary evolution. Geological Society of America Special Paper 293, Boulder, pp 51–59

Claeys P, Kiessling W, Alvarez W (2002) Distribution of Chicxulub ejecta at the Cretaceous–Tertiary boundary. In: Koeberl C, MacLeod KG (eds) Catastrophic events and mass extinctions: impacts and beyond. Geological Society of America Special Paper 356, Boulder, pp 55–68

Clarisse JM, Newman JN, Ursell F (1995) Integrals with a large parameter: water waves on finite depth due to an impulse. Proc R Soc Ser A450:67–87

Clausen JA, Gabrielsen RH, Johnsen E, Korstgård J (2003) Fault architecture and clay smear distribution. Examples from field studies and drained ringshear experiments. Norwegian J Geol 83:131–146

Collins GS, Turtle EP, Melosh HJ (2003) Numerical simulations of Silverpit crater collapse [abs] Lunar Planet Sci Conf 34, abs #2115, CD-ROM

Collins GS, Wünnemann K (2005) How big was the Chesapeake Bay impact? Insight from numerical modeling. Geology 33:925–928

Coney L, Gibson RL, Reimold WU, Koeberl C (2007a) Lithostratigraphic and petrographic analysis of ICDP drill core LB-07A, Bosumtwi impact structure, Ghana. Meteorit Planet Sci 42:569–589

Coney L, Reimold WU, Gibson RL, Koeberl C (2007b) Geochemistry of impactites and basement lithologies from ICDP borehole LB-07A, Bosumtwi impact structure, Ghana. Meteorit Planet Sci 42:667–688

Cordier P, Gratz AJ (1995) TEM study of shock metamorphism in quartz from the Sedan nuclear test site. Earth Planet Sci Lett 129:163–170

Crawford DA, Barnouin-Jha OS (2004) Computational investigations of the Chesapeake Bay impact structure [abs]. Lunar Planet Sci 35, abs #1757, CD-ROM

Croft SK (1985) The scaling of complex craters. Proc 15th Lunar Planet Sci Conf Part 2 J Geophys Res 90:C828–C842

Dahl JM, Schultz PH (2000) Strain rate measurements in vertical and oblique projectile impact experiments [abs]. Lunar Planet Sci 31, abs# 1901, CD-ROM

Dahl JM, Schultz PH (2001) Measurements of stress wave asymmetries in hypervelocity projectile impact experiments. Int J Impact Eng 26:145–155

Dallmann WK, Gjelberg JG, Harland WB, Johannessen EP, Keilen HB, Lønøy A, Nilsson I, Worsley D (1999) Upper Palaeozoic lithostratigraphy. In: Dallmann WK (ed) Lithostratigraphic Lexicon of Svalbard. Review and recommendations for nomenclature use. Upper Palaeozoic to Quaternary bedrock. Norsk Polarinstitutt, Tromsø, pp 127–214

DeClercq J, Dypvik H, Aagaard P, Jahren J, Ferrell RE Jr, Horton JW Jr (2009) Experimental alteration of artificial and natural impact melt rock from the Chesapeake Bay impact structure. In: Gohn G, Koeberl C, Miller KG, Reimold WO (eds) The ICDP – deep drilling project in the Chesapeake Bay impact structure: results from the Eyreville Core Holes. Geol Soc Am Spec Paper 458:559–570

Dienes JK, Walsh JM (1970) Theory of impact: some general principles and the method of Eulerian codes. In: Kinslow R (ed) High-velocity impact phenomena. Academic Press, New York, pp 46–104

Dimakis P, Braathen BI, Faleide JI, Elverhøi A, Gudlaugsson ST (1998) Cenozoic erosion and the preglacial uplift of the Svalbard–Barents Sea region. Tectonophysics 300:311–327

Donofrio RR (1981) Impact craters: implications for basement hydrocarbon production. J Pet Geol 3:279–302

Donofrio RR (1998) North American impact structures hold giant field potential. Oil Gas J 96:69–83

Doré AG, Jensen LN (1996) The impact of late Cenozoic uplift and erosion on hydrocarbon exploration: offshore Norway and some other uplifted basins. In: Solheim A et al (ed) Impact of glaciations on basin evolution: data and models from the Norwegian margin and adjacent areas. Glob Planet Change 12:415–436

Doyle P, Kelly SRA (1988) The Jurassic and Cretaceous belemnites of Kong Karls Land. Nor Polarinstitutt Skrifter 189:77

Dressler BO, Sharpton VL (1997) Breccia formation at a complex impact crater: Slate Islands, Lake Superior, Ontario, Canada. Tectonophysics 275:285–311

Dypvik H (1980) The sedimentology of the Middle to Upper Jurassic/Lower Cretaceous Janusfjellet Formation, Agardhfjellet and Sassenfjorden areas, Central Spitsbergen. Nor Polar Institutt Skrifter 172:79–134

Dypvik H (1993) Natural gamma activity – a possible aid in sedimentological field work. Nor Geologisk Tidsskrift 73:58–62

Dypvik H, Attrep M Jr (1999) Geochemical signals of the late Jurassic, marine Mjølnir impact. Meteorit Planet Sci 34:393–406

Dypvik H, Burchell MJ, Claeys P (2004a) Impacts into marine and icy environments. In: Dypvik H, Burchell M, Claeys P (eds) Cratering in marine environments and on ice. Springer, Berlin-Heidelberg, pp 1–20

Dypvik H, Claeys P, Deutsch A, Kyte FT, Matsui T, Smelror M (2008a) Marine impacts and environmental consequences – drilling of the Mjølnir structure, the Barents Sea. Sci Drill 6:55–57

Dypvik H, Eriksen DØ (1983) Natural radioactivity of clastic sediments and the contributions of U, Th and K. J Pet Geol 5:409–416

Dypvik H, Ferrell RE Jr (1998) Clay mineral alteration associated with a meteoric impact in the marine environment (Barents Sea). Clay Miner 33:51–64

Dypvik H, Ferrell RE Jr, Sandbakken PT (2003) The clay mineralogy of sediments related to the marine Mjølnir impact Crater. Meteorit Planet Sci 38:1437–1450

Dypvik H, Gudlaugsson ST, Tsikalas F, Attrep M Jr, Ferrell RE Jr, Krinsley DH, Mørk A, Faleide JI, Nagy J (1996) The Mjølnir structure – an impact crater in the Barents Sea. Geology 24:779–782

Dypvik H, Håkansson E, Heinberg C (2002) Jurassic and Cretaceous paleogeography and stratigraphic comparison in the North Greenland–Svalbard regions. Polar Res 21:91–108

Dypvik H, Harris NB (2001) Geochemical facies analysis of finegrained siliciclastics using Th/U, Zr/Rb, and (Zr+Rb)/Sr ratios. Chem Geol 181:131–146

Dypvik H, Jansa L (2003) Sedimentary signatures and processes during marine bolide impacts: a review. Sedimentary Geol 161:309–337

Dypvik H, Kalleson E (2010) Marine impacts–mechanisms of early post impact crater sedimentation. In Gibson R, Reimold WU (eds) Large meteorite impacts and planetary evolution IV. Geological Society of America Special Paper, in press

Dypvik H, Mørk A, Smelror M, Sandbakken PT, Tsikalas F, Vigran JO, Bremer GMA, Nagy J, Gabrielsen RH, Faleide JI, Bahiru M, Weiss H (2004b) Impact breccia and ejecta from Mjølnir crater in the Barents Sea – The Ragnarok Formation and Sindre Bed. Nor Geologisk Tidsskrift 84:143–167

Dypvik H, Nagy J, Eikeland TA, Backer-Owe K, Andresen A, Haremo P, Bjærke T, Johansen H, Elverhøi A (1991a) The Janusfjellet Subgroup (Bathonian to Hauterivian) on Central Spitsbergen: a revised lithostratigraphy. Polar Res 9(1):21–43

Dypvik H, Nagy J, Eikeland TA, Backer-Owe K, Johansen H (1991b) Depositional conditions of the Bathonian to Hauterivian Janusfjellet Subgroup, Spitsbergen. Sediment Geol 72:55–78

Dypvik H, Nagy J, Krinsley DH (1992) Origin of the Myklegardfjellet Bed, a basal Cretaceous marker in Svalbard. Polar Res 10:21–31

Dypvik H, Sandbakken PT, Postma G, Mørk A (2004c) Early postimpact sedimentation around the central high of the Mjølnir impact crater (Barents Sea, Late Jurassic). Sediment Geol 168:227–247

Dypvik H, Smelror M, Sandbakken PT, Salvigsen O, Kalleson E (2006) Traces of the marine Mjølnir impact event. Palaeogeogr Palaeoclimatol Palaeoecol 241:621–634

Dypvik H, Wolbach WS, Shuvalov V, Weaver SLW (2008b) Did the Mjølnir asteroid impact ignite Barents Sea hydrocarbon source rocks? In: Evans KR, Horton JW Jr, King DT Jr, Morrow JR (eds) The Sedimentary record of meteorite impacts. Geological Society of America Special Paper 437, Boulder, pp 65–72

Dypvik H, Zakharov V (2010) Mechanisms of late synimpact to early postimpact crater sedimentation in marine-target impact structures. In: Gibson RL, Reimold WU (eds) Large meteorite impacts and planetary evolution IV. Geological Society of America Special Paper 465: 301–318

Døssing A, Dahl-Jensen T, Thybo H, Mjelde R, Nishimura Y (2008) East Greenland Ridge in the North Atlantic Ocean: an integrated geophysical study of a continental sliver in a boundary transform fault setting. J Geophys Res 113:B10107, doi:10.1029/2007JB005536

Earth Impact Data Base (2010); http://www.unb.ca/passc/ImpactDatabase/

Egorov AY, Mørk A (2000) The east Siberian and Svalbard Triassic successions and their sequence stratigraphical relationships. Zentralblatt für Geologie und Paläontologie Teil 1:1377–1430

Eidvin T, Jansen E, Riis F (1993) Chronology of Tertiary fan deposits off the western Barents Sea: implications for the uplift and erosion history of the Barents Shelf. Mar Geol 112:109–131

Ekholm AG, Melosh JH (2001) Crater features diagnostic of oblique impacts: the size and position of the central peak. Geophys Res Lett 28:623–626

Eldholm O, Myhre AM, Thiede J (1994) Cenozoic tectonomagmatic events in the North Atlantic: potential paleoenvironmental implications. In: Boulter MC, Fischer HC (eds) Cenozoic plants and climates of the arctic, NATO ASI Series, 127. Springer, Heidelberg, pp 35–55

Eldholm O, Tsikalas F, Faleide JI (2002) The continental margin off Norway 62–75°N: Palaeogene tectonomagmatic segmentation and sedimentation. In: Jolley DW, Bell BR (eds) The North Atlantic Igneous Province: stratigraphy, tectonics, volcanic and magmatic processes. Geological Society, London, Special Publications 197, pp 39–68

Embry AF (1989) Correlation of Upper Palaeozoic and Mesozoic sequences between Svalbard, Canadian arctic archipelago, and northern Alaska. In: Collinson JD (ed) Correlation in hydrocarbon exploration. Norwegian Petroleum Society, Graham & Trotman, London, pp 89–98

Embry AF (1991) Mesozoic history of the Arctic Islands. In: Trettin HP (ed) Geology of the innuitian orogen and arctic platform of Canada and Greenland. Geol Surv Canada Geol Canada 3:371–433

Embry AF, Beauchamp B (2008) Sverdrup Basin. In: Sedimentary basins of the world, vol 5, Chapter 13. Elsevier, Amsterdam, pp 451–471

Engen Ø, Faleide JI, Dyreng TK (2008) Opening of the fram strait gateway: a review of plate tectonic constraints. Tectonophysics 450:51–69

Espindola JM, Mena M, de La Fuente M, Campos-Enriquez JO (1995) A model of the Chicxulub impact structure (Yucatan, Mexico) based on its gravity and magnetic signatures. Phys Earth Planet Int 92:271–278

Espitalie J, Laporte JL, Madec M, Marquis F, Leplat P, Paulet J, Boutefeu A (1977) Méyhode rapide de caractèrisation des roches mères. De leur potential pétrolier et de leur degree d'évolution. Revue del' Institute Frances Pétrole 32:23–42

Faleide JI, Tsikalas F, Breivik AJ, Mjelde R, Ritzmann O, Engen Ø, Wilson J, Eldholm O (2008) Structure and evolution of the continental margin off Norway and the Barents Sea. In: Gee D, Ladenberger A (eds) Nordic geoscience and 33rd IGC 2008, Episodes, Special Issue 31:82–91

Faleide JI, Vågnes E, Gudlaugsson ST (1993) Late Mesozoic-Cenozoic evolution of the southwestern Barents Sea in a rift shear tectonic setting. Mar Pet Geol 10:186–214

Fernandes MB, Skjemstad JO, Johnson BB, Wells JD, Brooks P (2003) Characterization of carbonaceous combustion residues I. Morphological, elemental and spectroscopic features. Chemosphere 51:785–795

Fertl WH, Rieke HH (1980) Gamma Ray spectral evaluation techniques identify fractured shale reservoirs and source rock characteristics. J Pet Technol 32(11):2053–2062

Fjeldskaar W, ter Voorde M, Johansen H, Christiansson P, Faleide JI, Cloetingh SAPL (2004) Numerical simulation of rifting in the northern Viking Graben; the mutual effect of modelling parameters. Tectonophysics 382:189–212

Fjæran T, Spencer AM (1991) Proven hydrocarbon plays, offshore Norway. In: Spencer AM (ed) Generation, accumulation and production of Europe's hydrocarbons. European Association of Petroleum Geoscientists Special Publication No 1. Oxford University Press, Oxford, pp 25–48

French BM (1998) Traces of catastrophe – a handbook of shockmetamorphic effects in terrestrial meteorite impact structures. Lunar and Planetary Iinstitute Contribution No 954, Lunar and Planetary Institute, Houston, p 120

Gabrielsen RH (1984) Longlived fault zones and their influence on the development of the southwestern Barents Sea. J Geol Soc Lond 141:651–662

Gabrielsen RH, Færseth RB, Jensen LN, Kalheim JE, Riis F (1990) Structural elements of the Norwegian Continental Shelf. Part I: The Barents Sea Region. Norwegian Pet Directorate Bull 6:33

Gabrielsen RH, Grunnaleite I, Ottesen S (1992a) Reactivation of fault complexes in the Loppa High area, southwestern Barents Sea. In: Vorren TO, Bergsager E, Dahl-Stamnes ØA, Holter E, Johansen B, Lie E, Lund TB (eds) Arctic geology and petroleum potential. Norwegian Pet Soc Spec Publ 2:631–641

Gabrielsen RH, Grunnaleite I, Rasmussen E (1997) Cretaceous and Tertiary inversion in the Bjørnøyrenna Fault Complex, south-western Barents Sea. Mar Pet Geol 14(2):165–178

Gabrielsen RH, Kløvjan OS (1997) Late Jurassic-early Cretaceous caprocks of the southwestern Barents Sea: Fracture systems and rock mechanical properties. In: Møller-Pedersen P, Koestler AG (eds) Hydrocarbon seals – importance for exploration and production. Norwegian Petroleum Society Special Publication 7, Elsevier, Amsterdam, pp 73–89

Gabrielsen RH, Kløvjan OS, Rasmussen A, Stølan T (1992b) Interaction between halokinesis and faulting: structuring of the margins of the Nordkapp Basin, Barents Sea region. In: Larsen RM, Brekke H, Larsen BT, Talleraas E (eds) Structural and tectonic modelling and its application to petroleum geology. Norwegian Pet Soc Spec Publ 1:121–131

Gardner GHF, Gardner LW, Gregory AR (1974) Formation velocity and density – the diagnostic basics for stratigraphic traps. Geophysics 39:770–780

Gault DE, Sonett CP (1982) Laboratory simulation of pelagic asteroidal impact: Atmospheric injection, benthic topography and the surface wave radiation field. In: French BM, Schultz PH (eds) Geological implications of impacts of large asteroids and comtes on the Earth. Geol Soc Am Spec Paper 190:69–92

Gault DE, Wedekind JA (1978) Experimental studies of oblique impact. Proc Lunar Planet Sci 9:3843–3875

Gavshin VM, Zakharov VA (1996) Bazhenovites on the Norwegian continental shelf. Sov Geol Geophys 32:52–59

Gee DG, Tebenkov AM (2004) Svalbard a fragment of the Laurentian margin. In: Gee DG and Pease V (eds) The Neoproterozoic Timanide Orogen and Eastern Baltica. Geol Soc Mem 30:101–206, 255

Gérard J, Buhrig, C (1990) Seismic facies of the permian section of the barents shelf: analysis and interpretation. Mar Pet Geol 7:234–252

Gerhard LC, Anderson SB, Lefever JA, Carlson CG (1982) Geological development, origin, and energy mineral resources of Williston Basin, North Dakota. Am Assoc Pet Geol Bull 66: 989–1020

Gersonde R, Deutsch A (2000) New field of impact research looks to the oceans. Eos Trans Am Geophys Union 81(20):221

Gersonde R, Kyte FT, Bleil U, Diekmann B, Flores JA, Gohl K, Grahl G, Hagen R, Kuhn G, Sierro FJ, Völker D, Abelmann A, Bostwick JA (1997) Geological record and reconstruction of the late Pliocene impact of the Eltanin asteroid in the Southern Ocean. Nature 390:357–363

Gisler GR, Weaver RP, Mader CL, Gittings ML (2004) Two- and three-dimensional asteroid impact simulations. Comput Sci Eng 6:46–55

Gjelberg J, Dreyer T, Høie A, Tjelland T, Lilleng T (1987) Late Triassic to Mid-Jurassic sandbody development on the Barents and Mid-Norwegian shelf. In: Brooks J, Glennie K (eds) Petroleum geology of North West Europe. Graham and Trotman, London, pp 1105–1129

Gjelberg J, Steel RJ (1995) Helvetiafjellet Formation (Barremian-Aptian), Spitsbergen: characteristics of transgressive succession. In: Steel RJ, Felt VL, Johannesen E, Mathieu C (eds) Sequence stratigraphy on the North west European margin. Norwegian Petroleum Society Special Publication 5. Elsevier, Amsterdam, pp 571–593

Glasstone S, Dolan PJ (1977) The effects of nuclear weapons, 3rd edn. US Department of Defense and US Department of Energy, US Government Printing Office, Washington, p 653

Glimsdal S, Pedersen G, Shuvalov VV, Dypvik H, Langtangen HP, Kristiansen Ø (2005) Tsunami generated by the Mjølnir impact [abs]. Lunar Planet Sci Conf 34, abs #1287

Glimsdal S, Pedersen GK, Atakan K, Harbitz CB, Langtangen HP, Løvholt, F (2006) Propagation of the December 26, 2004 Indian Ocean Tsunami: effects of dispersion and source characteristics. Int J Fluid Mech Res 33(1):15–43

Glimsdal S, Pedersen GK, Langtangen HP, Shuvalov V, Dypvik H (2007) Tsunami generation and propagation from the Mjølnir asteroid impact. Meteorit Planet Sci 42:1473–1493

Gohn GS, Koeberl C, Miller KG, Reimold WU, Cockell CS, Horton JW, Sanford WE, Voytek MA (2006) Chesapeake Bay impact structure drilled. Eos Trans Am Geophys Union 87:349, 355

Gohn GS, Koeberl C, Miller KG, Reimold WU, Browning JV, Cockell CS, Horton JW Jr, Kenkmann T, Kulpecz AA, Powars DS, Sanford WE, Voytek MA (2008). Deep drilling into the Chesapeake Bay impact structure. Science 320:1740–1745

Gradstein F, Ogg J, Smith A (2004) A geological time scale. Cambridge University Press, Cambridge, p 589

Gradstein FM, Agterberg FP, Ogg JG, Hardenbol J, van Veen P, Thierry J, Huang Z (1994) A Mesozoic time scale. J Geophys Res 99(B12):24051–24074

Gradstein FM, Agterberg FP, Ogg JG, Hardenbol J, Bäckström S (1999) On the Cretaceous time scale. Neues Jahrb Geol Palaontol Abh 212:3–14

Grant FS, West GF (1965) Interpretation theory in applied geophysics. McGraw-Hill, New York, p 584

Grantz A, Pease VL, Willard DA, Phillips RL, Clark DL (2001) Bedrock cores from 89° North: implications for the geologic framework and Neogene paleo-oceanography of the Lomonosov Ridge and a tie to the Barents shelf. Geol Soc Am Bull 113:1272–1281

Grieve RAF (1987) Terrestrial impact structures. Ann Rev Earth Planet Sci 15:245–270

Grieve RAF (1991) Terrestrial impact: the record in the rocks. Meteoritics 26:175–194

Grieve RAF, Langenhorst F, Stöffler D (1996) Shock metamorphism of quartz in nature and experiment: II. Significance in geoscience. Meteorit Planet Sci 31:6–35

Grieve RAF, Masaitis VL (1994) The economic potential of terrestrial impact craters. Int Geol Rev 36:105–151

Grieve RAF, Pesonen LJ (1992) The terrestrial impact record. Tectonophysics 216:1–30

Grieve RAF, Pesonen LJ (1996) Terrestrial impact craters: their spatial and temporal distribution and impacting bodies. Earth Moon Planet 72:357–376

Grieve RAF, Therriault A (2000) Vredefort, Sudbury, Chicxulub: Three of a kind? Annu Rev Earth Planet Sci 28:305–338

Grogan P, Nyberg K, Fotland B, Mylkebust R, Dahlgren S, Riis F (1998) Cretaceous magmatism south and east of Svalbard: evidence from seismic reflection and magnetic data. Polarforschung 68:25–34

Grue J, Pelinovsky EN, Fructus D, Talipova T, Kharif C (2008) Formation of undular bores and solitary waves in the Strait of Malacca caused by the 26. December 2004 Indian Ocean tsunami. J Geophys Res 113:C05008, doi:10.1029/2007JC004343

Grung-Olsen R, Hanssen OK (1987) Askeladd. In: Spencer AM et al (eds) Geology of Norwegian oil and gas fields. Graham and Trotman, London, pp 420–428

Gudlaugsson ST (1993) Large impact crater in the Barents Sea. Geology 21:291–294

Gudlaugsson ST, Faleide JI, Johansen SE, Breivik AJ (1998) Late Palaeozoic structural development of the south-western Barents Sea. Mar Pet Geol 15:73–102

Gustavsen FB (1995) Maringeologiske undersøkelser av berggrunnen i det nordlige Barentshavet (mellom Svalbard og Franz Josef Land). Cand Scient (candidatus scientarium) Thesis. University of Oslo, Oslo, p 156

Halvorsen E (1989) A paleomagnetic pole position of late Jurassic/Early Cretaceous dolerites from Hinlopenstretet, Svalbard and its tectonic implications. Earth Planet Sci Lett 94:398–408

Halvorsen E, Løvlie R, Andresen A (1996) Evidence of complete Tertiary remagnetization of Early Cretaceous dolerite dikes and sills from Spitsbergen, Svalbard. European Geophysical Society, 21st General Assembly, The Hague, C:138 [abs]

Hammer S (1963) Deep gravity interpretation by stripping. Geophysics 28:369

Hardenbol J, Thierry J, Farley MB et al (1998) Mesozoic and Cenozoic sequence chronostratigraphic framework of European basins. In: de Graciansky P-C, Hardenbol J, Jacquin T, Vail PR (eds) Mesozoic Cenozoic sequence stratigraphy of European basins. Soc Econ Paleontol Miner Spec Publ 60:3–13, 763–781

Harland WB (1969) Contribution of Spitsbergen to understanding of the tectonic eveolution of the North Atlantic region. In: Kay M (ed) North Atlanticgeology and continental drift. Am Assoc Pet Geol Mem 12:817–851

Harland WB (1971) Tectonic transpression in Caledonian Spits-bergen. Geol Mag 108:27–42

Hartung JB, Anderson RR (1996) A brief history on investigations of the Manson impact structure. In: Koeberl C, Anderson RR (eds) The Manson impact structure, Iowa: Anatomy of an impact crater. Geological Society of America Special Paper 302, Boulder, pp 31–43

Hayden T, Kominz M, Powars DS, Edwards LE, Miller KG, Browning JV, Kulpecz AA (2008) Impact effects and regional tectonic insights: Backstripping the Chesapeake Bay impact structure. Geology 36:327–330

Hayhurst CJ, Ranson HJ, Gardner DJ, Birnbaum NK (1995) Modelling of micro-article hypervelocity oblique impacts on thick targets. Int J Impact Eng 17:375–386

Heafford AP, Kelly SRA (1988) Carboniferous through Cretaceous panarctic tectonic events. In: Harland WB, Dowdeswell EK (eds) Geological evolution of the Barents Shelf region. Graham and Trotman, London, pp 19–32

Heard HC, Carter NL (1968) Experimentally induced 'natural' intragranular flow in quartz and quartzite. Am J Sci 266:1–42

Heinberg C, Håkansson E (1994) Late Jurassic-Early Cretaceous stratigraphy and depositional environment. Wandel Sea Basin, basin analyses. Scientific report 12, University of Copenhagen, Denmark, p 8

Henkel H, Reimold WU, Koeberl C (2002) Magnetic and gravity model of the Morokweng impact structure. J Appl Geophys 49:129–147

Henriet JP, De Batist M, Verschuren M (1991) Early fracturing of Palaeogene clays, southernmost North Sea: relevance to mechanisms of primary hydrocarbon migration. In: Spencer AM (ed) Generation, accumulation, and production of Europes hydrocarbons. Special Publication of the European Association of Petroleum Geoscientists, Oxford University Press, Oxford, pp 217–227

Hildebrand AR, Pilkington M, Ortiz-Aleman C, Chavez R, Urrutia-Fucugauchi J, Connors M, Graniel-Castro E, Camara-Zi A, Halpenny J, Niehaus D (1998) Mapping Chicxulub crater structure with gravity and seismic reflection data. In: Grady MM, Hutchison R, McCall GJH, Rotherby DA (eds) Meteorites: flux with time and impact effects. Geological Society of London Special Publication 140, London, pp 155–176

Horton JW Jr, Aleinikoff JN, Kunk MJ, Gohn GS, Edwards LE, Self-Trail JM, Powars DS, Izett GA (2005b) Recent research on the Chesapeake Bay impact structure, USA – Impact debris and reworked ejecta. In: Kenkmann T, Hörz F, Deutsch A (eds) Large meteorite impact III. Geological Society of America Special Paper 384, Boulder, pp 147–170

Horton JW Jr, Powars DS, Gohn GS (2005a) Studies of the Chesapeake Bay Impact Structure – The USGS-NASA Langley corehole, Hampton, Virginia, and related coreholes and geophysical surveys. US Geological Survey. USGS Professional Paper 1688, Reston, Virginia

Hovland M, Jodd AG (1988) Seabed pockmarks and seepages: impact on geology, biology and the marine environment. Graham and Trotman, London, p 293

Håkansson E, Birkelund T, Heinberg C, Hjort C, Mølgaard P, Pedersen SAS (1993) The Kilen Expedition 1985. Bull Geol Soc Denmark 40:9–32

Håkansson E, Heinberg C, Madsen L, Mølgaard S, Pedersen SAS, Piasecki S, Rasmussen JA, Stemmerik L, Zinck-Jørgensen K (1994) Wandel Sea Basin: basin analyses – Project summary. Scientific Report 1, University of Copenhagen, Denmark, p 11

Håkansson E, Heinberg C, Stemmerik L (1981) The Wandel sea basin from Holm Land to Lockwood Ø, Eastern North Greenland. Bull Grønlands Geol Undersøgelse 106:47–63

Håkansson E, Heinberg C, Stemmerik L (1991) Mesozoic and Cenozoic history of the Wandel Sea basin area, North Greenland. Bull Grønland Geol Undersøgelse 160:153–164

Ivanov BA, Artemieva NA (2002) Numerical modelling of the formation of large impact craters. In: Koeberl C, MacLeod KG (eds) Catastrophic events and mass extinctions: impact and beyond. Geological Society of America Special Paper 356, Boulder, pp 619–630

Ivanov BA, Turtle EP (2001) Modeling impact crater collapse acoustic fluidization implemented into a hydrocode [abs]. Lunar Planet Sci Conf 32, abs#1284, CD-ROM

Izett GA, Cobban WA, Obradovich JD, Kunk MD (1993) The Manson impact structure: ^{40}Ar/^{39}Ar age and its distal impact ejecta in the Pierre Shale in southeastern South Dakota. Science 262:729–732

Jackson HR, Johnson GL, Sundvor E, Myhre AM (1984) The Yermak Plateau: Formed at a triple junction. J Geophys Res 89:3223–3232

Jakobsson M, Grantz A, Kristoffersen Y, MacNab R (2003) Physiographic provinces of the Arctic Ocean seafloor. Geol Soc Am Bull 115:1443–1455

Jansa LF (1993) Cometary impacts into ocean: their recognition and the threshold constraint for biological extinctions. Palaeogeogr Palaeoclimatol Palaeoecol 104:271–286

Jansa LF, Pe-Piper G, Robertson BP, Friedenreich O (1989) Montagnais: A submarine impact structure on the Scotian shelf, eastern Canada. Geol Soc Am Bull 101:450–463

Jensen LN, Sørensen K (1992) Tectonic framework and halokinesis of the Nordkapp Basin. In: Larsen RM, Brekke H, Larsen BT, Talleraas E (eds) Structural and tectonic modelling and its application to petroleum geology. Norwegian Pet Soc Spec Publ 1:109–120

Johansen SE, Ostisty BK, Birkeland Ø, Fedorovsky YF, Martirosjan VN, Bruun Christensen O, Cheredeev SI, Ignatenko EA, Margulis LS (1993) Hydrocarbon potential in the Barents Sea region: Play distribution and potential. In: Vorren TO, Bergsager E, Dahl-Stamnes ØA, Holter E, Johansen B, Lie E, Lund TB (eds) Arctic geology and petroleum potential. Norwegian Pet Soc Spec Publ 2:273–320

Johnsen SO, Mørk A, Dypvik H, Nagy J (2001) Outline of the geology of Svalbard. NGF Abstr Proc Norwegian Geol Soc 1:93–112

Juhlin C, Pedersen LB (1987) Reflection seismic investigations of the Siljan Impact Structure, Sweden. J Geophys Res 92:14113–14122

Karp T, Milkereit B, Janle P, Danuor SK, Pohl J, Berckhemer H, Scholz CA (2002) Seismic investigation of the Lake Bosumtwi impact crater: preliminary results. Planet Space Sci 50:735–743

Kataoka T, Tsutahara M (2004) Transverse instability of surface solitary waves. J Fluid Mech 512:211–221

Keiswetter D, Black R, Steeples D (1996) Seismic reflection analysis of the Manson impact structure, Iowa. J Geophys Res 101:5823–5834

Kelly SRA (1988) Jurassic through Cretaceous stratigraphy of the Barents Shelf. In: Harland WB, Dowdeswell EK (eds) Geological evolution of the Barents Shelf region. Graham and Trotman, London, pp 109–130

Kenkmann T (2002) Folding within seconds. Geology 30:231–234

Kenkmann T (2003) Dike formation, cataclastic flow and rock fluidization during impact cratering: an example from the Upheaval Dome structure, Utah. Earth Planet Sci Lett 214:43–58

Kenkmann T, Jahn A, Scherler D, Ivanov BA (2005) Structure and formation of a central uplift: A case study at the Upheaval Dome impact crater, Utah. In: Kenkmann T., Hörz F, Deutsch A (eds) Large meteorite impacts III. Geol Soc Am Spec Paper 384:85–115

Kenkmann T, Kiebach F, Rosenau M, Raschke U, Pigowske A, Mittelhaus K, Eue D (2007) Coupled effects of impact and orogeny: Is the marine Lockne crater, Sweden, pristine? Meteorit Planet Sci 42:1995–2012

Kenkmann T, Wittmann A, Scherler D (2004) Structure and impact indicators of the Cretaceous sequence of the ICDP drill core Yaxcopoil-1, Chicxulub impact crater, Mexico. Meteorit Planet Sci 39:1069–1088

Kennedy AB, Chen Q, Kirby JT, Dalrymple RA (2000) Boussinesq modeling of wave transformation, breaking and run-up. Part I 1D J Waterway Port Coast Ocean Eng 126(1):39–47

Kieffer SW, Simonds CH (1980) The role of volatiles and lithology in the impact cratering processes. Rev Geophys 18:143–181

Kirschner CE, Grantz A, Mullen MW (1992) Impact origin of the Avak structure, Arctic Alaska, and genesis of the Barrow gas fields. Am Assoc Pet Geol Bull 76:651–679

Koeberl C (1989) New estimates of area and mass of the North American tektite strewn field. Proceedings of the 19th Lunar and Planetary Science Conference. Houston, Cambridge University Press, pp 745–751

Koeberl C (2007) The geochemistry and cosmochemistry of impacts. In: Davis A (ed) Treatise on Geochemistry, vol 1. Elsevier, Pergamon, pp. 1.28.1–1.28.52

Koeberl C, Armstong RA, Reimold WU (1997a). Morokweng, South Africa: A large impact structure of Jurassic-Cretaceous boundary age. Geology 25:731–734

Koeberl C, Brandstätter F, Glass BP, Hecht L, Mader D, Reimold WU (2007b) Uppermost impact fallback layer in the Bosumtwi crater (Ghana): mineralogy, geochemistry, and comparison with Ivory Coast tektites. Meteorit Planet Sci 42:709–729

Koeberl C, Milkereit B, Overpeck JT, Scholz CA, Amoako PYO, Boamah D, Danuor S, Karp T, Kueck J, Hecky RE, King JW, Peck JA (2007a) An international and multidisciplinary drilling project into a young complex impact structure: The 2004 ICDP Bosumtwi Crater Drilling Project-An overview. Meteorit Planet Sci 42:483–511

Korycansky DG, Asphaug E, Ward SN (2003) Impact tsunami calculations hydrodynamical simulations vs linear theory[abs]. Lunar Planet Sci Conf 34, abs #1195, CD ROM

Korycansky DG, Lynett PJ (2005) Offshore breaking of impact tsunami: the Van Dorn effect revisited. Geophys Res Lett 32:L10608

Koyi H, Talbot CJ, Tørudbakken BO (1993) Salt diapirs of the southwest Nordkapp Basin: Analogue modelling. Tectonophysics 228:167–187

Kriens BJ, Shoemaker EM, Herkenhoff KE (1999) Geology of the Upheaval Dome impact structure, southeast Utah. J Geophys Res 194:18867–18887, doi: 10.1029/1998JE000587

Kuznetsov NM (1965) Thermodynamic functions and shock adiabats for air at high temperatures. Mashinostroyenie, Moscow, p 464 (in Russian)

Kyte FT (2002). Unmelted meteoritic debris collected from Eltanin ejecta in Polarstern cores from expedition ANT XII/4. In: Deutsch A, Ivanov BA, Kyte FT, Gersonde R (eds) Oceanic impact; mechanisms and environmental perturbations. Deep Sea Res Part II Top Stud Oceanogr 49(6):1063–1071

Kyte FT, Langenhorst F, Tepley FJ III (2000) The Eltanin meteorite: Large messenger from the HED or mesosiderite parent body? [abs]. Lunar Planet Sci Conf 31, abs #1811, CD-ROM

Langenhorst F, Deutsch A, Hornemann U, Stöffler D (1992) Effect of temperature on shock metamorphism of single crystal quartz. Nature 356:507–509

Langenhorst F, Dypvik H (1996) Microstructural characteristics of shocked quartz from ejecta of the submarine Mjølnir impact structure, Barents Sea [abs]. Lunar Planet Sci Conf 27, abs # 1364, CD ROM

Langrock U, Stein R (2004) Origin of marine petroleum source rocks from the Late Jurassic to Early Cretaceous Norwegian Greenland seaway – evidence for stagnation and upwelling. Mar Pet Geol 21:157–176

Larsen RM, Fjæran T, Skarpnes O (1992) Hydrocarbon potential of the Norwegian Barents Sea based on recent well results. In: Vorren TO, Bergsager E, Dahl-Stamnes ØA, Holter E, Johansen B, Lie E, Lund TB (eds) Arctic geology and petroleum potential. Norwegian Pet Soc Spec Publ 2:321–331

Larssen GB, Elvebakk G, Henriksen LB, Kristensen S-E, Nilsson I, Samuelsberg T, Svanå TA, Stemmerik L, Worsley D (2005) Upper Paleozoic lithostratigraphy of the southern part of the Norwegian Barents Sea. Nor Geologiske Undersøkelse Bull 444:43

Lawver LA, Gahagan LM, Campbell DA, Brozena JM, Childers V (1999) Mid-Jurassic to Recent tectonic evolution of the Arctic region (*Powerpoint* animation, using the *PLATES'* animation software) [abs]. In: Lawver LA, Brozena JM, Kovacs LC, Childers V (eds) Compilations in the Canada Basin, Aerogeophysical Anomalies. Eos, Transactions, American Geophysical Union, Fall Meeting 1999, San Francisco, 80(46), p 1000

Lawver LA, Müller RD, Srivastava SP, Roest W (1990) The opening of the Arctic Ocean. In: Bleil U, Thiede J (eds) Geological history of the polar ocean. Klüwer Academic Publishers, Netherlands, pp 29–62

Leith TL, Weiss HM, Mørk A, Århus N, Elvebakk G, Embry AF, Brooks PW, Stewart KR, Pchelina TM, Bro EG, Verba ML, Danyushevskaya A, Borisov AV (1992) Mesozoic hydrocarbon sourcerocks of the Arctic region. In: Vorren TO et al (eds) Arctic geology and petroleum potential. Nor Pet Forening, Spec Publ 2:1–25

Lindström M, Shuvalov V, Ivanov B (2005) Lockne crater as a result of marine-target oblique impact. Planet Space Sci 53:803–815

Lindström M, Sturkell EFF, Törnberg R, Ormö J (1996) The marine impact crater at Lockne, central Sweden. Geologiska Fören Förhandlinger 118:193–206

Linjordet A, Grung-Olsen R (1992) The Jurassic Snøhvit gas field, Hammerfest Basin, offshore northern Norway. Am Assoc Pet Geol Mem 54:349–370

Ludwig JW, Nafe JE, Drake CL (1970) Seismic refraction. In: Maxwell AE (ed) The sea, vol 4. Wiley, New York, pp 53–84

Lundborg N (1968) Strength of rock-like materials. Int J Rock Mech Min Sci 5:427–454

Mackenzie GD, Maguire PKH, Denton P, Morgan JV, Warner M (2001) Shallow seismic velocity structure of the Chicxulub impact crater from modelling of Rg dispersion using a genetic algorithm. Tectonophysics 338:97–112

Maher HD Jr. (2001) Manifestations of the Cretaceous high arctic large igneous province in Svalbard. J Geol 109:91–104

Manum S, Throndsen T (1978) Dispersed organic matter in the Spitsbergen Tertiary. Nor Polarinst. Årbok 1977:179–187

Matsui T, Imamura F, Tajika E, Nakano Y, Fujisawa Y (2002) Generation and propagation of a tsunami from the Cretaceous-Tertiary impact event. In: Koerbel C, MacLeod KG (eds) Catastrophic events and mass extinctions: Impacts and Beyond. Geological Society of America Special Paper 356, Boulder, pp 69–77

Mauring E, Kihle O (2006) Leveling aerogeophysical data using a moving differential median filter. Geophysics 71:L5–L11

McGetchin TR, Settle M, Head JW (1973) Radial thickness variation in impact crater ejecta: implications for lunar basin deposits. Earth Planet Sci Lett 20:226–236

McGlaun JM, Thompson SL, Elrick MG (1990) CTH: a three-dimensional shock wave physics code. Int J Impact Eng 10:351–360

McKinnon WB (1982) Impact into the Earth's ocean floor: Preliminary experiments, a planetary model, and possibilities for detection. In: Silver LT, Schultz PH (eds) Geological implications of impacts of large asteroids and comets on Earth. Geological Society of America Special Paper 190, Boulder, pp 129–142

Mei CC (1989) Applied dynamics of ocean waves. Advanced series on ocean engineering, vol 1, 2nd edn. World Scientific, London, p 768

Melosh HJ (1989) Impact cratering: a geologic process. Oxford University Press & Clarendon Press, Oxford, p 245

Melosh HJ (2003) Impact-generated tsunamis: An over-rated hazard [abs]. Lunar Planet Sci Conf 34, abs #2013, CD-ROM

Melosh HJ, Ivanov BA (1999) Impact crater collapse. Ann Rev Earth Planet Sci 27:385–425

Merrill RB, Schultz PH (1981) Proceedings of the conference on multiring basins; formation and evolution. Pergamon Press, New York, p 289

Midtkandal I, Nystuen JP, Nagy J (2007) Paralic sedimentation on an epicontinental ramp shelf during a full cycle of relative sealevel fluctuations; the Helvetiafjellet Formation in Nordenskiøld land, Spitsbergen. Norwegian J Geol 87:343–359

Miles JW (1977) Diffraction of solitary waves. Zeitschrift für angewandte Mathematik und Physik 28:889–902

Miles JW (1980) Solitary waves. Ann Rev Fluid Mech 12:11–43

Milkereit B, Green A, Wu J, White D, Adams E (1994) Integrated seismic and borehole geophysical study of the Sudbury igneous complex. Geophys Res Lett 21:931–934

Milton DJ, Glikson AY, Brett R (1996) Gosses Bluff – a latest Jurassic impact structure, central Australia. Part 1: Geological structure, stratigraphy and origin. J Aust Geol Geophys 16: 453–486

Montanari A, Koeberl C (2000) Impact stratigraphy-The Italian record. Springer, Berlin-Heidelberg, p 364

Morgan JV, Christeson GL, Zelt CA (2002a) Testing the resolution tomogram across the Chicxulub crater. Tectonophysics 335:215–226

Morgan JV, Warner MR (1999) The third dimension of a multiring impact basin. Geology 26: 407–410

Morgan JV, Warner MR, Chicxulub working group (1997) Size and morphology of the Chicxulub impact crater. Nature 390:472–476

Morgan JV, Warner MR, Collins GS, Melosh HJ, Christeson GL (2000) Peak-ring formation in large impact craters: geophysical constraints from Chicxulub. Earth Planet Sci Lett 183: 347–354

Morgan JV, Warner MR, Grieve R (2002b) Geophysical constraints on the size and structure of the Chicxulub impact crater. In: Koeberl C, MacLeod KG (eds) Catastrophic events and mass extinctions: impacts and beyond. Geological Society of America Special Paper 356, Boulder, pp 49–46

Mørk A, Dallmann WK, Dypvik H, Johannessen EP, Larssen GB, Nagy J, Nøttvedt A, Olaussen S, Pcelina TP, Worsley D (1999) Mesozoic lithostratigraphy. In: Dallmann WK (ed) Lithostratigraphic Lexicon of Svalbard. Review and recommendations for nomenclature use. Upper Paleozoic to Tertiary. Norsk Polarinstitutt, Tromsø, pp 127–214

Mørk A, Elvebakk G (1999) Lithological description of subcropping Lower and Middle Triassic rocks from the Svalis Dome, Barents Sea. Polar Res 18(1):83–104

Mørk A, Knarud R, Worsley D (1982) Depositional and diagenetic environments of the Triassic and Lower Jurassic succession of Svalbard. In: Embry AF, Balkwill HR (eds) Arctic geology and geophysics: Proceedings of the 3rd international symposium on Arctic geology. Canadian Society of Petroleum Geologists, Calgary, pp 371–398

Mørk A, Smelror M (2001) Correlation and noncorrelation of high order circumarctic Mesozoic sequences. Polarforschung 69:65–72

Movshovich EV, Milyavsky AE (1990) Morphology and inner structure of Kamensk and Gusev astroblemes. In: Masaitis VL (ed) Impact craters of the Mesozoic-Cenozoic boundary. Nauka Press, Leningrad, pp 110–146

Myhre AM, Eldholm O, Sundvor E (1982) The margin between Senja and Spitsbergen fracture zones: implications from plate tectonics. Tectonophysics 89(1–3):33–50

Nagy J, Basov VA (1998) Revised foraminiferal taxa and biostratigraphy of Bathonian to Ryazanian deposits in Spitsbergen. Micropaleontology 44:217–255

Nagy J, Løfaldli M, Bäckström SA, Johansen H (1990) Agglutinated foraminiferal stratigraphy of Middle Jurassic to basal Cretaceous shales, Central Spitsbergen. In: Hemleben C (ed) Palecology, biostratigraphy, paleoceanography and taxonomy of agglutinated foraminifera. Kluwer Academic Press, Dordrecht, pp 969–1015

Nagy J, Løfaldli M, Bomstad K (1988) Aspects of foraminiferal distribution and depositional conditions in Middle Jurassic to Early Cretaceous shales in eastern Spitsbergen. Abhandlungen der Geologischen Bundesanstalt Wien 41:287–299

Nardin TR, Røssland KG (1992) Restoration of the eroded section in the western Barents Sea. In: Vorren TO, Bergsager E, Dahl-Stamnes ØA, Holter E, Johansen B, Lie E, Lund TB (eds) Arctic geology and petroleum potential. Norwegian Pet Soc Spec Publ 2:607–618

Naumov MV (2002) Impact generated hydrothermal systems: data from Popigai, Kara, and Puchezh-Katunki impact structures. In: Plado J, Pesonen LJ (eds) Impacts in Precambrian shields. Springer, Berlin-Heidelberg, pp 307–321

Nettleton LL (1976) Gravity and magnetics in oil prospecting. McGraw-Hill, New York, p 464

Northwest Geophysical Associates Inc (1991) GM-SYS version 1.88 software package

Nøttvedt A, Cecchi M, Gjelbjerg JG, Kristensen SE, Lønøy A, Rasmussen A, Rasmussen E, Skott PH, van Veen PM (1993) Svalbard-Barents Sea correlation: A short review. In: Vorren TO, Bergsager E, Dahl-Stamnes ØA, Holter E, Johansen B, Lie E, Lund TB (eds) Arctic geology and petroleum potential. Norwegian Pet Soc Spec Publ 2:363–375

Nyland B, Jensen LN, Skagen J, Skarpnes P, Vorren TO (1992) Tertiary uplift and erosion in the Barents Sea: Magnitude, timing, and consequences. In: Structural and tectonic modeling and its application to the petroleum geology. Norwegian Pet Soc Spec Publ 1:153–162

Ohm, SE, Karlsen DA, Austen TJF (2008) Geochemically driven exploration models in uplifted areas. Examples from the Norwegian Barents Sea. Am Assoc Pet Geol Bull 92:1191–1223

O'Keefe JD, Ahrens TJ (1982) The interaction of the Cretaceous/Tertiary Extinction Bolide with atmosphere, ocean and solid Earth. In: Silver LT, Schultz PH (eds) Geological implications of large asteroids and comets on the Earth. Geol Soc Am Spec Paper 190:103–120

O'Keefe JD, Ahrens TJ (1985) Sampling of planetary material by oblique impact jet entrainment [abs]. Lunar Planet Sci 16:629–630

O'Keefe JD, Ahrens TJ (1986) Oblique impact – A process for obtaining meteorite samples from other planets. Science 234:346–349

O'Keefe JD, Ahrens TJ (1999) Complex crater: relationship of stratigraphy and rings to impact conditions. J Geophys Res 104(E11):27091–27104

O'Keefe JD, Stewart ST, Ahrens TJ (2001a) Impact on comets and asteroids [abs]. Lunar Planet Sci 32, abstract #2002

O'Keefe JD, Stewart ST, Lainhart ME, Ahrens TJ (2001b) Damage and rockvolatile mixture effects on impact crater formation. Int J Impact Eng 26:543–553

Okubu CH, Schultz RA (2007) Compactional deformation bands in Wingate Sandstone; additional evidence of an impact origin of Upheaval Dome, Utah. Earth Planet Sci Lett 256:169–181

Ormö J, Lindström M (2000) When a cosmic impact strikes the seabed. Geol Mag 137:67–80

Ormö J, Shuvalov V, Lindström M (2002) Numerical modeling for target water depth estimation of marine target impact craters. J Geophys Res 107:3-1–3-9, doi:10.1029/2002JE001865

Otha Y (1994) Caledonian and Precambrian history of Svalbard: A review and an implication of excape tectonics. Tectonophysics 231:183–194

Parker JR (1967) The Jurassic and Cretaceous sequence in Spitsbergen. Geol Mag 104:487–505

Pascal C, Gabrielsen RH (2001) Numerical modelling of Cenozoic stress patterns in the mid Norwegian margin and the northern North Sea. Tectonics 20:585–599

Pedersen G (1994) Nonlinear modulations of solitary waves. J Fluid Mech 267:83–108

Peregrine DH (1966) Calculations of the development of an undular bore. J Fluid Mech 25: 321–330

Peregrine DH (1976) Interaction of water waves and currents. Adv Appl Mech 16:10–117

Pesonen LJ, Lehtinen M, Deutsch A, Elo S, Lukkarinen H (1996) New geophysical and petrographic results of the Suvasvesi N impact structure, Finland. Lunar Planet Sci 27:1021

Peters JL (1949) The direct approach to magnetic interpretation and its practical application. Geophysics 14:290–320

Pierazzo E, Artemieva N, Ivanov B (2005) Starting conditions for hydrothermal systems underneath Martian craters: Hydrocode modeling. In: Kenkmann T, Hörz F, Deutsch A (eds) Large meteorite impacts III. Geological Society of America Special Paper 384, Boulder, pp 443–457

Pierazzo E, Kring DA, Melosh HJ (1998) Hydrocode simulations of the Chicxulub impact event and the production of climatically active gasses. J Geophys Res 103:28607–28626

Pierazzo E, Melosh HJ (1999) Hydrocode modeling of Chicxulub as an oblique impact even. Earth Planet Sci Lett 165:163–176

Pierazzo E, Melosh HJ (2000) Understanding oblique impacts from experiments, observations, and modeling. Ann Rev Earth Planet Sci 28:141–167

Pierazzo E, Vickery AM, Melosh HJ (1997) A reevaluation of impact melt production. Icarus 127:408–423

Pilkington M, Grieve RAF (1992) The geophysical signature of terrestrial impact craters. Rev Geophys 30:161–181

Pilkington M, Hildebrand A, Ortiz-Aleman C (1994) Gravity and magnetic field modelling and structure of the Chicxulub crater, Mexico. J Geophys Res 99:13147–13162

Pilkington M, Jansa LF, Grieve RAF (1995) Geophysical studies of the Montagnais impact crater, Canada. Meteoritics 30:446–450

Plescia JB (1996) Gravity investigation of the Manson impact structure, Iowa. In: Koeberl C, Anderson RR (eds) The Manson impact structure, Iowa: Anatomy of an Impact Crater. Geological Society of America Special Paper 302, Boulder, pp 89–104

Plescia JB, Shoemaker EM, Shoemaker CS (1994) Gravity survey of the Mount Toondina impact structure, South Australia. J Geophys Res 99:13167–13179

Poag CW (1996) Structural outer rim of Chesapeake Bay impact crater: Seismic and borehole evidence. Meteorit Planet Sci 31:218–226

Poag CW, Koeberl C, Reimold WU (2004) The Chesapeake Bay crater: Geology and geophysics of a Late Eocene submarine impact structure. Springer, Berlin-Heidelberg, p 522

Poag CW, Poppe LJ (1998) The Toms Canyon structure, New Jersey outer continental shelf: A possible late Eocene impact crater. Mar Geol 145:23–60

Poag CW, Powars DS, Poppe LJ, Mixon RB (1994) Meteoroid mayhem in the Ole Virginny: source of the North American tektite strewn field. Geology 22:691–694

Pohl J, Stöffler D, Gall H, Ernstson K (1977) The Ries impact crater. In: Roddy DJ, Pepin RO, Merrill RB (eds) Impact and explosion cratering. Pergamon Press, Tarrytown, pp 343–404

Popov Y, Romushkevich R, Bayuk I, Korobkov D, Mayr S, Burkhardt H, Wilhelm H (2004) Physical properties of the rocks from the upper part of the Yaxcopoil-1 drill hole, Chicxulub crater. Meteorit Planet Sci 39:799–812

Powars DS, Bruce TS (1999) The effects of the Chesapeake Bay impact crater on the geologic framework and the correlation of hydrogeologic units of southeastern Virginia, south of the James River. US Geological Survey. US Geological Survey Professional Paper 1612, Reston, p 82

Puura V, Suuroja K (1992) Ordovician impact crater at Kärdla, Hiiumaa Island, Estonia. Tectonophysics 216:143–156

Quaide WL, Oberbeck VR (1968) Thickness determinations of the lunar surface layer from lunar impact craters. J Geophys Res 73:5247–5270

Rawson PF, Hoedemaker PJ, Abuirre-Urreta MB, Avram F, Ettachfini M, Kelly SRA, Klein J, Kotetishvili EV, Owen HG, Ropolo P, Thompson MRA, Wippich M, Vasicek Z (1999) Report on the 4th International Workshop of the Cretaceous cephalopod team (ICPG-Project 362). Scr Geol 3:3–13

Reid AB, Allsop JM, Granser H, Millet AJ, Somerton IW (1990) Magnetic interpretation in three dimensions using Euler deconvolution. Geophysics 55:80–91

Richardsen G, Vorren TO, Tørudbakken BO (1993) Post-Early Cretaceous uplift and erosion in the southern Barents Sea: a discussion based on analysis of seismic interval velocities. Nor Geologisk Tidsskrift 73:3–20

Riches P, Traub-Sobott I, Zimmerle W, Zinkernagel U (1986) Diagenetic peculiarities of potential Lower Jurassic reservoir sandstones, Troms 1 area, off northern Norway, and their tectonic significance. Clay Miner 21:565–584

Rider GMH (1991) The geological interpretation of well logs. Whittles Publishing, Caithness, p 175

Riding JB, Thomas JE (1992) Dinoflagellate cysts of the Jurassic system. In: Powell AJ (ed) A stratigraphic index of Dinoflagellate Cysts. British Micropalaeontological Society Publication Series. Chapman and Hall, London, pp 7–97

Robin E, Molina E (2006) Chronostratigraphy, composition, and origin of Ni-rich spinel from the Late Eocene Fuente Caldera section in Spain: one impact or more? Meteorit Planet Sci 41:1231–1248

Robin E, Rocchia R, Siret D, Dypvik H (2001) Discovery of nickel iron particles in the ejecta bearing strata of the latest Jurassic Mjølnir meteorite impact (Barents Sea). Norwegian Geol Soc Abstr Ser 1:67–68

Roddy DJ (1977) Largescale impact and explosion craters: comparisons of morphological and structural analogs. In: Roddy DJ et al (eds) Impact and explosion cratering. Pergamon Press, Tarrytown, pp 185–246

Roddy DJ, Schuster SH, Rosenblatt M, Grant LB, Hassig PJ, Kreyenhagen KN (1987) Computer simulations of large asteroid impacts into oceanic and continental sites – preliminary results on atmospheric, cratering and ejecta dynamics. Int J Impact Eng 5:525–541

Rokoengen K, Mørk A, Mørk MBE, Smelror M (2005) The irregular base Cretaceous reflector offshore Mid Norway: A possible result of the Mjølnir impact in the Barents Sea? Geol Surv Norway Bull 443:19–27

Russell JS (1845) Report on waves. Proceedings of the 14th meeting of the British Association for the Advancement of Science, York, pp 311–390, Plates XLVII–LVII

Sandbakken PT (2002) A geological investigation of the Mjølnir crater core (7329/03-U-01), with emphasis on shock metamorphosed quartz. Cand Scient (candidatus scietarium) thesis. University of Oslo, Oslo, p 142

Sandbakken P, Langenhorst F, Dypvik H (2005) Shock metamorphism of quartz at the submarine Mjølnir impact crater, Barents Sea. Meteorit Planet Sci 40:1363–1375

Sawatzky HB (1977) Buried impact craters in the Williston Basin and adjacent area. In: Roddy DJ, Pepin RO, Merrill RB (eds) Impact and explosion cratering. Pergamon Press, New York, pp 461–480

Schmidt RM, Holsapple KA (1982) Estimates of crater size for large body impact: Gravity scaling results, In: Silver LT, Schultz PH (eds) Geological implications of impacts of large asteroids and comets on Earth. Geological Society of America Special Paper 190, Boulder, pp 93–102

Schmoker JW (1981) Determination of organicmatter content of Appalachian Devonian shales from gammaray logs. Am Assoc Pet Geol Bull 65:1285–1298

Scholz CA, Karp T, Brooks KM, Milkereit B, Amoako PYO, Arko JA (2002) Pronounced central uplift identified in the Bosumtwi impact structure, Ghana, using multichannel seismic reflection data. Geology 30:939–942

Scholz CA, Karp T, Lyons RP (2007) Structure and morphology of the Bosumtwi impact structure from seismic reflection data. Meteorit Planet Sci 42:549–560

Schultz PH (1992) Atmospheric effects on ejecta emplacement and crater formation on Venus from Magellan. J Geophys Res 97:16183–16248

Schultz PH (1996) Effect of impact angle on vaporization. J Geophys Res 100:21117–21135

Schultz PH (1999) Ejecta distributions from oblique impacts into particulate targets [abs]. Lunar Planet Sci Conf 30, abs #1919

Schultz PH, Anderson RR (1996) Asymmetry of the Manson impact structure: evidence for impact angle and direction. In: Koeberl C, Anderson RR (eds) The Manson impact structure, Iowa: anatomy of an impact crater. Geological Society of America Special Paper 302, Boulder, pp 397–417

Schultz PH, D'Hondt S (1996) Cretaceous-Tertiary Chicxulub impact angle and its consequences. Geology 24:963–967

Schultz PH, Gault DE (1990) Prolonged global catastrophes from oblique impacts. In: Sharpton VL, Ward PD (eds) Global catastrophes in Earth history; An interdisciplinary conference on impacts, volcanism and mass mortality. Geological Society of America Special Paper 247, Boulder, pp 239–261

Schultz PH, Gault DE (1992) Recognizing impact signatures in the planetary record [abs] International conference on large meteorite impacts and planetary evolution, Sudbury, Canada. Lunar and Planetary Institute, Houston, Texas, Contribution

Sclater JG, Christie RAF (1980) Continental stretching: An explanation of the postmid-Cretaceous subsidence of the central North Sea basin. J Geophys Res 85:3711–3739

Sharpton VL, Burke K, Camargo ZA, Hall SA, Lee S, Marin LE, Suarez RG, Quezada MJM, Spudis PD, Urrutia FJ (1993) Chicxulub multiring impact basin: Size and other characteristics derived from gravity analysis. Science 261:1564–1567

Sharpton VL, Grieve RAF (1990) Meteorite impact, cryptoexplosion and shock metamorphism; A perspective on the evidence at the K/T boundary. In: Sharpton VL, Ward PD (eds) Global catastrophies in Earth history; An interdisciplinary conference on impacts, volcanism, and mass mortality. Geological Society of America Special Paper 247, Boulder, pp 301–318

Sharpton VL, Marin LE, Carney JL, Lee S, Ryder G, Schuraytz BC, Sikora P, Spudis PD (1996) Model of the Chicxulub impact basin. In: Ryder G, Fastovsky D, Gartner S (eds) The Cretaceous Tertiary events and other catastrophes in Earth history. Geological Society of America Special Paper 307, Boulder, pp 55–74

Shoemaker EM (1962) Interpretation of lunar craters. In: Kopal Z (ed) Physics and astronomy of the Moon. Academic Press, New York, pp 283–351

Shoemaker EM, Wolfe RF, Shoemaker CS (1990) Asteroid and comet flux in the neighborhood of Earth. In: Sharpton VL, Ward PD (eds) Global catastrophes in earth history; An interdisciplinary conference on impacts, volcanism and mass mortality. Geological Society of America Special Paper 247, Boulder, pp 155–170

Shuto N (1985) The Nihonkai-Chubu earthquake tsunami on the north Akita Coast. Coast Eng Jpn 28:255–264

Shuvalov VV (1999) Multidimensional hydrodynamic code SOVA for interfacial flows: Application to thermal layer effect. Shock Waves 9:381–390

Shuvalov VV (2002a) Numerical model of dust ejection induced by meteoroid impacts. Int J Impact Eng 27:377–385

Shuvalov VV (2002b) Numerical modeling of the impacts into shallow sea. In: Plado J, Pesonen LJ (eds) Impacts in Precambrian shields, impact studies. Springer, Berlin-Heidelberg, pp 323–336

Shuvalov VV (2003a) Mechanisms of Tsunami generation by impacts [abs] Large Meteorite Impacts 3, Nördlingen, August 2003

Shuvalov VV (2003b) Numerical modeling of the Eltanin impact. Meteorit Planet Sci 38, [abs], # 5149

Shuvalov VV, Dypvik H (2004) Ejecta formation and crater development of the Mjølnir impact. Meteorit Planet Sci 39:467–479

Shuvalov V, Dypvik H, Tsikalas F (2002) Numerical simulations of the Mjølnir marine impact crater. J Geophys Res 107:doi 10.1029/2001JE001698

Simonson BM, Davies D, Wallace M, Reeves S, Hassler SW (1998) Iridium anomaly but no shocked quartz from Late Archean microkrystite layer: Oceanic impact ejecta? Geology 26:195–198

Skagen JI (1992) Effects of hydrocarbon potential caused by Tertiary uplift and erosion in the Barents Sea. In: Vorren TO, Bergsager E, Dahl-Stamnes ØA, Holter E, Johansen B, Lie E, Lund, TB (eds) Arctic geology and petroleum potential. Norwegian Pet Soc Spec Publ 2:711–719

Skilbrei JR (1993a) An evaluation of magnetic basement determinations from the southwestern Barents Sea. In: Skilbrei JR (ed) Interpretation of geophysical data from the northwestern Barents Sea and Spitsbergen. Dr Ing Thesis. University of Trondheim, Norway, p. 160

Skilbrei JR (1993b) Short note: The straight slope method for basement depth determination revisited. Geophysics 58:593–595

Skogseid J, Planke S, Faleide JI, Pedersen T, Eldholm O, Neverdal F (2000) NE Atlantic continental rifting and volcanic margin formation. In: Nøttvedt A, Larsen BT (eds) Dynamics of the Norwegian Margin. Geol Soc Spec Publ 167:295–326

Smelror M (1994) Jurassic stratigraphy of the western Barents Sea Region: A review. Geobios 17:441–451

Smelror M, Dypvik H (2005) Dinoflagellate cyst and prasinophyte biostratigraphy of the Volgian-Ryazanian boundary strata, western Barents Shelf. Nor Geologiske Undersøkelse Bull 443: 61–69

Smelror M, Dypvik H (2006) The sweet aftermath: Environmental changes and biotic restoration following the marine Mjølnir impact (Volgian-Ryazanian boundary strata, Barents Shelf). In: Cockell C, Koeberl C, Gilmour I (eds) Biological processes associated with impact events. Springer, Berlin-Heidelberg, pp 143–178

Smelror M, Dypvik H, Tsikalas F (eds) (2001) 7th Workshop of the ESF Impact Programme. Submarine craters and ejecta-crater correlations, and icy impacts and icy targets. Abstract and proceedings of the Norwegian Geological Society, vol 1, p 114

Smelror M, Dypvik H, Mørk A (2002) Phytoplankton blooms in the Jurassic Cretaceous boundary beds of the Barents Sea possibly induced by the Mjølnir impact. In Buffetaut E, Koeberl C (eds) Geological and biological effects of impact events. Lecture notes in Earth Sciences, Impact Studies. Springer, Berlin-Heidelberg, pp 69–81

Smelror M, Kelly SRA, Dypvik H, Mørk A, Nagy J, Tsikalas F (2001a) Mjølnir (Barents Sea) meteorite impact offers a Volgian-Ryazanian boundary marker. Newsl Stratigr 38:129–140

Smelror M, Mørk A, Monteil E, Rutledge D, Leereveld H (1998) The Klippfisk Formation – a new lithostratigraphic unit of lower Cretaceous platform carbonates on the Western Barents Shelf. Polar Res 17(2):181–202

Smelror M, Mørk MBE, Mørk A, Løseth H, Weiss HM (2001b) Middle Jurassic-Lower Cretaceous transgressive-regressive sequences and facies distribution off Troms, northern Norway. In: Martinsen OJ, Dreyer T (eds) Sedimentary environments offshore Norway – Palaeozoic to recent. Norwegian Pet Soc Spec Publ 10:211–232

Smelror M, Petrov OV, Larssen GB, Werner SC (2009) Geological history of the Barents sea. Geological survey of Norway, Trondheim, p 135

Smit J (1999) The global stratigraphy of the Cretaceous-Tertiary boundary impact ejecta. Ann Rev Earth Planet Sci 27:75–113

Smith DG, Harland WB, Hughes NF, Pickton CAG (1976) The geology of Kong Karls Land, Svalbard. Geol Mag 113:193–304

Snyder DB, Hobbs RW (1999a) Deep seismic reflection profiles across the Chicxulub crater. In: Dressler B, Sharpton VL (eds) Large Meteorite Impacts and Planetary Evolution II. Geological Society of America Special Paper 339, Boulder, pp 269–279

Snyder DB, Hobbs RW (1999b) The BIRPS Atlas II: A second decade of deep seismic reflection profiling. Lond Geol Soc Lond 3:CD-ROMs

Solheim A, Andersen ES, Elverhøi A, Fiedler A (1996) Late Cenozoic depositionnal history of the western Svalbard continental shelf, controlled by subsidence and climate. Glob Planet Change 12:135–148

Sonett CP, Pearce SJ, Gault DE (1991) The oceanic impact of large objects. Adv Space Res 11: 77–86

Spudis PD (1993) The geology of multiring impact basins. Cambridge University Press, Cambridge, p 263

Steel RJ, Gjelberg J, Helland-Hansen W, Kleinsphen K, Nødtvedt A, Rye Larsen M (1985) The Tertiary strikeslip basins and orogenic belt of Spitsbergen. In: Biddle KT, Christie-Blick N (eds) Strikeslip deformation, basin formation and sedimentation. Soc Econ Paleontol Mineral Spec Publ 37:339–359

Steinberger B, Torsvik TH (2008) Absolute plate motions and true polar wander in the absence of hotspot tracks. Nature 452:620–623

Stewart DJ, Berge K, Bowlin B (1995) Exploration trends in the Southern Barents Sea. In: Hanslien S (ed) Petroleum exploration and exploitation in Norway. Norwegian Pet Soc Spec Publ 4: 253–276.

Stewart SA, Allen PJ (2002) A 20-km diameter multiringed impact structure in the North Sea. Nature 418:520–523

Stewart SA, Allen PJ (2005) 3D seismic reflection mapping of the Silverpit multiringed crater, North Sea. Geol Soc Am Bull 117:354–368

Stöffler D, Ewald U, Ostertag R, Reimold WU (1977) Research drilling Nördlingen 1973, Ries: composition and texture of polymictic impact breccias. Geol Bavarica 75:163–190

Stöffler D, Langenhorst F (1994) Shock metamorphism of quartz in nature and experiment: I. Basic observation and theory. Meteoritics 29:155–181

Sturkell E, Lindström M (2004) The target peneplain of the Lockne impact. Meteorit Planet Sci 39:1721–1731

Sugita S, Schultz PH (2002) Initiation of run-out flows on Venus by oblique impacts. Icarus 155:265–284

Supernaw IR, McCoy AD, Link AJ (1978) Method for insitu evaluation of the source rock potential of earth formations: US Patent 4,071,744, January 31

Surlyk F, Zakharov VA (1982) Buchiid bivalves from the upper Jurassic and lower Cretaceous of East Greenland. Palaeontology 25:727–753

Suuroja K, Suuroja S, All T, Floden T (2002) Kärdla (Hiiumaa Island, Estonia) – the buried and well-preserved Ordovician marine impact structure. Deep Sea Res Part II 49:1121–1144

Sættem J, Bugge T, Fanavoll S, Goll RM, Mørk A, Mørk MBE, Smelror M, Verdenius J (1994) Cenozoic margin development and erosion of the Barents Sea: Core evidence from southwest of Bjørnøya. Mar Geol 118:257–281

Talwani M, Eldholm O (1972) Continental margin off Norway: a geophysical study. Geol Soc Am Bull 83:3575–3606

Talwani M, Heirtzler JR (1964) Computation of magnetic anomalies caused by twodimensional bodies of arbitrary shape. In: Parks GA (ed) Computers in the mineral industries, Part 1. Stanford University Publication. Geol Sci 9:464–480

Talwani M, Worzel JL, Landisman M (1959) Rapid gravity computations for twodimensional bodies with application to the Mendocino submarine fracture zone. J Geophys Res 64:49–59

Tanaka M (1986) The stability of solitary waves. Phys Fluids 29(3):650–655

Telford WM, Geldart LP, Sheriff RE (1990) Applied geophysics. Cambridge University Press, Cambridge, p 770

Thompson SL, Lauson HS (1972) Improvements in the chart D radiation-hydrodynamic CODE III: Revised analytic equations of state. Report SC-RR-71 0714. Sandia National Laboratory, Albuquerque, p 119

Tillotson JH (1962) Metallic equations of state for hypervelocity impact. General Atomic Report GA-3216:137

Torsvik TH, Olesen O (1988) Petrophysics and Palaeomagnetism initial report of the Norwegian Geological Survey Laboratory, Norges Geologiske Uundersøkelse report no. 88.171

Torsvik T, Mosar J, Eide EA (2001) Cretaceous – Tertiary geodynamics; a North Atlantic exercise. Geophys J Int 146:850–866

Torsvik T, Van der Voo R, Redfield TF (2002) Relative hotspot motions versus true polar wander. Earth Planet Sci Lett 202:185–200

Tsikalas F (1992) A study of seismic velocity, density and porosity in Barents Sea wells (N-Norway). Master thesis. University of Oslo, Oslo, p 169

Tsikalas F (2005) Mjølnir Ccater as a result of oblique impact: Asymmetry evidence constrains impact direction and angle. In: Koeberl C, Henkel H (eds) Impact tectonism. Impact Studies. Springer, Berlin-Heidelberg, pp 285–306

Tsikalas F, Faleide JI (2004) Nearfield erosional features at the Mjølnir impact crater: The role of marine sedimentary target. In: Dypvik H, Burchell M, Claeys P (eds) Cratering in marine environments and on ice. Impact Studies. Springer, Berlin-Heidelberg, pp 39–55

Tsikalas F, Faleide JI (2007) Postimpact structural crater modification due to sediment loading: An overlooked process. Meteorit Planet Sci 42:2013–2029

Tsikalas F, Faleide JI, Eldholm O, Dypvik H (2002b) Seismic correlation of the Mjølnir marine impact crater to shallow boreholes. In: Plado J, Pesonen LJ (eds) Impacts in Precambrian Shields. Impact Studies. Springer, Berlin-Heidelberg, pp 307–321

Tsikalas F, Gudlaugsson ST, Eldholm O, Faleide JI (1998c) Integrated geophysical analysis supporting the impact origin of the Mjølnir Structure, Barents Sea. Tectonophysics 289:257–280

Tsikalas F, Gudlaugsson ST, Faleide JI (1998a) Collapse, infilling, and postimpact deformation at the Mjølnir impact structure, Barents Sea. Geol Soc Am Bull 110:537–552

Tsikalas F, Gudlaugsson ST, Faleide JI (1998b) The anatomy of a buried complex impact structure: The Mjølnir Structure, Barents Sea. J Geophys Res 103:30469–30484

Tsikalas F, Gudlaugsson ST, Faleide JI, Eldholm O (1999) Mjølnir Structure, Barents Sea: A marine impact crater laboratory. In: Dressler B, Sharpton VL (eds) Large meteorite impacts and planetary evolution II. Geological Society of America Special Paper 339, Boulder, pp 193–204

Tsikalas F, Gudlaugsson ST, Faleide JI, Eldholm O (2002a) The Mjølnir marine impact crater porosity anomaly. Deep Sea Res Part II 49:1103–1120

Turtle EP, Pierazzo E, Collins GS, Osinski GR, Melosh HJ, Morgan JV, Reimold WU (2005) Impact structures: What does crater diameter mean? In: Kenkmann T, Hörz F, Deutsch A (eds) Large meteorite impacts III. Geological Society of America Special Paper 384, Boulder, pp 1–24

Ugalde H, Artemieva N, Milkereit B (2005) Magnetization on impact structures – constraints from numerical modelling and petrophysics. In: Kenkmann T, Hörz F, Deutsch A (eds) Large meteorite impacts III. Geological Society of America Special Paper 384, Boulder, pp 25–42

Ugalde H, Danuor SK, Milkereit B (2007) Integrated 3D model from gravity and petrophysical data at Lake Bosumtwi impact crater, Ghana. Meteorit Planet Sci 42:859–866

Urrutia-Fucugauchi J, Morgan J, Stöffler D, Claeys P (2004) The Chicxulub Scientific Drilling Project (CSDP). Meteorit Planet Sci 39:787–790

Urrutia-Fucugauchi J, Morán Zenteno D, Sharpton VL, Buffler R, Stöffler D, Smit J (2001) The Chicxulub Scientific Drilling Project. Infrastructura científica y desarrollo tecnológico 3.Mexico City: Insituto de Geofisica, Universidad Nacional Autónoma de México, p 45

Vågnes E, Faleide JI, Gudlaugsson ST (1992) Glacial erosion and tectonic uplift in the Barents Sea. Nor Geologisk Tidsskrift 72:333–338

Vermeesch PM, Morgan JV (2004) Chicxulub central crater structure: initial results from physical property measurements and combined velocity and gravity modelling. Meteorit Planet Sci 39:1019–1034

Vigran JO, Mangerud G, Mørk A, Bugge T, Weitschat W (1998) Biostratigraphy and sequence stratigraphy of the Lower and Middle Triassic deposits from the Svalis Dome, Central Barents Sea, Norway. Palynology 22:89–141

Vogt PR, Taylor PT, Kovacs LC, Johnson GL (1979) Detailed aeromagnetic investigation of the Arctic Basin. J Geophys Res 84:1071–1089

von Dalwigk I, Ormö J (2001) Formation of resurge gullies at impacts at sea: The Lockne crater, Sweden. Meteorit Planet Sci 36:359–369 von Engelhardt W (1990) Distribution, petrography and shock metamorphism of the ejecta of the Ries crater in Germany: A review. Tectonophysics 171:259–273

von Engelhardt WV, Bertsch W (1969) Shock induced planar deformation structures in quartz from the Ries Crater, Germany. Contrib Mineral Petrol 20:203–234

Vorren TO, Laberg JS (2001) Late Quaternary sedimentary processes and environment on the Norwegian-Greenland Sea continental margins. In: Martinsen OJ, Dreyer T (eds) Sedimentary evironments offshore Norway – Paleozoic to Recent. Norsk Petroleumsforening Special Publication pp 451–456

Walderhaug O (1992) Magnitude of uplift of the Stø and Nordmela Formations in the Hammerfest Basin – a diagenetic approach. Nor Geologisk Tidsskrift 72:321–323

Ward SN (2001) Landslide tsunami. J Geophys Res 106:11201–11215

Ward SN, Asphaug E (2000) Asteroid impact tsunami: A probabilistic hazard assessment. Icarus 145:64–78

Ward SN, Asphaug E (2002) Impact tsunami-Eltanin. Deep Sea Res II 49:1073–1080

Ward SN, Asphaug E (2003) Asteroid Impact Tsunami of 2880 March 16. Geophys J Int 153:F6–F10

Weiss R, Wünnemann K, Bahlburg H (2003) Oceanic impacts, tsunamis, and the influence of the water depth on the quantity and characteristics of the generated waves [abs] Large Meteorite Impacts 3, abs #4081, Nördlingen, CD ROM

Wessel P, Smith WHF (1998) New improved version of generic mapping tools released, EOS Trans Am Geophys Union 79(47):579

Wierzbowski A, Smelror M, Mørk A (2002) Ammonites and dinoflagellates in the Upper Oxfordian and Kimmeridgian of the northeastern Norwegian Sea (Nordland VII offshore area): Biostratigraphical and biogeographical signifycance. Neues Jahrb Geol Palaeontol Abh 226:145–164

Wignall PB, Hallam A (1991) Biofacies, stratigraphic distribution and depositional models of British onshore Jurassic black shales. Geol Soc Lond Spec Publ 58:291–309, doi:10.1144/GSL.SP.1991.058.01.19

Winzer SR (1972) The Steen River astrobleme, Alberta, Canada. Proc 24th Int Geol Congr Sect 15:148–156

Wolbach WS, Widicus S, Dypvik H (2001) A preliminary search for evidence of impact-related burning near the Mjølnir impact structure, Barents Sea [abs] Lunar Planet Sci Conf 32, abs #1332, CD-ROM

Won IJ, Bevis M (1987) Computing the gravitational and magnetic anomalies due to a polygon: Algorithms and fortran subroutines. Geophysics 52:232–238

Worsley D (2008) The post-Caledonian development of Svalbard and the western Barents Sea. Polar Res 27:298–317

Worsley D, Aga O (1986) Evolution of an arctic archipelago – The geological history of Svalbard. Statoil, Norway, p 121

Worsley D, Johansen R, Kristensen SE (1988) The Mesozoic and Cenozoic succession of Tromsøflaket. In: Dalland A, Worsley D, Ofstad K (eds) A lithostratigraphic scheme for the Mesozoic and Cenozoic succession offshore mid- and northern Norway. Norwegian Pet Directorate Bull 4:42–65

Wünnemann K, Ivanov BA (2003) Numerical modeling of the impact craterdepth parameter dependence in an acustically fluidized target. Planet Space Sci 51:831–845, doi:10.1016/J.PSS.2003.08,001

Wyllie MRJ, Gregory AR, Gardner GHF (1956) Elastic wave velocities in heterogeneous and porous media. Geophysics 21:41–70

Wyllie MRJ, Gregory AR, Gardner GHF (1958) An experimental investigation of factors affecting elastic wave velocities in porous media. Geophysics 23:459–493

Yershova YS (1983) Ob'yasnitel'naya zapiska k biostratigraficheskoy skheme yurskikh i nizhnemelovykh otozheniy arkihpelaga shpitsbergen. PGO Sevmorgeologiya 62:1–50 (in Russian)

Zakharov VA (1981) Buchiidy i biostratigraphiya boreal'noy verkhney yuri I neocoma. Tr Inst Geologii I Geofiziki Novosibirsk 458:1–271 (in Russian)

Zakharov VA (1983) Oxfordian at Kheta. In: Bobolepov KV (ed) Paleogeography of the North of USSR during Jurassic. Nauka Press, Novosibirsk, p 190 (in Russian)
Zakharov VA, Lapukhoy AS, Shenfil OV (1993) Iridium anomaly at the Jurassic-Cretaceous boundary in northern Siberia. Russian J Geol Geophys 34:83–90
Zakharov VA, Pruner P, Rogov MA (2007) Advances in correlation of the Jurassic-Cretaceous boundary interval of Arctic and south Europe based on magne- to- and biostratigraphy. The Arctic Conference Days. Abstr Proc Geol Soc Norway 2:315–316
Zakharov VA, Rogov MA (2003) Boreal-Tethyan mollusk migrations at the Jurassic- Cretaceous boundary time and biogeographic ecotone position in the Northern Hemisphere. Stratigr Geol Correl 11:152–171
Zakharov VA, Shurygin BN, Levchuk MA, Pinous OV, Sahagian D (1998) Eustatic signals in the Jurassic and Lower Cretaceous (Neocomian) deposits of the West-Siberian sedimentary basin. Russian Geol Geophys 39(11):1493–1504 (Translated from Geologiya i Geofizika 39, © 1999 by Allerton Press, Inc)
Zakharov VA, Shurygin BN, Kurushin NI, Meledina SV, Nikitenko BL (2002) A Mesozoic ocean in the Arctic: paleontological evidence. Russian Geol Geophys 43(2):155–181 (Translated from Geologiya i Geofizika 43, © 2002 by Allerton Press, Inc)
Zakharov VA, Surlyk F, Dalland A (1981) Upper Jurassic - Lower Cretaceous buchia from Andøy, Northern Norway. Nor Geologisk Tidsskr 61:261–269
Zamyshlyaev BV, Evterev LS (1990) Models of dynamic deforming and failure for ground media. Nauka Press, Moscow, p 215 (in Russian)
Zuber M, Smith DE, Lemoine FG, Neumann GA (1995) The shape and internal structure of the moon from the clementine mission. Science 266:1839–1843
Åm K (1975) Magnetic profiling over Svalbard and surrounding shelf areas. Nor Polarinstitutt Årbok 1973:87–89
Århus N (1991) Dinoflagellate cyst stratigraphy of some Aptian and Albian sections from northern Greenland, southeast Spitsbergen and Barents Sea. Cretaceous Res 12:209–225
Århus N, Kelley SRA, Collins JSH, Sandy MR (1990) Systematic palaeontology and biostratigraphy of two Lower Cretaceous condensed sections from the Barents sea. Polar Res 8(2):165–194

Index

E

F

G

H

I

N

O

P

Q

R

T

MIX
Papier aus verantwortungsvollen Quellen
Paper from responsible sources
FSC® C105338

If you have any concerns about our products,
you can contact us on
ProductSafety@springernature.com

In case Publisher is established outside the EU,
the EU authorized representative is:
Springer Nature Customer Service Center GmbH
Europaplatz 3, 69115 Heidelberg, Germany

Printed by Libri Plureos GmbH
in Hamburg, Germany